Muddling
Through

Muddling Through

Pursuing Science
and Truths in the
21st Century

Mike Fortun

Herbert J. Bernstein

*To David,
With great
admiration & fondness,
love
Herb*

COUNTERPOINT
WASHINGTON, D.C.

Library of Congress Cataloging-in-Publication Data
Fortun, Michael.
  Muddling through : pursuing science and truths in the twenty-first century / Mike Fortun and Hebert J. Bernstein.
      p.   cm.
  Includes bibliographical refernces and index.
  ISBN 1-887178-48-1 (alk. paper)
    1. Science.   2. Science—Social aspects.   3. Research—Social aspects.   I. Bernstein, Hebert J.   II. Title.
Q172.F67   1998
509'.05—dc21   98-38517
                                                                         CIP

Jacket and book design by Amy Evans McClure
Composition by Westview Press

Printed in the United States of America on acid-free paper that meets the American National Standards Institute Z39-48 Standard.

COUNTERPOINT
P.O. Box 65793
Washington, D.C. 20035–5793

Counterpoint is a member of the Perseus Books Group.

10 9 8 7 6 5 4 3 2 1

Contents

Acknowledgments

Thanks first to all of our colleagues at the Institute for Science and Interdisciplinary Studies (ISIS), who over the years have helped make it a "House of Experiment" in the best sense of that seventeenth-century concept: Jeff Green, Karen Sutherland, Elizabeth Motyka, Kelly Erwin, Risa Silverman, Jane Benbow, Jim Oldham, Scott Tundermann, Erich Schienke, Lily Louie, and Abby Drake. David Gruber has been a true, steadfast, and wise patron and cotheoretician; it's a simple fact that ISIS would not exist without him. Greg Prince, Nina Shandler, Everett Mendelsohn, Michelle Murrain, William Nugent, and Sean Decatur have shared the responsibility, work, and occasional worry of Board service. We thank all our colleagues at Hampshire college and the Five College Consortium who provide the institutional and intellectual arena in which ISIS has flourished.

Thanks to Marcus Raskin, who started us on the path to this book. It's hard to say exactly when that was—a collaborator and mentor for almost twenty years, he has helped to develop and understand those forms of knowledge that go beyond the story of science-as-usual—what Marc originally called reconstructive knowledge. He was a principal coauthor of *New Ways of Knowing: Science, Society, and Reconstructive Knowledge*, a precursor to this book. This book itself owes its start to Marc's "Paths to the Twenty-First Century" project, which initiated a series of volumes on knowledge and its public purposes. Our own intellectual and social projects in the democratic reconstruction of the sciences bear an enormous debt to his example and wisdom.

Thanks to all the people, too many to name, who have shared with us their knowledge in and of the sciences, their challenges, and their life situations. Their resilience and inventiveness are truly inspiring. They include the neighbors of military bases, wanting to know if toxic sites at "their" installation endanger health or environment; the indigenous people of the Ecuadorian rainforest, contending daily with the incursions of oil exploration into their lives; and our students who teach as well as learn, wanting to know the intricacies of quantum mechanics and other sciences, or the cultural analyses of the sciences. And thanks to our many other teachers, the professional colleagues in the sciences, the humanities, and all the disciplines in between—that is, *all* the disciplines. There are more of them than we could directly reference in the following pages. Whether of atoms or words, from the laboratory or the library, their inventions have powered and shaped our own.

Thanks to another group who form a sort of invisible college of support. Their work and deeds show that action and care can be joined in and for life: Harry Saal, Danny Greenspun, Michael Ubell, Cora Weiss, Peter Weiss, Paula Hawthorne, Thomas Ewing, Jock Herron, Jaenet Guggenheim, Gae Eisenhardt, Howard Eisenberg, Arlene Eisenberg, Adele Simmons, Timi Joukowsky, Woody Wickham, Harriet Barlow, Michael Shandler, Lucy McFadden, Michael Mann, Carol Salzman, Scott Nadel, Kate Downes, David Wiener, Sam Wiener, and Nick Seamon.

Less invisibly: thanks to the National Science Foundation for its grant (SBR–9601757) in support of Fortun's work. Thanks also to the John D. and Catherine T. MacArthur Foundation, the Samuel Rubin Foundation, the Institute for Scientific Interchange, Hampshire College, and Five Colleges Inc.

Thanks to Rich Doyle for his affirmations of multiplicity. His early enthusiastic reading of the entire manuscript kept us and it going through the initial uncertainties. Thanks to Jack Shoemaker for both his patience and encouragement, and also to Trish Hoard and the rest of the staff at Counterpoint Press. It was our unbelievable good fortune to work with a publisher so committed to beautiful and thoughtful books. We weren't sure what a "development editor" was when Nancy Heneson was assigned to us in this capacity; now we know the standard by which the field should be judged. She broke us of our worst habits, taking the mass of words we had heaved up and helping us turn it into a book. Whatever ambiguities remain were either beyond the reach of even her extraordinary talents, or intentional on our part. In either case, we are solely responsible for any residing confusions, errors, omissions, and excesses in the text.

Lastly, our deepest thanks to those for whom the word "thanks" seems especially frail but who—lucky us—know our frailties all too well, and love us anyway: Kim Fortun, Mary Mayers Bernstein, Carolyn Joy Bernstein, and Laila Jael Bernstein. It's to them, and to all of our parents—Alice and Raymond Fortun, Edith and Harry Bernstein, and Lillian Bernstein—that we dedicate this book.

Prologue

Just because we are finite beings, located, situated, embodied, we can, and can only, muddle through. . . . Scientists muddle through with staggering success. Only their success is rather different than they imagine. It depends not on any possibility of translating thought into action, but on the conjoining practices of a colluding community of common language speakers. Our task . . . is to make sense of the successes of science in terms of the particular linguistic and material conventions that scientists have forged for their own sorts of muddling through.

Evelyn Fox Keller[1]

For the scientists who work within them daily, and for the people who avidly tune in to their work through the media, the sciences are an endless source of exhilaration, insight, and invention. The sciences demand some of our best thinking and most dextrous manipulations. In terms that seem utterly dependable and authoritatively final, the sciences tell us about what's truly happening in the genes, cells, and organs of our bodies; about the evolution of life; about how our brains and minds really work, as well as about the brains and (perhaps) minds of computers; about the lawful and awe-ful realities of subatomic particles and vast cosmological stretches of time and space. The ingenuity of scientists connects to the intricacies of nature, and the resulting combination of the these forces feeds us, heals us, transports us, inspires us.

Yet few things are more unsettling than working in or observing the sciences today. Yesterday's truths are quickly forgotten, made obsolete by the truer truths just released. Medical breakthroughs soon show serious limitations or disturbing side effects. The next cosmic discovery will cost taxpayers a billion dollars more than the previous one. Many scientists with great ideas see those ideas go unsupported or undeveloped, and they encounter a public that is often uninterested, ill-informed, or hostile. Urgent controversies over health, behavior, and environment resist consensus and often seem only to splinter into bitter disagreements *among* scientists. Nature seems not only more and more complex but also opaque and even downright ornery. And so the problems pile up, and distrust and disillusionment set in.

These are contradictions, to say the least. This book will do its work not by resolving them, but by asking the questions that fall in between. Making sense of these contradictions is perhaps the hardest challenge for democracy in the twenty-first century. Of course we need to cultivate scientific literacy, but first we need to ask some basic questions about how the sciences are produced, applied, and understood.

The sciences of today create the worlds of tomorrow. They show their effects on our bodies, our conceptions of self, and our polity. Yet the sciences continue in large part to be viewed as a resource of absolutely certain, objective answers gleaned from a pure and exact "scientific method." This is, to put it bluntly, wrong, and all the really great (and some not so great) scientists know it. They know that the relationship between creative scientific inquiry and the requirements of a pluralist democratic society is anything but the straightforward application of supposedly neutral facts to social problems. They know that today we are living "after the fact," in a world in which science can no longer be regarded as the oracle of those cultural and material reforms necessary to a just society.

This world demands a kind of literacy that makes sense of the sciences not in terms of infinity and transcendence, but of finitude and location; not in terms of awesome translations from the real to the ideal, but of complex conjunctions and collusions among things, words, and deeds; not in terms of a book of nature discovered and decoded by a small group of experts, but of an ongoing essay written and spoken by many, in a shifting, generative language. In short, we have to engage with the sciences as the kind of activities they have always been, the *only* kind of activities they can be: muddling through.

Muddling Through is a book about the sciences in the late twentieth century and about the kind of sciences we need for the twenty-first. It is a book about how the sciences make sense of the world and provide sense to the world. Think of *Muddling Through* as the basic text for a different kind of science literacy project, a project to reimagine and then enact the sciences as operations of language and thought and as attempts, trials, *limited experiments* involving things, ideas, and just about everything in between.

This is also a book about politics (not policy) and culture—that is, about how the sciences are made through arduous and diverse political processes. This book is about how the sciences affect politics not only through technological invention but by generating the images and metaphors that we apply to every situation and phenomenon we encounter, and by providing the blueprints we use to make and legitimate crucial social decisions. The connections between the sciences and democratic pluralism need to be revitalized, through both new concepts and innovative social forms.

We use the plural "connections" deliberately to illustrate that the sciences and democracy must link up at many levels, from the policy panels in Washington and Brussels to the workings of curiosity and inquiry in each of us. Democratic society needs pluralism and participation not only in the application of science, but even more importantly in its production. In the words of François Jacob: "An age or a culture is characterized less by the extent of its knowledge than by the nature of the questions it puts forward."[2] We want *more* people asking *more*, and different, kinds of questions about what's "really" the case, and we want them in the laboratories, in the field, in the hospital wards, in the classrooms, and on the funding panels.

For this kind of social innovation to happen, and to happen in a way that helps produce good science rather than being simply prohibitive, people need an understanding of the sciences that is more complex than conventional accounts provide. Such accounts often hinge on the image and imagination of the hero, a "great scientist" like Einstein or Newton. But the notion of greatness making history, and as the proper basis for writing history, is passé. Still, there will be many places in this book where we will invoke the words of people from the high culture of science. In doing so we play a risky double game: using their authority and heroic stature not to undermine that authority or stature but to call them into question. Where, precisely, does the force of their ideas come from? Was it wrested from nature? Did it spring full-blown from their minds? Or are other things, other processes, and other people involved? Indeed, many of the scientists we discuss have often been most skilled at calling their own authority, as conventionally understood, into question. At their best, the sciences themselves put both the world, and their own processes of questioning that world, into the picture, the frame of inquiry. We also use recent historical and cultural explorations of some of the great figures of science—Copernicus, Galileo, Darwin—to work toward a more complex rendering of what the sciences actually are, and why they are often so successful, as well as sometimes risky and destructive.

Even so, the sciences are far more than what these noble figures represent. So many books today take that focus, purveying the conventional picture of science as an exalted mode of wondrous discovery, intellectual adventure, and abstract theorizing, undertaken by a chosen few. We choose to include stories about lesser-known scientists, with an emphasis on the sciences as practice, an activity that is socially complicated as well as intellectually complex. Our stories come not just from the realm of high culture and theory but from the laboratory, the courtroom, the toxic waste site, the popular television program.

These stories are more frequently included, and at greater length, than the reader may be used to finding in books of this type. We do this because it is necessary both to convey the complexity of the issues involved, the enormous amount of detail and the subtlety and difficulty of the questions, and to provide a better sense of just what kind of remarkable achievement the sciences in fact represent. We mix the old and the new, jumping across time, disciplines, and cultures. We range all over the scientific territory, displaying fragments that don't always assemble into an overall lesson, a moral, a sense of certainty, or some other comforting whole. We want these brief, detailed, but incomplete looks into particular episodes to spark skeptical interest and further inquiry. And while we certainly want to convince and persuade, we do not want to oversimplify. Above all, we want to upset faith in, and move far beyond, the usual accounts of the sciences.

The familiar story goes like this: The "scientific method" always starts from stable, given facts—observations, measurements in the form of numbers, isolated and purified substances that are part of an unchanging, solid reality. Logic then compels the assembling (a process carefully controlled by existing theory) of these indisputable pieces of the real world into a theory which literally re-presents that world: a perfect match that, when done correctly, admits no doubt. The theory is checked by the fact, by the real world, and underwritten by the rigor and purity of the scientific method. Together, hard fact and reliable method provide the sciences with a unique and powerful tool for self-correction that eliminates (in the long run) all forms of bias and error, yielding a neutral and objective progressive approach to equally neutral and objective final truths. Fact and method thus answer what we call the "really?" questions: Does this particular chemical really cause cancer? Is intelligence really genetic? Is the physical universe really made up of quarks and leptons, held together by bosons? Is biodiversity really important for the survival of the planet? Is homosexuality really rooted in biology? And so on. These answers, arrived at free from the disturbing influences of culture, society, language, or political power, can then be applied with the utmost assurance in the larger spheres of society and politics. Because of their faithful objectivity and scrupulous neutrality, the sciences can be a social resource—solving problems of health, hunger, and communication—precisely because they are immune to social or political influences. Their hard-won apparent neutrality, paradoxically, makes the sciences socially powerful.

So much for conventional accounts. We can no longer excuse the errors and simplifications on which such popular tales are based, no longer afford to have this view of the sciences circulate in the social and political realms. The stakes are too high.

If there is a single concept central to the errors, simplifications, and negative social effects of this conventional account of the sciences, it is purity. And against this concept's family of terms—purity, pure, purists—we will run our own family of terms: muddied, muddled, middled, messy, complex, hybridized, and many others. We are not particularly fond of such stark oppositions as pure/muddled, as the reader will see over the course of this book, but occasionally they come in handy. It is better to think about the sciences as muddled rather than pure; to imagine the borders between the sciences and the worlds of language, culture, and politics as muddied rather than clear and distinct; to know scientists as complex hybrid figures rather than rarefied heroes; to see the work of the sciences as a complicated interaction with a messy world, an exchange involving tools, words, things, and even more nebulous entities, rather than a methodical, pristine encounter between mind and nature.

Our point is *not* to drag the sciences through the mud, nor to dismiss the potential of science for social reform. Nothing we write here should obscure the passion we have, and the drive that the "great scientists" embody, to create truly embracing, challenging, and productive encounters with the world. Pursuing the sciences can be an amazing enterprise of rigorous thinking and subtle guessing, creative manufacturing and respectful listening, exercised will and inflicted surprise, hard work and even harder play. We need the sciences now more than ever, but we have to have them reimagined and re-formed—re-formed by attention to their own history, by overt attempts to enact them differently, and by new protocols for questioning and judging the sciences as they develop.

In these times when, as Avital Ronell phrases it, "America is being emptied of the desire to know,"[3] anyone who critically questions the sciences runs the heavy risk of being labeled "antiscience," a charge leveled as liberally as "anti-American" once was. If we have to make such silly generalizations, our preference is to say that we are simply "pro-inquiry." Each of us is a committed practitioner of the craft in which we were originally trained, yet we cross the lines as well. The physicist (Bernstein) appreciates the work of scholars in the fields of science studies, and considers it, at its best, to be just as robust, intellectually demanding, important, and valid—and just as fallible and culture-bound—as the sciences. The historian of science respects the work of scientists, knowing the power, creativity, efficacy, and legitimacy of the sciences—and their limitations and social embeddedness—through historical, philosophical, and cultural analysis. We believe that not only is such a dialogue between these endeavors possible, but that such collaboration and hybridization can produce both better science and better scholarship on the sciences, as well as more democratic social processes.

Expanding on Evelyn Fox Keller's imagery in the opening quote, we argue that the pursuit of both the sciences and of democracy is best imagined and enacted as "muddling through." Few things are more dangerous than unmuddled absolute faith in any answer or method, scientific or political. When it comes to the sciences, there are no simple answers like "just purify them," "just add values to them," "just keep them in their place," "just get rid of them," or even "just democratize them." They can't be pure, they already have values, they're everywhere, we can't get rid of them even if enough of us were stupid enough to want to, and democratizing them is an experiment, not an answer. We reject these and all similar imprecise, grand formulations: that the sciences disenchant the world; that they contribute to the loss of our souls; that they mechanize and reduce an organic, holistic cosmos; that they are essentially a violent way of knowing; that they destroy community; and so many more. Very high-minded, and very unhelpful— which is hardly surprising, since these expressions of the problem depend on concepts that are just as pure and idealized (souls, wholes, communities, and values) as their counterparts in the sciences which they aim to oppose.

Throughout this book we will show the way such paired sets of opposites recur both in and around the sciences. Dichotomies such as science/antiscience, mechanical/organic, fact/value (and countless others) structure the way we do and think about the sciences. While such polar oppositions are in some sense inescapable, new scientific literacies will depend on getting *in between* them. It is in the in-between, the muddled middle, where change happens, where creativity can be found, where the new emerges, where abundance dances—where the sciences *are* the sciences.

The in-between is the place scientists have in fact sought out and worked in for hundreds of years, and which they still seek today. While their own public representations of their work and the ways in which we hear about it in the media emphasize the seemingly miraculous, wondrous, and powerfully theoretical aspects of scientific inquiry, another part of the story that is usually (but not always) subsumed is the trial-and-error method: days, months, and years of what many scientists call "tinkering"—trying to get a piece of equipment to work properly, interpreting messy data, separating signal from noise, articulating new theory and explanations for new phenomena. Any decent scientist knows that results and explanations are always open to revision—indeed, *must* be revised if those results and explanations are to mean or work for anything. Answers are always only temporary, one-off, close but no cigar. They are guaranteed to make themselves obsolete by virtue of their own inescapable insufficiency.

The middle is also where uncertainty, risk, chance, and error can be found, and where one is therefore best advised to "muddle through." "Muddling through" is by no means a perfect principle—how could it be, centered as it is on imperfection?

The political scientist Charles Lindblom wrote an important article in 1959 titled "The Science of Muddling Through." Lindblom was working in a particular historical context in which operations research and systems theory were becoming quite powerful in organizational theory and in government policy. These disciplines claimed to be comprehensive and strictly rational, taking every possibility into account, optimizing material outcomes and maximizing efficiency while minimizing social conflict, providing the best solution to differences of values and goals. Lindblom called them "root methods," and while they were rarely practiced as precisely as they said they could be, and just as rarely yielded their intended results, these root methods, these total sciences were formalized as the best theory to be taught in professional schools and to use policy circles.

To this Lindblom opposed the "branch method," the "muddling through" practiced in the real world by administrators and policy makers, but hardly ever acknowledged as systematic and knowledge-based. "Muddling through" worked with incomplete information, lack of time and resources for full analysis of all factors and options, irreducible conflicts of values and political choices, the pressure of outside interests, the inertia of past decisions, and so on. Branch methods were better, in Lindblom's articulation, because they acknowledged finitude, the necessity and value of compromise, and they admitted that the lack of a system could itself be systematic in its own way. There was no sense of finality, an analysis accomplished and set in motion; you always had to go back again and again. Moreover, even when people claimed to be doing totally rational and systemic analysis, they were actually muddling through—they *had* to, given the complexity of the problems and systems with which they were dealing. Abstract root methods may have been the idealized theory, but muddling branch methods were the actual, empirical practice.[4]

We argue many of the same points for the sciences in general. Acknowledging that things are muddled seems to make it difficult if not impossible to render the kinds of judgments about the sciences that we say our society sorely needs. Judging muddles is difficult, as we"ll see; it is *not* impossible. Still, there is another twist: imperial "muddling through," whether British or Austro-Hungarian, relied on the shared assumptions and prejudices of a gentry class whose education so set their thinking that improvised and muddled solutions—creative as they otherwise might be—nevertheless resulted in solutions that preserved class privilege. We have to be on the lookout for these kinds of effects, the way that the terms and frames of inquiry can sneak back up and surprise or contradict us.

We ask the reader to be patient, to let our argument emerge, in true muddling fashion, over the course of the book. While we will return again and again to the domain of the in-between, to the middle zone between opposed terms or viewpoints, to the place of compromise and negotiation, this does not mean that muddling through amounts simply to a desire for a happy medium, in political or intellectual terms. Instead, it entails a commitment to the "unhappy middle." Far from being indecisive, noncommittal, or blandly middle-of-the-road, muddling through is marked by perseverance: a dogged pursuit and relentless enactment of both pointed inquiry ("How do you know that?") and thoughtful social and political work ("Given *that*, let's try this . . . "). As a result, it does not imply muddle-headedness, but in fact the opposite. Through precise description of the complex ways that the work of our heads, our hearts, and our tongues intersects in the sciences, and how those intersections are in turn inextricably linked to events and movements in culture and politics, muddling through represents the only inquiry adequate to the monstrous, rapidly changing worlds of both nature and society. Recognizing its own limits, it suggests only actions that are cautious and responsible. And if it lacks unwavering belief in timeless principles, it makes up for it in vigilance, inventiveness, and democratic spirit.

How, then, do we use this book to turn muddling into muddling through, to steer through the bountiful and treacherous waters of the sciences today? Chapter by chapter, Section I introduces and elaborates on four navigational tactics that together define the method of muddling through. They are not only definitions, however; they are also responses to some hoary assumptions about the way science is thought about and conducted, namely, that facts are found, that theory and language mirror the world, and that science is a politically and culturally neutral tool.

First, *Facts are not found, but made*. The scientific method does not discover truth, it produces it. Chapter 1 thus focuses on *experimenting*: it is at this middle level that the muddle between facts and theories in the sciences is most easily located. We avoid unquestioned theoretical abstractions, grounding our inquiries instead in the realm of human activities, where flesh-and-blood people negotiate with cranky equipment, murky concepts, and an evasive "nature." One of the most important stories told here concerns the particular social function played by facts and the experimental production of facts in seventeenth-century England, the time and place in which the sciences first became truly experimental. All of the stories in this chapter help us see how facts are made, without being made up, and how facts should always be subject to extensive inquiry.

Second, *Theory and language refract the world, not reflect it*. Chapter 2 turns to *articulating*, focusing on the array of activities that produce what are conventionally referred to as scientific theories, as well as the broader narratives, world views, and interpretations that supplement their meanings. The notion of articulating steers us away from conceptions of theory as mirrorings of a world composed of atomistic facts, and toward an understanding of (many kinds of) theory whose status as "truth" depends less on faithful reflection of a preexistent world, than on the viability, strength, or robustness of a tangle of connections, or articulations. We employ at least two usages of the word "articulation": One, we argue that pursuing sciences requires a better understanding of how, and where, language works, that the sciences involve a struggle to articulate something that has never been said before, an attempt to put new things into new words (and new words into new things). Two, we use "articulation" to refer to the way the sciences are coupled and jointed: a vast connection of words, things, instruments, social trends, and funding sources. These articulations reach down into the level of experimenting and the experimental production of facts, across to other articulations from the same or other fields of thought, and up into the domain of culture and its narratives, and their embeddedness in social institutions. Our challenge is to demonstrate how the appeal and effectiveness of the sciences come not from their mirroring of nature, but from the density and peculiar strengths of these webbed linkages.

Which brings us to our response to the third assumption: *Science is never neutral, but always charged, moving in a field of cultural and political forces.* The sciences are not tools to be wielded for good or evil by the powers that be, but inquiry infrastructures composed not only of instruments, theories, and language but of larger institutions and their material and cultural resources. Thus, instead of reinforcing the usual distinctions between knowledge and power, reason and force, we introduce in Chapter 3 the analytic of *powering/knowing*. Using Galileo and Darwin as central examples, we show how the conventional ideals of knowledge and reason unsullied by baser considerations of power and resources do indeed require rethinking. But that doesn't mean simply that might makes right. Good science has always depended on a variety of power sources, and the sciences have always been an active, charged matrix remarkably sensitive to the pushes and pulls of seemingly distant ideas, institutions, people, culture, and, of course, capital. But the "charges" between the sciences and their historical and social contexts are contingent rather than determined; the affinities among the sciences, politics, and cultures are sometimes coarse and commanding, but just as often supple, subtle, delicate, and indirect. In any case, these contingent affinities are quite real; they shape and shade what we know, what we call truth, reason, nature, and justice.

If the purity and objectivity of the sciences were once guaranteed by their freedom from the corrupting influences of power and their faithful mirroring of a real world, what upholds and legitimates a system built out of experimenting, articulating, and powering? If the sciences are geared less toward faithful, objective representations of a primordial reality, and more toward the production of novel effects and entities, new social possibilities and unheard-of ideas, does this mean that anything goes? Can we construct facts or theories according to personal or political whim? We take up such knotty issues in chapter 4, and suggest that the demanding and difficult process of *judging* should be installed near the center of the complex webs spun through the sciences. We discuss notorious historical episodes such as the legitimation of eugenics in Germany and the United States, and Lysenkoism in the Soviet Union, and equally tangled current dilemmas posed by toxic torts and scientific fraud, to show how ever-present ambiguity and the volatile mix of the political and the scientific demand subtle, thoughtful, yet ultimately risky acts of judgment, every step of the way.

Thus, in Section I we are committed to muddling *up*: messing with conventional understandings of the sciences, blurring the boundaries between supposedly distinct things like facts and theories, hybridizing categories like "cultural" and "scientific," complexifying the figures of famous (and not-so-famous) scientists and the variety of forces which they drew upon and unleashed. Scientists (at least the really good ones) have always been adept at such muddling up. It has been crucial to their success, even if they weren't always aware of it, and we have much to learn from it. Scientists tweak their instruments and experimental setups to break up current theories. They imagine and articulate new ideas to order previously disorderly and nonsensical experimental results. They judge, they guess, they leap logical gaps, they combine rigor and risk, they cobble together money, time, people, ideas, and a host of other things. Section I explores these unmethodological methods, confirming that close inquiry into the practices of the sciences can help us understand the vital role they have to play in our future.

If Section I is about muddling up, Section II can be thought of as accounts of muddling *in*: getting one's hands dirty, running new experiments, creating new institutional resources, organizing communities. If Section I develops one meaning of "after the fact"—developing the tools and questions for living in a spacetime where facts are no longer as straight as we liked to think they were—then Section II interprets "after the fact" in another way: as the active pursuing of the sciences, chasing after them, *wanting* them.

Chapters 5 through 8 detail, respectively, efforts to clean up the military's toxic wastes at Westover Air Reserve Base in Massachusetts, an emergent illness that has been dubbed multiple chemical sensitivities (MCS), the articulations between health and "human nature" produced in the fields of molecular biology and biotechnology, and current work at the theoretical and experimental frontiers of quantum mechanics.

These chapters vary somewhat in voice and tone among themselves and from the rest of the book. The mixture of you-are-there reportage and more detached accounts reflects our differing degrees of direct, personal involvement in the stories told. But all of the accounts detail the critical tensions involved in work within the sciences and the many ways in which culture and the sciences both collude and collide. They show that the potential for pluralized democratic engagement with technical problems *does* exist, as does the urgent need for new ways to think about, and take, responsibility within the sciences.

After muddling up and muddling in, we reiterate some of the processes, promises, and problems of muddling *through* in Section III, an essay on the guiding principles and methods for the scientific literacies we hope to encourage.

Reimagining and reenacting the sciences in a democracy is a demanding project. We will all have to develop a stomach for contradictions and ambiguity and find ways to ask yet another question. Because every insight we gain will be accompanied by a certain blindness, we will have to keep experimenting—with new substances and machines, of course, but also with new habits of thought, new languages, new practices, and new colluding communities.

All of which is a fairly direct introduction to the content, questions, and themes of the book. But pursuing science is never so direct (for which we should be grateful). So let us proceed as the sciences proceed in their modes of inquiry and action—by indirection as well as direction, by the meander as well as the beeline, by uncontrollable excess conjoined with careful delimitation, by trial and error . . .

PART I

"...practicing a rationality..."

I think that the central issue of philosophy and critical thought since the eighteenth century has always been, still is, and will, I hope, remain the question: what is this Reason that we use? What are its historical effects? What are its limits, and what are its dangers? How can we exist as rational beings, fortunately committed to practicing a rationality that is unfortunately criss-crossed by intrinsic dangers? One should remain as close to this question as possible, keeping in mind that it is both central and extremely difficult to resolve. In addition, if it is extremely dangerous to say that Reason is the enemy that should be eliminated, it is just as dangerous to say that any critical questioning of this rationality risks sending us into irrationality.

—Michel Foucault

The effort really to see and really to represent is no idle business in face of the *constant* force that makes for muddlement. The great thing is indeed that the muddled state too is one of the very sharpest of the realities, that it also has color and form and character. . . .

—Henry James, *What Maisie Knew*

Chapter 1

Experimenting

Invisible Lines of Force

The most primitive things in conventional accounts of the sciences are facts. Facts are supposed to be "brute": stubborn, unchangeable features of the world which serve as the building blocks of science and its theoretical representations. That the word "brute" is usually attached to facts signals their primal, even violent, nature: If you ignore the facts, you're going to get hurt.

When something is brute, it's beyond—or *beneath*—argument. Hence the persistence in the memory of many scientists, and in their diatribes against antirealists and social constructionists, of stories like the eighteenth-century English author and dictionary-maker Dr. Samuel Johnson (faithfully reported by the Johnson-fact-obsessed Boswell) kicking a stone, thus refuting idealist philosopher George Berkeley: *POW! Berkeley's talking trash, so what's the point of listening or debating?* Or the even more apocryphal story of Galileo leaving his Vatican trial, having agreed to recant his heliocentric teachings, and stomping his foot: *CLOMP! And yet it moves!* Or today, when a sociologist challenges the naive realism of scientists, the most frequent response that a scientist makes in defense: *You're a social constructionist? Try stepping out of an airplane: AAAAAIIIEEEEEEE. . . . ..WHOMP!* Almost before it begins, the argument always ends with some kind of thud. The whole thing starts to look like the comic book version of science that it is, superheroes slugging it out with villains in confrontations that require only minimal and guttural texts in balloons.

Matters of fact—or questions of what facts are and do— demand much more subtle treatments. Surely there are ways to approach facts, ask questions about them, that fall between these polemical caricatures. To begin with, you could ask what facts are supposed to do according to conventional philosophy of science. That is: what are facts, ideally? Then you could go on to ask how (and why) facts have assumed that role historically. How do people in various situations and in different historical periods go about deciding what is or is not a fact? In other words: what are facts, in fact?

The most important question, however, is whether, to paraphrase Foucault, this is really the best place to start staying close to the questions of the sciences.

In fact, it isn't. As the title of this chapter indicates, the better (but by no means perfect or essential) place to start is with *experimenting*, which, for the moment, we take as the emblematic activity of the sciences. Starting an inquiry into the sciences with a focus on experimenting immediately puts you in the muddle of things, where people build, write, question and requestion, blunder and triumph, are surprised and disappointed. Experimenting is a fantastic, puzzling, productive, messy complex of practices—including the practice of theorizing. Experimenting involves encounters with a . . . world. (Later in this chapter, we'll introduce what we think is a better term—real*itty*—but "world" will have to do for now.) Such encounters are indeed forceful, but hardly characterized by slapstick violence.

Still, "the fact" exerts a gravitational force that seems to draw all discussions of the sciences inexorably toward it. Conventional views of the sciences are all about grounding, and facts are the solid ground on which the building process starts. (That's why the foot and the ground, and all that kicking and stomping and falling, keep turning up in those stories defending realism's honor and virtue.) You have to start with the facts, and *only* with the facts. And so, as much as we would like to start this book elsewhere, we can't.

Earlier in this century, the dream of those philosophers and scientists associated with the Vienna Circle and its various brands of logical positivism, was that the sciences would be firmly grounded in these brute facts, once one delineated the strict rules by which facts could be built up into theory. Facts were to be the fundamental building blocks of the sciences, the only solid foundation on which to build knowledge and society. Despite decades of critique, this remains a widely accepted and widely deployed picture of science: Scientists make observations of an unchanging world, develop hypotheses on the basis of those observations, submit these to logical and empirical testing, to finally arrive at a mirroring, a literal re-presentation of the world.

But all that is so ideal as to be dull, and far too general and vague. No scientist actually works that way, even if they write textbooks or popular articles that say they do. It's much better to look at specific matters of fact in specific cases, to see what they do and how they do it. We now have ample evidence that the distinction between what the world is in fact, and our theoretical representations of it, is a quite muddled affair indeed. But before someone starts kicking a stone or thumping a table, the point of muddling this distinction is not to deny that something called "the real world" or "nature" matters, or that we are free to choose between relativized representations, none truer than any other. That also would be far too general and vague. The histories of the sciences, and their continual daily practice, are chock-full of unexpected encounters, strange new results, and daily confrontation with a material world that continually surprises and forces inquiry to begin again.

For the time being, we'll accept the conventional conception of facts as the solid basis that assists in the scientific project of representing a world. It's not that facts as building blocks aren't capable of providing solidity; they can provide a solid basis for scientific work, but always as part of an architectural strategy, and after a lot of skilled and sometimes not-so-skilled work. The reconstruction of our representations of the world is a neverending, unavoidable process. Looking at the interplay between solid facts and shifting strategies of representation allows us to ask a different set of questions: What muddled world-representation structures are nevertheless stable—at least temporarily, and for specific purposes? What makes them stable, or unstable? What will further experimenting accomplish?

Our views are close to those developed as part of the pragmatist tradition in philosophy, often distilled down to the "Big Three" figures of Charles Sanders Peirce, William James, and John Dewey. That tradition worked in a middle space between idealism and realism, trying to avoid the simple collapses to which philosophy was so prone, everything getting reduced to either a subject or an object, a foot or a stone. Words Dewey wrote in 1916, to introduce his provocatively titled *Essays in Experimental Logic*, could just as well describe our position:

> The position taken in these essays is frankly realistic in acknowledging that certain brute existences, detected or laid bare by thinking but in no way constituted out of thought or any mental process, set every problem for reflection and hence serve to test its otherwise merely speculative results. It is simply insisted that as a matter of fact these brute existences are equivalent neither to the objective content of the situations, technological or artistic or social, in which thinking originates, nor to the things to be known—of the objects of knowledge.[1]

Writing in an era of rapid industrialization, Dewey and other pragmatists often employed metaphors of production to make the conceptual distinctions they thought important. Dewey went on in this essay to compare brute existences to mineral rock or raw ore "in its undisturbed place in nature," a "brute datum" to "the metal undergoing extraction from raw ore for the sake of being wrought into a useful thing," and the object(s) of knowledge to the final "manufactured object." It is a scheme that depends not on two opposing terms (fact/theory, object/subject) but on *three* terms, with "brute datum" in the middle, between nature and human commerce. Existence, purpose, and knowledge would be another way to express this indissoluble triad. The most important characteristic of Dewey's kind of fact is that it was made (but not made up) to continue the productive process.

Simplistic charges of subjectivism were to Dewey a "depressing revelation" of the traps people were prone to falling into when talking about knowledge—idiotic traps involving stepping out of airplanes and such. "To stumble on a stone need not be a process of knowledge; to hit it with a hammer, to pour acid on it, to put pieces in the crucible, to subject things to heat and pressure to see if a similar stone can be made, *are* processes of knowledge."[2] The sciences, as a form of inquiry, depend on the third, middle term highlighted by Dewey. They also represent a kind of third term themselves: the sciences are neither subjective musings circling inside one's head, nor random collisions with a brutal world, but something in the middle— something that always involved *signs*:

> In every case, it is a matter of fixing some given physical existence as a sign of some other existences not given in the same way as is that which serves as a sign. These words of Mill might well be made the motto of every logic: "To draw inferences has been said to be the great business of life. Everyone has daily, hourly, and momentary need of ascertaining facts which he has not directly observed. . . . It is the only occupation in which the mind never ceases to be engaged." Such being the case, the indispensable condition of doing the business well is the careful determination of the sign-force of specific things in experience. And this condition can never be fulfilled as long as the thing is presented to us, so to say, in bulk.[3]

What follows in this chapter are some stories in which facts figure centrally, but not in bulk. Facts are not something from the bulk bins of raw nature, but the carefully packaged items neatly displayed on the supermarket shelves of the sciences. The following are stories *about* facts, less in the sense of giving direct definitions to this term, and more in the sense of trying to show what goes on about and around facts. They should illustrate how, if you consider facts not as inert things but as a "sign-force," you find yourself not just bumping into them, but being pushed or pulled in certain directions. Like one magnet approaching another, approaching the topic of fact can involve overwhelming attraction or repulsion, sudden reversals of polarity which turn things upside down or inside out, and the seemingly magical ordering of random filings from a far-flung territory into patterns showing the invisible lines of force.

Getting Centered

The Copernican Revolution is the paradigmatic example of a scientific revolution, when the old ways of seeing, knowing, and doing things with the natural world suddenly—or not so suddenly, as the case may be—change into new perceptions, conceptions, and activities. The Copernican Revolution, the story goes, established one of the most basic, entrenched, and universally accepted facts that has become practically a test of sanity: Everyone knows that the earth goes around the sun; you'd have to be crazy, or maybe just a skepticism-infected postmodern social constructionist, to think otherwise.

Since no event more marks Europe's emergence from the Dark Ages, when we are in effect told that religion enforced insanity, it's worthwhile looking closely at who Copernicus was, what he did and didn't do, and how exactly he did it.

Let's quickly get some standard misconceptions out of the way—misconceptions that nevertheless carry a lot of weight within scientific communities and the general public today. After a lot of work by a lot of historians and philosophers, dedicated to describing how and why this scientific revolution turned out the way it did, there is little doubt that:

1. Copernicus's heliocentric system was not *simpler* than the earth-centered, geocentric system inherited from Ptolemy; it was full of baroque mechanisms, some of which were more contrived and complicated than Ptolemy's. (Technically, it wasn't even heliocentric—Copernicus had to put the center of the universe at an abstract point *near* the center of the sun.)

2. For a long time, the Copernican system was no better at *predicting* astronomical events than the Ptolemaic.

3. Copernicus didn't build his system up *inductively* from the "facts" of observation; as Johannes Kepler would point out a bit later, both Copernican and Ptolemaic systems at times contradicted observations (which were not altogether reliable to begin with).

4. The Ptolemaic system was not *falsified*, i.e., demonstrated to be wrong, while the Copernican theory continually held up under trial.[4] Here things become a bit more complicated, but it seems clear that even if such disproof is considered essential, that kind of falsification didn't really occur until the early nineteenth century (with the introduction of stellar parallax observations in 1838). In which case you have a very long, and very muddy, revolution—not to mention a lot of explaining to do as to why all of the great scientists until then were totally, "rationally" committed to heliocentrism.[5]

These criteria of simplicity, increased capacity for prediction, reliance on induction rather than metaphysical hypotheses, and withstanding tests to prove it false, were for much of the twentieth century thought to be the hallmarks of rational progress in scientific theories. Little wonder, then, that when historians and philosophers showed that Copernicus and his heliocentric system didn't exhibit these characteristics, Thomas Kuhn would write in 1959 that "to astronomers the initial choice between Copernicus's system and Ptolemy's could only be a matter of taste, and matters of taste are the most difficult of all to define or debate."

But you needn't be a zealous science purist to feel that this phrase "matter of taste" could use a little more precision or specificity. Since Kuhn, historians and philosophers of science have become more adept at defining and debating what "matters of taste" are and how they operate. And by looking at some of their work, we can better understand how those matters of taste meld with matters of fact.

It's not that Copernican scholars have stopped looking at what historians of science sometimes call the "internal" complexities and achievements of sixteenth-century astronomy. (Particularly in the 1960s and 1970s, historians of science thought it was their job to distinguish between "internal" factors [mathematics, logic, etc.—the "real stuff"] that pushed the sciences to progress, and the "external" factors [cultural beliefs, social institutions, etc.] which could only hold science back or distort it. Now many of them know that the most interesting things happen in between these categories.) The conventionally scientific part of Copernicanism is still a salutary and inexhaustible topic, and you can find library shelves packed with volumes of articles crammed with detailed accounts of Copernicus's observational data, intricate geometric demonstrations, and other technical issues considered internal to the sciences, and exclusively constitutive of them. This aspect of the Copernican achievement remains compelling to historians and astronomers alike.

But not quite compelling enough. *Why* and *how* the Copernican Revolution happened, and why it happened so slowly as to make "revolution" a misnomer, remain nagging questions for many. Simple stories have an easy appeal—and the story, "the revolution happened because Copernicus got the facts right," is as simple as they come. But to really understand such a profound scientific and cultural change requires more than attention to the force of logic and factual evidence alone.

To begin complicating this story, we could ask whether Copernicus *really* believed that the sun was at (or near) the center of the universe, or just thought that this was a convenient and satisfying model. Moreover, whatever Copernicus really believed in the privacy of his own brain, how did his book instruct readers, then and now, to think about his system?

Copernicus's *De revolutionibus* was prefaced by a letter from his editor, Andreas Osiander, a letter that has generated four hundred years of controversy and much ill will toward editors in general. Osiander has been called everything from stupid to unethical, from an obstructionist and obsequious theologian and an enemy of science to a daring religious and scientific heretic. What has been at issue is the view which he put at the front of Copernicus's book, that "it is quite clear that the cause of the apparent unequal motions are completely and simply unknown to this art [astronomy]. And if any causes are devised by the imagination, as indeed very many are, they are not put forward to convince anyone that they are true, but merely to provide a correct basis for calculations." The insertion was anonymous, and thus ambiguous to readers.

Did Copernicus himself subscribe to such a view, that his system was just a better model for "saving the appearances" and possibly improving the calendar, but did not describe how the universe really was constructed? Almost certainly not. But such a view—that philosophy (and natural philosophy) could provide plausible, probable explanatory models but never the ultimate truth of God's causal mechanisms—was not unusual in either astronomy or philosophy more generally in this period. (This conceptual image of a nature which exceeds our capacity to represent it perfectly is one which will come up a number of times in the following pages, and we will be taking it quite seriously later.) Most scholars agree that in addition to misrepresenting Copernicus's actual convictions, Osiander's anonymous introductory letter served a strategic social function: it gave some protection to Copernicus's work in a time of profound religious and political upheaval. If this apparently anonymous introduction had not been included, all that was genuinely new and useful in Copernicus's book might have been almost entirely overlooked or immediately dismissed as simple heresy. The dissembling introduction at least forestalled and ameliorated that fate.

And in fact, for decades after its publication, most astronomers and other scholars used De revolutionibus for primarily the reason Osiander suggested: to make better calculations. The astronomer Praetorius used it to improve the Ptolemaic, earth-centered system, and did so quite effectively. The Danish nobleman-astronomer Tycho Brahe employed it similarly, devising a hybrid system which made excellent predictions, in which the earth remained motionless at the center of the universe where it had always been, with the sun going around it, while all the planets wheeled around the sun.[6]

While not exactly in the Middle Ages, Copernicus is nevertheless best thought of as "in the middle"—the middle between the Middle Ages and modernity. His attachment to the circle as the most virtuous of the geometric forms, a heavy hangover from the ancient Greeks, made him cling to baroque mechanisms like the orbit-on-orbit epicycle, the deferent, and the eccentric, and even more baroque combinations of these. Such devices made his system seem physically absurd to contemporaries not only in the Vatican, but in the scientific community as well. He held on tightly to the idea of crystalline heavenly spheres. He worked with old beliefs, sketchy data, and "irrational" commitments to both old and new disciplines of knowledge—and he did great work.

We will come back to Copernicus and his middled position in a later chapter. For now, we only wanted to show the dependence of a fact on a larger framework. To say: "It's a matter of fact that the earth goes around the sun," actually excludes the crucial parenthetical remark: "(Within the Copernican system of calculation, observation, and theorization) it's a matter of fact that the earth goes around the sun." This fact is no longer a fact within the modern theory of General Relativity, which allows us to specify *any* frame of reference for centering our factual measurements: earth, sun, center of the galaxy, or any arbitrary point in space. Within the current Einsteinian system of General Relativity, it's a matter of fact that the movements of earth and sun correspond to the curvatures of a four-dimensional construct called spacetime.

Nevertheless, there are stable relationships between the movements and observed positions of the planets, the sun, and surrounding stars—a "brute existence" or, in another of Dewey's terms, "original *res* of experience." But those relationships aren't at all meaningful or purposeful, and thus for Dewey—and for us—do not make a fact. They still must be refined into a "sign-force," which is what the Ptolemaic, Copernican, Tychoan, or Einsteinian system does. Within any of those systems, what was a jumble of observations and relationships gets packaged into a "useful thing." With it, you can build a Global Positioning System of satellites as an aid to navigation—a thoroughly Ptolemaic technology. You can predict the appearances and even the impacts of comets—the latter on other planets like Jupiter, we hope. Useful, reliable, even "mechanical" relationships and interactions? Definitely. Universal truth? An unnecessary and immodest, albeit reassuring, metaphysical claim.

Thus, ironically, the people and computers in the NASA control rooms worked within an earth-centered Tychoan system to launch inhabitants of the twentieth century to the moon, the first of the heavenly spheres. You could see this briefly in the hit movie *Apollo 13*, where the historically obsolete but pragmatically efficient view of the earth-as-calculating-center-of-the-cosmos flashed briefly across the screen. But such muddling within an outmoded theoretical framework was largely behind the scenes, while the main narrative centered on a different brand of muddling through. The earth-bound engineers and space-cast astronauts together did some quick-and-dirty calculations, and hastily rigged together components intended for other jobs into an air purifier that would get them home again. Such "kludge jobs" (see Chapter 2), whether cosmological or air-cleaning systems, can be remarkably effective, flexible, and heroic, even in the face of disaster.

Merging Beliefs and Facts to Make Experimental Science

You might think that, given a corpus of alchemical investigations and writings that was far more extensive than his famous work in natural philosophy and mathematics, we would present Isaac Newton as a study in muddling contradictions.

Instead, we take a brief look at the lesser-known but nevertheless important figure of Robert Boyle. Boyle's work in natural philosophy was of the kind we would now anachronistically refer to as "chemistry," the less glamorous status of which might go some way toward explaining why fewer people would recognize his name than would recognize Newton's. (Boyle's Law states that the volume of a gas increases proportionately as its pressure decreases at constant temperature. Most often written as part of the "gas law," $pV = nRT$, it describes the reliable relationship between the pressure (p), volume (V), and temperature (T) of a specific amount (n moles) of gas.) But Boyle, as one of the key figures in the founding of experimental science and of one of the earliest scientific institutions, England's Royal Society, deserves at least as much attention as Newton.

With *Leviathan and the Air-Pump: Hobbes, Boyle, and the Experimental Life*, the historians of science Steven Shapin and Simon Schaffer have produced one of the most empirically and thematically rich and suggestive works on the origins of modern experimental science in seventeenth-century England. We can't consider all of their themes here in the detail they deserve—how science was both entertaining spectacle as well as an engine for facts; how experiments had to be publicly witnessed, and how new literary devices and genres for such witnessing were instituted (and still operate powerfully today); how the laboratory became a separate and privileged social space; how the problem of replication haunted all enactments of experiment; and many others. What we focus on here is their explication of how facts became centrally important to science, and how objectivity and truth came to be defined in terms of these facts.

Until the middle of the seventeenth century, "knowledge" and "science" were kept markedly distinct from matters of "opinion." The former adhered to the absolute certainty of "demonstrative sciences" like logic and geometry. Natural philosophers—whom it wouldn't be terribly wrong to think of as physical scientists—emulated these demonstrative sciences, and so produced "the kind of certainty that compelled absolute assent." But over the second half of the century, experimentalists like Robert Boyle and others associated with the newly established Royal Society came to redefine our expectations of science in terms of what was *probably* (that middling, gray area between knowledge and opinion) the case about nature. "Physical hypotheses were provisional and revisable; assent to them was not obligatory, as it was to mathematical demonstrations; and physical science was, to varying degrees, removed from the realm of the demonstrative. The probabilistic conception of physical knowledge was not regarded by its proponents as a regrettable retreat from more ambitious goals; it was celebrated as a wise rejection of a failed project. By the adoption of a probabilistic view of knowledge one could attain to an *appropriate* certainty and aim to secure *legitimate* assent to knowledge claims."[7]

It was the "matter of fact" that would undergird this new project of probable hypotheses, because the fact was what could provide the greatest degree of assurance, or moral certainty. Like Copernicus—or at least like Copernicus's preface-writer and, as we'll see in a later chapter, like Galileo—Boyle and the English experimentalists believed that God could produce similar effects through a variety of mechanisms.

How to get to a "matter of fact," then? According to Boyle, by aggregating individual *beliefs*. Here is an interesting muddle at the origins of experimental science: Solid, foundational matters of fact had themselves to be founded on the somewhat shakier matter of belief. To be more precise: The beliefs to be aggregated into facts would be those of individuals who could be trusted, whose eyes had been properly trained, and who would report faithfully on what they witnessed—in a word, gentlemen. Experimental science would have to be a noble pursuit.

(We're simplifying a bit here, and also excluding much interesting detail. One effect of the "gentlemen" argument at the time was to exclude those people, including women, associated with alchemy and other traditions of knowledge, keeping them out of the historical process of defining what the sciences would become. And things would get even more complicated later, as experimental science came to be celebrated as the best route out of the lower or middle classes, and into the more refined strata of society.)

Shapin and Schaffer point to what they call three "technologies" that allowed Boyle and others to produce matters of fact: a material technology of experimentation, in this case the air-pump; a literary technology that made "virtual witnessing" possible for a wider community that couldn't be squeezed into the Royal Society's demonstration chambers for every meeting; and a social technology that established certain conventions for how experimentalists should deal with each other and with each other's knowledge-claims. Each of these three technologies, or knowledge-producing tools, "embedded the others," and none could be said to be more fundamental than the others. The material technology of the air-pump required specific social organizations, such as the distinction between common mechanics, technicians, and demonstrators, and the gentlemen scientists, as well as the creation of a privileged laboratory space set apart from the rest of the world (social organizations which remain operative today). The literary technology of the new scientific report extolled the value of those social arrangements, embodied in its rhetoric the new social values of modesty and probabilism, and opened the new experimental findings to wider questioning or affirmation. Indeed, running the machine of the literary technology—i.e., reading—was in some sense equivalent to actually running the machinery of the experiment.

The air-pump, as historian of science A. Rupert Hall has put it, was "the cyclotron of the age." Relatively few of them existed, as they required the patronage of a state, royalty, or elite scientific society to be built. They were cranky, prone to malfunction and requiring constant tinkering and maintenance. This work was not always pretty and, as with all the sciences today, depended on a lot of craft knowledge and unseen, uncredited technicians. Although Boyle wanted a larger air-pump than the thirty-quart one he used, he reported that his nameless "glass-men" were at the limits of their abilities. As sealants and lubricants, Boyle had recipes (which he did not always write down) that included "sallad [sic] oil," "melted pitch, rosin, and wood-ashes," and for fixing small cracks, linen spread with a mixture of quicklime, cheese scrapings, and water ground to a paste "to have a strong and stinking smell."

The central problem against which all this effort was directed was the controversial Torricellian phenomenon. First performed in 1644, the phenomenon occurs when you take a long glass tube, closed at one end, and place it in a tub of mercury. The tube fills completely with mercury, and is then inverted so the open end remains in the tub and the closed end is at the top. You then see that the mercury no longer fills the tube, but leaves the top empty—or looking empty, anyway. We call it a barometer today, and regard it as a good measure of air pressure. For the seventeenth century, it might in retrospect be called a scandalometer, because it was a good measure of the scandalous variety of doctrines and opinions about what actually was, or what possibly could be, in that seemingly empty space at the top of the tube. Torricelli, Pascal, Roberval, and Descartes were just a few of the natural philosophers who weighed in on this controversy. The empty space at the top of the tube couldn't be isolated from other social and philosophical controversies; what was or wasn't at the top of the tube was a question inextricably bound up with disturbing differences between conservative Scholastics and radical empiricists, and even more disturbing associations among the concepts of "subtle matter" and spirit, and the practices of witchcraft. It was a seemingly empty space that sparked a seemingly endless series of weighty disputes.

Here's where Boyle thought his material-literary-social technology of experimental science could show its greatest value: it could stop, or at least sidestep, all these endless vituperative arguments that were so easily joined with religious, national, and political differences, and restrict the discussion (and therefore what would count as real knowledge) to matters of fact. In terms that we'll elaborate more on in Chapter 2: where others continually *articulated* the phenomenon at the top of the mercury column with other philosophical and religious beliefs or articulations, Boyle sought to disarticulate or disentangle questions and statements about nature from religious or political ones, and rearticulate those statements within the new material-literary-social technology of experimental science. Thus, in his writing Boyle decided to "speak so doubtingly, and use so often, *perhaps, it seems, it is not improbable*, and such other expressions, as argue a diffidence of the truth of the opinions I incline to, and that I should be so shy of laying down principles, and sometimes of so much as venturing at explications."[8] That decision led to an extremely powerful and persistent "literary technology": the often strained rhetoric of this contrived modesty (as well as what Boyle called his "prolixity") still characterizes scientific papers today.

This modesty was not a disadvantage or deferral, but an asset. Is there or isn't there a vacuum—whose side are you on?: Boyle would adroitly sidestep "so nice a question" and would not "dare to take upon me to determine so difficult a controversy." Doing so would "make the controversy about a vacuum rather a metaphysical, than a physiological question; which therefore we shall no longer debate here. . . ." As Shapin and Schaffer comment:

> The significance of this move must be stressed. Boyle was not "a vacuist" nor did he undertake his *New Experiments* to prove a vacuum. Neither was he "a plenist," and he mobilized powerful arguments against the mechanical and nonmechanical principles adduced by those who maintained that a vacuum was impossible. What he was endeavoring to create was a natural philosophical discourse in which such questions were inadmissible. The air-pump could not decide whether or not a "metaphysical" vacuum existed. This was not a failing of the pump; instead, it was one of its strengths. Experimental practices were to rule out of court those problems that bred dispute and divisiveness among philosophers, and they were to substitute those questions that could generate matters of fact upon which philosophers might agree.[9]

Unlike the alchemists, the experimentalists practiced their trade in public, in the assembly room of the Royal Society (although critics like Thomas Hobbes countered that the exclusive Royal Society hardly counted as "public"). There the experiments were witnessed and, in language drawn from the legal world, the results were *judged* by many observers of the trial. (We'll return to the inescapable inexactness of judging in the sciences in Chapter 4.) But Boyle, perhaps to maximize the number of people who could be recruited to the new experimentalist ways, showed a more inclusive attitude toward alchemy and the alchemists than did some of his contemporaries, such as Isaac Newton. Newton believed that alchemy and the new science should be kept as distinct practices and forms of knowledge, and so engaged in both while keeping his extensive alchemical pursuits secret. Newton criticized Boyle for simply publishing work that dealt (albeit critically) with the alchemical tradition. Working his new separation between matters of fact and the language of theory, Boyle's set of social conventions incorporated some (although not enormous) tolerance for social difference: "Let his opinions be never so false, his experiments being true, I am not obliged to believe the former, and am left at liberty to benefit myself by the latter."[10] Taken seriously, that's not only a laudable social convention for the sciences, but a productive one as well. It's an excellent legacy from this period which needs to be better preserved.

Boyle's sociotechnical innovation is easy to "naturalize": experimental practices were simply better at producing a better kind of knowledge, as subsequent history has shown. But as we'll see, just what "better" meant at that time, in that social context, was hotly debated. That's where the work of the new "social technology" was important, since it redefined conventions of proper discourse and what would count as knowledge. Boyle argued for and established new *social conventions* concerning how knowledge was or was not to be produced: disputes were to be about findings and not about the character of the investigator; ad hominem attacks were out. Indeed, the status of the individual investigator or philosopher, so prone to dogmatism or "enthusiasm," was supposed to shrink in comparison to the communal endeavor. And the community was obliged not to believe in each other's opinion, but only to assent to what they believed they saw in nature, apparent through the matters of fact produced by experiment.

Again, these are in many ways admirable ideals: public witnessing, the necessity of judging, and an agreement to at least *defer* social, political, and philosophical differences—to leave them aside and read only the natural world as it was presented through experiments. They can, with some cleaning up and rearticulation, be salvaged as goals which the sciences can and should pursue today.

But how was such a consensus, based only on witnessing, possible? Nature itself might be the basis for brute matters of fact, but did that mean that it could serve as the basis of knowledge? What makes some beliefs more believable than others? And just what does "judging" involve?

Richard Feynman's Half-Assed Atoms

After all that ancient history, it might be nice to shift registers to something and someone a little more recent, a little more familiar, and perhaps a little stranger at the same time.

Richard Feynman is surely one of the twentieth century's most famous physicists, and rightly so. From his contributions to the Manhattan Project (employing the ancient abacus for calculations of nuclear physics) to his work on the commission investigating the space shuttle *Challenger* disaster (running a tabletop experiment with ice water and O-ring in a congressional hearing room), he has worked unconventionally yet productively on science in the most public spheres. His invention of "Feynman diagrams" was enormously important to the theory and practice of particle physics. He wrote thoughtful and deservedly influential physics textbooks, and was also adept in the more popular vein of the *Surely You're Joking, Mr. Feynman!*, isn't-physics-amazing-and-beautiful-even-if-it-is-a-bit-strange genre. There's no question that in this culture he rates as a *Genius*, the word that science writer James Gleick picked for the title of his Feynman biography.

But when Feynman was not so concerned with tying up his own thinking and the wonders of physics into tidy publishable packages, did he see the world any differently? Did he think about his own thinking in physics as something that was even stranger than his books represented—stranger because less accessible, more muddled?

Interviewed by the physicist and historian of science Silvan S. Schweber, Feynman tried to articulate how he worked and thought through visualization. Reading this, we get some insight into how Feynman might have actually experienced the raw, brute world of things:

> I cannot explain what goes on in my mind clearly because I am actively confusing it and I cannot introspect and know what's happening. But visualization in some form or other is a vital part of my thinking and it isn't necessary I make a diagram like that. The diagram is really, in a certain sense, the picture that comes from trying to clarify visualization, which is a half-assed kind of vague [sic], mixed with symbols. It's very difficult to explain, because it is not clear. My atom, for example, when I think of an electron spin in an atom, I see an atom and I see a vector and a Ψ [psi] written somewhere, sort of, or mixed with it somehow, and an amplitude all mixed up with xs. It is impossible to differentiate the symbols from the thing; but it is very visual. It is hard to believe it, but I see these things not as mathematical expressions but a mixture of mathematical expression wrapped into and around, in a vague way, around the object. . . . What I really am trying to do is to bring birth and clarity, which is really a half-assedly thought out pictorial semi-vision thing. OK?[11]

Now you might say that we shouldn't take such statements very seriously: they're uttered in an interview, where the speaker doesn't have time to think carefully, or come up with the right words, let alone better syntax or grammar. But it's actually not far in spirit from what Feynman might have written in one of his books. Feynman's brilliance in physics is unquestioned. But you could also say that when asked where that brilliance stems from, or what goes into it, Feynman finds exactly the right words—a language of persistent muddling. He *really can't* "explain . . . clearly," he *really is* "actively confusing" his thought process, he *really does* rely on a "half-assed kind of vague" thing clumsily called visualization, before he sets down a diagram or equation on paper.

When Feynman is doing the work that made him famous, when he's doing physics, he is working with a world that is a mixture, a concatenation of mathematical symbols written over, or in, or *somewhere* around—*somehow*—an object like an atom. It is impossible to differentiate these things, extract the symbols from the things-in-themselves. Feynman's brain isn't capable of it, and neither are ours or anyone else's. It's hard to believe, because it so goes against our familiar way of thinking. You couldn't have asked Feynman, "Is electron spin really a fact?," because as soon as the words "electron spin" were out of your mouth, he'd already have had some "half-assed" but nevertheless quite powerful vision of a psi-function and all the attendant mathematics, together with an arrowlike angular momentum vector. Electron spin, or the fact that electrons have spin, doesn't exist (meaning, in the root sense of "exist," it doesn't *stand out*) without the supporting mathematics wrapped, vaguely, around it. Don't think of electron spin as a brute existence; think of it as a "sign-force." There's not first an atom, and then some idea of spin, and then the formulas for that spin-state: it's all at once, and it's not just in Feynman's head, it's out there, loose in the world. And it exists totally before, within, and constitutive of saying that electron spin is "real." He is—we are—it is—off and running, pushing and pulling, exerting peculiar sign-forces. The fact is always after the fact.

Sort of.

OK?

Charged Electron, Committed Physicist

Electron spin is one of those weird quantum properties, you might argue, but what about the charge of the electron? Surely there's nothing half-assed about that: reliably charged electrons ran through the computers that this manuscript was created on, powered the printing presses and other machines that produced this book, generated the light by which you're now reading it—and even helped to mediate the neural processes by which you gave some meaning to the preceding paragraph.

Another historical episode shows how a focus on experimenting can complicate even something as reliable as electron charge. It also starts to suggest how an emphasis on muddling actually forces us to new kinds of precision—both practically and conceptually, and in the pursuits of science and history. Eventually, we'll compare two well-scrutinized episodes from two different periods in the history of science, and two different disciplines. One story is about the Nobel prize–winning American physicist Robert Millikan's experimental establishment of the fundamental charge of the electron early in this century, and how that became a fact via what could be termed, ungenerously, as fudging the data. That's the story told here. The second story is a more recent controversy, and centers on the molecular biologist (and Nobel laureate) David Baltimore's involvement in a series of immunological experiments, which is elaborated in Chapter 4. Both cases involve the proper care and handling of "the facts;" both show how ambiguity presents opportunities as well as dangers, elicits and demands creativity even as it allows for self-deception or fraud. In the end, precision and ambiguity end up in an uneasy co-existence, requiring a series of judgments.

Millikan's oil-drop experiment is famous within both science and the history of science, a fame secured in part by the analysis provided by the physicist and historian of physics Gerald Holton.[12] High-school students now try to replicate Millikan's efforts in their physics classes. What they learn is that even with much better equipment than Millikan had, and even knowing how the experiment is supposed to turn out (what the data are supposed to look like), it remains a maddeningly quirky, delicate, and difficult task.

The question facing Millikan and others in the opening decades of this century was whether the newest fundamental particle, the electron, had a precise unit of electric charge associated with it, or if the charge varied continuously from one electron to another. Other physicists had already established that the charge-to-mass ratio (e/m) of what they called "cathode rays" was constant, and that these cathode rays might be particles better called electrons. If the electrons all carried the same charge, they would also all have the same mass, and would qualify as "atoms" of electricity, in the original meaning of that word as an indivisible unit.

Working experimentally, Millikan set about his inquiries by constructing an apparatus consisting of two slightly separated, electrically charged horizontal plates enclosed in a box. He sprayed small droplets of oil between these plates, a process which gave most of the oil droplets some electric charge. By rapidly and carefully adjusting the voltage on the plates, while looking at the droplets through a small telescope, Millikan tried to get one droplet after another to hang perfectly suspended between the plates. A series of by no means trivial or easy observations, assumptions, and calculations allowed Millikan to judge the weight of each suspended droplet, and to then equate the force of gravity trying to pull it down with the electric force from the charged plates which kept it hanging, suspended in the telescope's sights. Knowing the voltage on the plates, knowing the weight of the drops, and knowing the force of gravity on them, Millikan could calculate the only remaining unknown in the equations: the charge on the droplet.

We write that Millikan was working experimentally, but that's not quite true. Yes, he had ingeniously dreamed up and built this apparatus, and disciplined his body to sit in front of it with one eye fixed on the telescope, hand twiddling the voltage dial and rapidly writing down observations. But if he had been working only experimentally, he would have ended up with what almost all the high school students today end up with: a data set that *suggests* that the charge on the oil droplets comes in precise units of one, two, three, or four electron charges, but doesn't quite *convince*, because the data set is muddled with observations and calculations that might also indicate a continuous range of charge values in between the neat set of integers, and especially some that seem to indicate a charge of one-half or one-third of the most common value.

Millikan was successful because he was working *theoretically* as well as experimentally. Millikan's notebooks and writings at the time reveal that he was *committed in advance* to the theory of unitary electron charge, and so knew which experimental outcomes to throw away. If he hadn't been so committed, he would have been awash in a sea of largely undifferentiated data (as was a competitor, the Austrian physicist Felix Ehrenhaft). Millikan was an excellent muddler. In the words of Peter Galison:

> Millikan supported atomic theory and in general a granular representation of nature. . . . By the strength of his convictions, he set aside some measurements not in accord with his atomistic hypothesis about electric charge. . . . [He] was faced with a choice where dogmatic pronouncements on "experimental method" would not help: to ignore all expectations leads to chaos, yet to adamantly follow prior beliefs can blind the experimenter to novelty.[13]

Millikan violated the prime directive that scientists are supposed to observe, harboring an irrational (in the sense of being before the fact) commitment to a certain theory. And his violation of that conventional expectation became the path to the fact. A kind of feeling for the apparatus was also part of doing good science for Millikan. He had an excellent sense of when the equipment he had built was working well and when it wasn't, when the complex and unquantifiable environmental conditions in the room were conducive to a good experimental run and when they weren't, when he had made good observations and when he hadn't. Being a good, precise experimentalist requires a much more imprecise ability, that of judging. (But when we turn to questions of judging and the case of David Baltimore in Chapter 4, we'll see that theoretical commitments in the face of experimental ambiguity, and a "feeling for the apparatus," don't always work out so well.)

Still, you might object, even if all that is true and Millikan didn't fit the conventional model of the orthodox scientist, he was able to do what he did because, in the end, nature *really* is that way, and the electron *really* does have a unitary charge of 1.602×10^{-19} coulombs. True—or shall we say, true enough. Thousands of physicists, electrical engineers, chemists, and other scientists make good use of that very stable fact every day. But not one of them could tell you what "charge" *really* is in the first place, what the name refers to—at least not without referring to an enormous series of other concepts, measuring devices, experimental effects, and disciplined procedures. As the eminent experimental British physicist who pioneered the study of electricity, Michael Faraday, wrote in 1839 about terms such as "ion," "electrode," "cathode" and others that he invented: "These terms being once well defined will, I hope, in their use enable me to avoid much periphrasis and ambiguity of expression. I do not mean to press them into service more frequently than required, for I am fully aware that names are one thing and science another."[14] Names and sciences don't comprise such easily separated spheres, but it is still good to be reminded that "the charge of the electron" is a shorthand term that helps us avoid the long, ambling, periphrastic description of a finely crafted network of technologies, techniques, equations, social infrastructure, and other names that together allow something called "the charge of the electron" to be produced—and produced reliably. If you really want to know why the charge of the electron is really 1.602×10^{-19} coulombs, you'd want to know the incredibly periphrastic history of electromagnetic theory and experiments, coupled with the technological and social history of electrical technologies.

Faraday, and later his compatriot James Clerk Maxwell, did brilliant theoretical and experimental work in electricity and electromagnetic theory. We tend to forget, and our science textbooks rarely reflect the fact, that they did this work in quite a different scientific culture from our own. One quality of that culture is hinted at in Faraday's statement above: these British physicists (the German physicists like Hertz and Helmholtz are quite another story) were always aware of that persistent gap between names and sciences, between our representations of a world and the world itself. As our earlier stories have suggested, this kind of modesty is an enduring philosophical strand in the sciences, and one well worth preserving. It's a modesty that redirects our attention to the middle space, to the terms and technologies that we put into that gap, stopgaps "pressed into service" to better manage both our ideas and our world.

But by the end of this chapter, we might say something else—something that tries to get around that most useful, natural, and totally outdated and inadequate stopgap phrase: "the world itself."

Ludwik Fleck and the Signal that Resists

Ludwik Fleck conducted noteworthy and sometimes remarkable work in the fields of bacteriology, serology, medical practice, and sociology and philosophy of science. Fleck was a man at a number of crossroads, subjected to a variety of forced migrations, and dedicated to scientific work in what were frequently less than ideal circumstances.

Fleck was born, educated, and began his biomedical career in Lvov, Poland. Beginning in 1928, he headed the bacteriological laboratory of the Social Sick Fund in Lvov, until anti-Jewish measures forced his dismissal in 1935. He continued his medical and research practice in private, where in addition to working on streptococci he devised a method to improve the Wassermann test for syphilis. When the Soviets took over Lvov in 1939, he became director of the City Microbiological Laboratory, head of the microbiology department of the State Bacteriological Institute, and held a number of other positions as well. Then when the Germans occupied Lvov in 1941, Fleck became director of the bacteriological laboratory of the Jewish Hospital. Prompted by the severe typhus epidemic that gripped the Jewish ghetto, Fleck muddled through with "primitive means" to use the urine of typhus patients in rapid diagnostic and vaccination procedures. His methods were so effective, in fact, that when the Germans arrested him in 1942, he was forced to put his skills to work producing the vaccine for German troops.

Defiant truth-telling was not the most effective survival strategy at the time. Asked by the Germans if his vaccine would be suitable for Aryans, Fleck responded: "Of course, provided that the vaccine is prepared from the urine of Aryans and not of Jews." Fleck and his staff were taken first to Auschwitz, where they were attached to the camp's hospital and forced to produce vaccine; Fleck was later taken to Buchenwald, where he continued this work. The Nazis killed Fleck's two sisters and their families in Poland, before he was freed when the U.S. army entered Buchenwald in April 1945. He resumed his scientific work in Lublin, holding a number of prestigious positions and receiving numerous awards, including election to the Polish Academy of Science in 1954. Fleck died in 1961, in Israel.

In addition to this extraordinary life and career, another of Fleck's enduring legacies is his book, *The Genesis and Development of a Scientific Fact*. Published in Switzerland in 1935 (as a Jew, he couldn't publish in Germany), the extent to which this book was read and discussed in Europe is only now beginning to become clear. Its greatest acknowledged effect would seem to come much later: at the very least, Fleck's book reassured Thomas Kuhn that many of his own ideas about paradigms and revolutions in the sciences were not crazy.

The book's seemingly paradoxical title struck Kuhn when, as he reports, Harvard president and (briefly) U.S. High Commissioner to Germany James Conant related "with glee" the remarks of a German associate: "How can such a book be? A fact is a fact. It has neither genesis nor development."[15] But the book in fact *is*, and it details the historical shifts in what was believed to really cause syphilis, and how that condition came to be diagnosed and defined.

Besides his work on syphilis (which is best read in his own words), another, briefer episode related in Fleck's book also conveys his views on scientific research, experimentation, and the status of "facts."

Fleck relates a story about how he and an assistant grew a streptococcus from the urine of a female patient. "Its unusually rapid and profuse growth attracted our attention, as did pigment formation, which is very rare with streptococci."

Fleck had never seen such colors in bacteria and "remembered only vaguely having read about them." Wanting to find out more about these germs, he planned a series of experiments—and then didn't do them. Instead—as he recalled and recounted after the fact—"the project turned largely into a study of variability." Such a thing undoubtedly happens all the time in science: An investigator starts out interested in one set of questions and phenomena, and ends up doing something completely different. But what doesn't usually happen is that an investigator asks the crucial reflexive question that Fleck then asked: How could this have happened?

Because he inquired about his own process of inquiry, wondering how it came to be that he did something he didn't plan to do, Fleck becomes one of our archetypal muddling, reflexive, admirable scientists. He had been studying the literature on variability in bacteria, and how they were classified according to various species, and because he knew that Streptococcus often reminded laboratory scientists of Staphylococcus, he also remembered something about staph colonies having different colors. So he suggested to his colleague that she determine if their strep colonies could be split into lighter and denser colonies. They could: There were a large number of "ordinary yellowish, transparent colonies," and there were a very few "small, white, and more opaque ones."

Now they really began trying to patch things and explanations together, unsure whether they could "even claim with any certainty and assurance that a real problem existed at all." Those first differences that they had noticed—between yellowish and white bacterial colonies—"all became unstable in subsequent generations." But that became a new difference to work with—the difference between the offspring of the few special colonies and that of the majority. That difference even seemed to increase as they kept creating new generations of the germs, which Fleck attributed to the "partly subconscious selection of the most divergent colonies." But again, "all attempts to formulate this difference had to be dropped right after the next inoculations. . . ." They continued working.

At this point, many scientists would have reconstructed their narrative into the "Eureka!" genre, where the truth suddenly becomes miraculously and immediately clear. But Fleck saw it differently: ". . . at last, after we had gained comprehensive experience, a formulation crystallized." The emphasis here is on experience rather than the flash of intuitive confirmation—the vague, "partly subconscious" yet powerful cumulative effects of habit, work, acquaintance via immersion, and half-remembered scientific literature. Of course, sudden insight is often an important characteristic of the sciences, but the frequent emphasis on insight in the sciences can easily lapse into a kind of mysticism.

For Fleck, the truth doesn't appear, the mind of the deity isn't revealed, the scientist's mind doesn't seize the wondrous fact, but a *formulation crystallizes* out of this viscous gel of experience. It wasn't the difference in color that was important, as they had first thought. It wasn't the difference in intensity of color in subsequent generations that was important. It now appeared that all the colonies had the same color, but differed in structure. And unlike the color, those differential structures—later called the "smooth" (type G) and the "curly" (type L)—"could be perpetuated through transfers." They had—discovered? made? crystallized?—a fact.

Fleck explained why he thought it important to subject the reader to this kind of account (which we have simplified immensely), one that followed the vicissitudes of the actual experimental encounters:

> It shows (1) the material offering itself by accident; (2) the psychological mood determining the direction of the investigation; (3) the associations motivated by collective psychology, that is, professional habits; (4) the irreproducible "initial" observation, which cannot be clearly seen in retrospect, constituting *a chaos*; (5) the slow and laborious revelation and awareness of "what one actually sees" or *the gaining of experience*; (6) that what has been revealed and concisely summarized in a scientific statement is an artificial structure, related but only genetically so, both to the original intention and to the substance of the "first" observation. The original observation need not even belong to the same class as that of the facts it led toward.[16]

Fleck tried to anticipate how many "theoreticians," trying to get this complex mess of process tamed into the usual ordered accounts of science, might come up with alternative formulations. They could say that Fleck and his colleague's initial observation, "devoid of any assumptions," was something like "Today one hundred large, yellowish, transparent and two smaller, lighter, more opaque colonies have appeared on the agar plate." But that seemingly raw statement of brute facts, Fleck points out, "anticipates a difference between the colonies, which could actually be established only at a later stage of a long series of experiments." Recall that the first observation was colors, shortly after muddled with the memory of some literature on color in staph colonies. All right, the theoreticians might respond: you would have to investigate all the colonies, and all their properties, and then you could see how there were one hundred of this first sort and two of the second sort. Well, writes Fleck, "it's altogether pointless to speak of *all* the characteristics" of something like bacteria, because you always begin from some combination of habit, professional experience, or some motivation that's not entirely conscious. A good scientist always begins, at least in some measure, after the fact. Why delude yourself into thinking the process is or should be random, arbitrary, or completely inclusive? "Observation without assumption," writes Fleck, "which psychologically is nonsense and logically a game, can therefore be dismissed."

But two types of observation, with variations along a transitional scale, appear definitely worth investigating: (1) *the vague initial visual perception*, and (2) *the developed direct visual perception of a form*. . . .

Direct perception of form requires being experienced in the relevant field of thought. The ability directly to perceive meaning, form, and self-contained unity is acquired only after much experience, perhaps with preliminary training. At the same time, of course, we lose the ability to see something that contradicts this form. . . . Visual perception of form therefore becomes a definite function of thought style. The concept of being experienced, with its hidden irrationality, acquires fundamental epistemological importance. . . .

By contrast, the vague, initial visual perception is unstyled. Confused partial themes in various styles are chaotically thrown together. Contradictory moods have a random influence upon undirected vision. . . . Nothing is factual or fixed. Things can be seen almost arbitrarily in this light or that. There is neither support, nor constraint, nor resistance and there is no "firm ground of facts."[17]

This beautiful passage is well worth lingering over. Unstyled, undirected, or purposeless perception—the kind of raw empiricism that we saw Boyle apparently aiming for—can't provide facts or solidity. Meaning requires discipline; all perception and thought, in order for it to count as science, has to be styled. And that means living with contradictions and trade-offs. Direct perception of form or meaning is only possible through the indirection of being trained in a "thought-style." And while you gain enormously from being so styled—indeed, it's a prerequisite for doing science—you lose at least some of your ability to see other, new things. The sciences drive a hard bargain.

Fleck was a scientist who knew that conventional ideas about facts and what happened in the processes called experimenting had to be reworked, rethought, and reinvented. And like scientists who continually invent new instruments and procedures, and then experiment with them, the analyst of the sciences has to invent new concepts like "thought-style" and "thought-collective"—and then see if they work or not, and for what ends. Fleck invented "thought-style" to draw our attention away from the nonsensical idea of observation without assumption, and the associated idea of brute facts drawn from an independent reality, and focus that attention instead on what some scholars now call the elements and processes of social construction: how habits of thought, long-standing disciplinary and cultural assumptions, personal history, professional education and training, and even moods and chance all work together to stylize perception and theorizing.

A thought-style is similar to what Kuhn would later call a paradigm; both terms have been ungenerously equated with "fashion." That's simply a very bad translation (often deliberately so). You don't change thought-styles like an old shirt or skirt; they are embedded, as we'll see in more detail over the next few chapters, in words, practices, funding patterns, our science curricula, and in many other things and relationships. They change slowly, and not simply through individual will or commercial promotion. Thought-styles (and there are usually more than one operating within a given field of science at any one time) are created and transmitted by thought-collectives, which isn't as totalitarian as it might sound. Narrowly construed, Fleck used thought-collective to refer to the scientific subcommunity that developed and tested new ideas, trained its newer members to see certain forms and ask certain types of questions, and acted in a variety of other ways to control the arbitrariness and chaos of unstylized reality.

Fleck, ruminating on his experience as a creative, successful muddler in the sciences, gives future scientists a very valuable gift. He gives to them, and to the rest of us as well, some training and stylizations to help us perceive the sciences and give meaning to them in a new way. He gives us a way to stop thinking about facts themselves, and to start thinking about "how a *fact* arises. *At first there is a signal of resistance in the chaotic initial thinking, then a definite thought constraint, and finally a form to be directly perceived.* A fact always occurs in the context of a history of thought and is always the result of a definite thought style. . . . The *fact* thus represents a *stylized signal of resistance in thinking*. Because the thought style is carried by the thought collective, this 'fact' can be designated in brief as the *signal of resistance by the thought collective*. . . . The fact thus defined as a 'signal of resistance by the thought collective' contains the entire scale of possible kinds of ascertainment, from a child's cry of pain after he has bumped into something hard, to a sick person's hallucinations, to the complex system of science."[18]

While it might seem that Fleck was something of an idealist, locating that resistance in thinking itself, a careful reading of that passage just cited shows that to be a poor interpretation. Fleck rearticulates the fact as "the signal of resistance by the thought collective." By mentioning the child's cry he connects the idea of a scientific thought-collective back to the naive realist's cartoon violence of feet, stones, and bodies falling out of airplanes. He even shows how preverbal utterances, truly brute cries of pain, have to be seen in social context: the cry is a signal to parents!

Scientists meet all kinds of resistance; they bump up against things all the time. That's what makes the sciences challenging, exhilarating, creative, and useful. But it's a mistake to think of reality, the material world, as either the first or only signals of resistance. Fleck and numerous scholars since him help us see how habit, tools, moods, money, professional training, metaphors, and other things that aggregate into a thought-style and thought-collective (fairly imprecise terms which we will try to specify in more detail later) have what we could call very *material* effects.[19]

When he says that there is no "firm ground of facts," Fleck isn't pulling the rug from under our feet. He's redirecting our attention again, because we've been looking for solidity in the wrong place—or rather, we've been looking for it in just *one* place: the facts. The solidity of facts, like the truth of scientific theories that we will be discussing in more detail over the next few chapters, isn't simply a consequence of their being "grounded in reality." Fleck directs our attention to the "signal that resists," a signal (or Dewey's sign-force) that comes not just from nature or reality, but from . . . where?

The short, imprecise answer is: from all over the place. A better answer comes in the next, final section of this chapter: from realitty.

Experimenting with Some Physicists . . . and Realitty

Later in the book we undertake a more extended exposition of some history of and current research in quantum physics. But it would be rather strange not to mention something about quantum physics in a chapter devoted to experimenting, since this particular branch of physics is most responsible for blurring the boundaries between experimental instruments, the acts of measuring and observing, the spinning of theory, and perhaps even the line between mind and reality.

Ever since the physicist Max Planck introduced the concept of the quantum into physics in 1900, these once distinct and stable categories have destabilized more and more. Planck (and many other physicists since) felt that the trembling extended beyond the boundaries of physics itself, and took on vast cultural significance. "We are living in a very singular moment of history," Planck wrote in his 1932 popular book, *Where Is Science Going?*—"a moment of crisis" that showed itself "not only in the actual state of public affairs but also in the general attitude towards fundamental values in personal and social life." There was a vague but growing unease in the world that Planck associated with "skeptical attack," a spirit that might have been appropriate to religion's "doctrinal and moral" systems, and even to art, but now the skeptic had "invaded the temple of science." Even within physics, the holiest of holy domains, "the spirit of confusion and contradiction has begun to be active." Established principles of causality, the independence of external reality from measurement and thought, were under suspicion from within, from physicists themselves.

Indeed, Planck was in no small part responsible for this situation. His invention in 1900 of the quantization of energy as a rather desperate means of reconciling disagreeable implications of classical theory with some reliable data on thermal radiation was the small event that snowballed into the quantum revolution that now disturbed him.

Where some believed this might mark the beginning of a "great renaissance," Planck's sympathies were more in line with those who saw the "tidings of a downfall to which our civilization is fatally destined." The "logical coherence" of science had once served as protection against this skeptical "contagion." Now even great scientists like Planck knew that it could no longer serve as the "firm foundation," the "rock of truth" on which we could "take our stand and feel sure that it is unassailable and that it will hold firm against the storm of skepticism raging around it."[20]

That's a rather fearsome, and fearful, vision, and one that is still felt and expressed by many scientists today. Concern about the pernicious cultural influence of postmodernists, social constructionists, and other skeptical types has almost reached the level of hysteria in some portions of the scientific community. It's something that has changed little between Planck's time, and the "Science Wars" of our day: elder statesmen of science extolling a quasireligious defense of the secure, solid, unassailable ground of scientific truth as the bedrock of civilization, feeling besieged by the skeptical and possibly immoral forces of unbelief and contradiction.[21]

Meanwhile, things in the quantum world seem only to have gotten more puzzling and entangled, and quantum physicists have been forced over and over again to acknowledge the unavoidable collusions that go on between their own experiments, observations, theories, and instruments, and the real outer world. Physicist John Wheeler has been one of the most original and audacious quantum theorists of the post–World War II era. (He also helped develop the A-bomb and the hydrogen bomb.) Recently, he has distilled most of his work into the pithy phrases "No question? No answer!" and "It from bit."

> Otherwise put, every *it*—every particle, every field of force, even the spacetime continuum itself—derives its function, its meaning, its very existence entirely, even if in some contexts indirectly—from the apparatus-elicited answers to yes-or-no questions, binary choices, *bits*.
>
> It from bit symbolizes the idea that every item of the physical world has at bottom—at very deep bottom, in most instances—an immaterial source and explanation; that which we call reality arises in the last analysis from the posing of yes–no questions and the registering of equipment evoked responses; in short, that all things physical are information-theoretic in origin and this is a *participatory universe*.[22]

"This is a participatory universe." It's a bold statement—even an overstatement, of the kind that has made quantum physics so prone to vastly overextended and overinterpreted conclusions about how uncertainty, indeterminacy, or the interpenetration of subject and object become definitive, all-embracing cosmic principles that illuminate everything from holistic medicine to enlightened business management strategies to the failures of justice in televised trials. (And if a feminist theorist of science were to write something similar, she or he would be met with much harsher criticism than Wheeler receives.) But Wheeler offers this formulation as a "tentative idea or working hypothesis" that might very well be wrong, but as "the policy of the engine inventor, John Kris, reassures us, 'Start her up and see why she don't go!'"

Moreover, a close reading of the passage above shows that Wheeler is careful to leave certain qualifications in place: the immateriality at the bottom of the material world may lie so deep as to be not worth asking after. The interactions between questions and things, instruments and objects, in some contexts other than the subatomic, quantum context may be so indirect as to be irrelevant. We have to try to be very modest and precise here: Photons are not immune systems, and the equipment for evoking a response from a subatomic particle is not the equipment for evoking a measurement of groundwater contaminants. The strange interactions in the muddled space between bit and it, question and answer, virtual and real, immaterial theory and material fact, are going to be different in different scientific pursuits and contexts. A recurring demand in this book will be to track the specific interactions that happen in the wildly varied situations in which the sciences are actively pursued.

Nevertheless, Wheeler's concept of a participatory universe drawn from quantum physics is an excellent working hypothesis to try to incorporate in all of the sciences, and in all of our thinking and experimenting with the sciences. The sciences have to start with the knowledge that there is always some path, however indirect, by which theory and fact participate with each other. All of our scientific pursuits have to somehow acknowledge that the real and the virtual partake of each other, that the most factual and solid signal that resists is always informed by seemingly immaterial thought-styles. That's a much better way to pursue the sciences than with our current conventional notions of solid, solitary facts, securely walled off from or unmuddled with our changing concepts and cultures.

To start seeing why, let's return to the world of Max Planck and his book *Where Is Science Going?* We've already seen how Planck was deeply concerned about the cracks in the foundations of science and reason, and the implications which these cracks of skepticism and "irrationalism" held for the wider culture. Trying to shore up those foundations, Planck set out two theorems that together formed what he called the "cardinal hinge" on which the physical sciences turn. They are theorems that many scientists today would agree with, and which form our common-sense, conventional view of the sciences: "(1) *There is a real outer world which exists independently of our act of knowing*, and (2) *The real outer world is not directly knowable*." [Planck's italics]

Planck, who had just been lamenting the spirit of contradiction abroad in the current culture, admitted that these theorems were themselves mutually contradictory—that is, he could not rationally justify their conjunction. "And this fact," he continued, "discloses the presence of an irrational or mystic element which adheres to the physical sciences as to every other branch of human knowledge. The knowable realities of nature cannot be exhaustively discovered by any branch of science. . . . We must accept this as a hard and fast irrefutable fact. And we cannot remove this fact by trying to fall back upon a basis which would restrict the scope of science from the very start merely to the description of sensory experiences. The aim of science is something more. It is an incessant struggle towards a goal which can never be reached. Because the goal is of its very nature unattainable. It is something that is essentially metaphysical. . . ."[23]

Somewhere along the line, Planck argued, the scientist has to jump the wall or make the leap, and introduce a "metaphysical hypothesis":

> [S]cience demands also the believing spirit. Anybody who has been seriously engaged in scientific work of any kind realizes that over the entrance to the gates of the temple of science are written the words: *Ye must have faith*. It is a quality which the scientist cannot dispense with. . . . The reasoning faculties alone will not help him forward a step, for no order can emerge from that chaos of elements unless there is the constructive quality of mind which builds up the order. . . . Again and again the imaginary plan on which one attempts to build up that order breaks down and then we must try another. This imaginative vision and faith in the ultimate success are indispensable. The pure rationalist has no place here.[24]

This conventional view of the sciences hinges on a series of contradictions, or juxtaposed oppositions: physics/metaphysics, reason/imagination, science/faith, and even laboratory/temple. While the public, common-sense view of the sciences almost exclusively privileges the term in front of the slash—putting its best face forward—in fact, the opposing term is always present. It *has* to be there, even if it generally has to be swept under the rug for public occasions. There's certainly nothing wrong with contradictions per se; they're a part of any system, and indeed, the sciences wouldn't be possible or nearly so generative without the engine of contradiction. But when contradiction is denied or swept under the rug, it becomes much harder to get a sense of the peculiar effects of that particular contradiction. More importantly, it becomes almost impossible to imagine another system of thinking that might work just as well, although harboring its own contradictions.

Planck's radically bifurcated world is utterly dependent on opposed terms like imagination and reason, faith and science—and fact and theory. Planck asks us to have faith in what cannot possibly make sense, that science approaches the truth ever closer, enacting a hyperbolic nonencounter with the absolutely separate— ever nearing reality, but never touching it. That can't make sense and *has to* be an article of faith, because if "nature" or "truth" is indeed radically separate in the sense that Planck postulates, then it's impossible to know if you are approaching it more closely or not. As kids, many of us have played the game where someone hides an object, and then gives verbal cues to someone else seeking it: *you're getting warmer, warmer, colder, oooohhhh you're burning up!* But for that game to be played successfully, someone has to know where the object is hidden. The game needs a God, in other words, who can be outside the system looking in. In order to judge distance and progressive approach, you have to know the final destination—which Planck has just said is humanly unknowable.

Furthermore, if it's true that scientists have to have this kind of faith, it also true that they can't put too much faith in faith. Passionate adherence to a set of theoretical beliefs is a recipe for bad science . . . sometimes. Because at the same time, scientists have to be skeptical about being consistently skeptical, since that is also likely to land them in trouble. Deciding when to have faith, and when to skeptically attack results and theories is an expert, subtle art or craft. There are no hard-and-fast rules to follow in this world of contradictions and experimenting, and the best scientists—like Copernicus, Millikan, Fleck, and even Planck himself—will be the ones who can muddle through.

But what makes us think we really need Planck's kind of faith anyway? Why is it that our culture, by and large, encourages and even demands that we believe that the sciences progressively approach and approximate a full and final truth? Why are we so often presented with the stark alternatives: a world of rock-solid, time-less facts-or completely adrift in a raging sea?

One possibility is that the sciences then gain the same exalted and authoritative status as religion has had, as Planck's use of the word "temple" should make clear. There are many ways in which the sciences, and our cultural attitudes toward them, are indeed like religion. But it would be a mistake to reduce and equate the sciences to religion, and to say that reason and the sciences are simply another form of faith. The world is much more complicated than that. Part of that complexity involves the fact that progressivism in politics has traditionally been connected to this progressivism in the sciences; eliminate (even if only in our epistemology) the real world which is approached progressively, and you eliminate any basis for making judgments not only about what is scientifically "better," but also morally or politically "better." We will question this linkage more closely in Chapter 4.

Our point here is more modest: It's *possible* to pursue another kind of thought-style regarding the sciences, the work of experimenting, and the status of facts. This other thought-style —"muddling through" is one way of putting it—is coherent, rational, productive, reliable, and critical of established truths, including its own. This way of pursuing science is no more (and no less) marked by its own internal contradictions than the conventional view, described succinctly in Planck's two axioms. The opening glimpses of that thought-style provided here by Fleck, Dewey, and others will be used as building blocks for the following chapters.

And in the end, the ways in which Planck and Fleck or Dewey view the nature of the world and the nature of the sciences might be very similar, or at least compatible. Each seems to be saying that "nature" or "reality" remains inaccessible, and that what we have access to is only a kind of imperfect simulation of it, in the form of what Fleck calls our "stylized knowledge," or what Planck calls our "constructive quality of mind."

And there is certainly plenty of room for common ground among these worlds, a middle space or interzone where the differences might be better tolerated or even suspended. Diplomatic negotiations in the science wars are not only possible, they're necessary, and they could be quite productive. Planck's image of the sciences as an "incessant struggle" toward an unattainable goal is very close in spirit to what we are calling "pursuing sciences." Throw in his notion of an inexhaustible outer world that can't be known directly, and we can find even more room for discussion. An inexhaustible reality, a nature so excessive that it can provide multiple changing answers to our multiple, changing, questioning experiments, may not be so far from a world in which facts are made and nature is produced—or what Donna Haraway, borrowing from Native American traditions, calls a "trickster nature": a shape-shifting, provocative world which can never quite be pinned down. There should be *some* room to work with here in the middle.

And it's in the middle, at the hinge of contradictions, where muddling through happens. Linger over *Coastline Measure*, the Mark Tansey painting which appears on the cover of this book. Planck's either/or choice—either the firm, factual rocks of solid truth, or the raging storm of skepticism—becomes a both/and here. Pursuing sciences isn't a matter of approaching an outside world ever more closely, but rather the difficult, slippery, often risky and usually admirable work at the limit, on the border of solid rock and stormy sea. When the surveying team returns next year, the coastline will still be a disjunctive juncture of stability and change, but its precise contours will have shifted into a new pattern whose measure must be taken again. Immersed in their work, concentrating on keeping their footing in this treacherous territory, they may lack the privileged perspective of the distant observer who can see the suggestion of an oddly repeating, fractal structure that penetrates to the limits of their world.

Is this what bedrock is—a layered, iterated pattern made solid by its own cross-referenced iterations?

Do the stories we've told above cohere into some kind of theory of facts? Not exactly. You wouldn't necessarily say that Tansey's painting is a theory of coastlines. But just as Tansey provides an image that prompts a series of speculations about rocky shores and the immense forces shaping them, we've tried to provide a collage of fact-images whose structure is meant to initiate a new set of responses to the question: what are the sciences *grounded* in?

In other words, if you think this chapter is supposed to answer one of our "really?" questions—"Are facts really discovered via experimenting, in a separate reality out there, or are they really manufactured, social constructions?"—you're going to be disappointed. That's probably what a good philosopher would call an undecidable question. At the very least, judging by the number of books already written on the subject, it's the kind of question that promises to engage philosophers and many others for a long time to come.

It's not the most productive question. What we're more interested in, and what we hope will be more productive in the long run, is inventing a new word that can help keep us close to the question without resolving it, to paraphrase Foucault—a word for further experiments on experimenting.

Why does it seem so naturally, intuitively true that, as Planck says, "there is a real outer world which exists independently of our act of knowing," but which our acts of knowing called the sciences can uncover and approach (but never meet)? Part of the answer certainly has to do with the experiences that scientists have every day in their laboratories or at their desks, where it seems to them undeniable that nature compels their hands and minds in powerful ways. Another reason has to do with the dominant culture of the sciences in which we've all grown up, including the way that the sciences are almost always taught in our schools. But another part of the answer lies still deeper, and has to do with language.

The ways in which most of us (including Planck) think about what the sciences really are and how they really work are threaded together with what language is and how it works. It's no accident that Planck's opposition between subjective act of knowing and objective world resonates so strongly with the subject-(verb)-object structure of language. Alternative styles for thinking about the sciences don't find nearly so comfortable a fit, and so lack that natural feeling of common sense: facts aren't discovered, but produced?! By what subject? Out of what object? *The signal that resists?!* Language cannot simply be overthrown or overcome.

Language can, however, be played, and the first step might be a new word. Throughout this chapter we've referred to this *something* that has been variously termed "a material world," "the world itself," or just "the world." "Reality" is the key word here. The word "reality" signifies the irrefutable thing, or the signified, that makes Johnson's toe hurt, makes Millikan's experiment eventually come out correctly, and flattens the social constructionist's body at the end of the airplane experiment. It's a word full of finality, for the brute reality of brute facts.

Can we come up with an alternative signifier? The thing, the signified that we're trying to point to here is what will emerge over the course of the next few chapters as "a muddle, or complex assemblage, of material, social, cultural, linguistic, technical and other forces—although those things are just our provisional names, too—that constitute what is most frequently called reality." *That's* the thing that makes pursuing the sciences so hard, so useful, so important, and so problematic.

That's the thing that we have to learn to recognize as the ground on which the sciences are pursued, and on which they depend—ground in the sense of the violent conjunction of rock/sea.

But that phrase in quotation marks, as you can see, is a clumsy and long construction that you would soon tire of seeing in print. "The signal of resistance by the thought collective" is certainly not without its problems, either—not the least of which is that it tends to alienate scientists who aren't already partly sympathetic. We need another term for "reality" that will work on our coastline, in the middle between realists and constructivists. It should be close to "reality," since it's an important word that we wouldn't want to get rid of all together. But the new word should be something which breaks "reality" up at the same time, just a bit. (And all we need is just a bit, since reality as a concept depends on absolute purity, zero leakage between facts and theories, things and words, objects and practices, land and sea.) So instead of reality, from now on we'll be writing (about) reality. When you speak it, you can't hear the difference. We can go around talking about reality, just as all the other good scientists do. You can only see the difference in writing, where it introduces a difference into the word-concept of reality in two ways. One is quite literal: reality is an anagram of "alterity," a word which signifies difference, otherness in general, in all its multiple forms. Alterity is "the state of being other or different; diversity, otherness." Like reality, then, reality is a kind of alien presence that can mock our wills and desires.

The second difference incorporated in reality is an injection of time, always represented in scientific equations by the lowercase italic t. Reality is what you get when you throw time into the equations of reality. If reality always stays the same, but we believe that we approach it ever more closely, reality always changes, and it's no longer a question of nearness and approach, but of successive experimental practices and their successive (and successful) thought-styles.

The sciences today are better seen as a matter of re-producing reality so that it can be worked on and experimented with, not simply re-presenting reality so that it can be thought about and understood. Why is science better seen that way? Why open a book with such stories? Or as Dewey asked eighty years ago: "Since instrumentalism admits that the table is really 'there,' why make such a fuss about whether it is there as a means or as an object of knowledge?"[25]

Dewey formulated a number of responses to that question, but the primary one was that the conventional view of science and the facts "commits us to a view that change is in some sense unreal, since ultimate and primary entities, being simple, do not permit of change."[26] Seeing entities like facts and things as complex rather than simple, as mediated and manufactured rather than ultimate and primary, at least holds open the possibility of change. Fleck's world is not divided in two; he doesn't need or want us to believe in a separate or future world of total truth that can be approached. He only wants us to pay attention to *this* world, to watch how we stylize our observations and thoughts, to think about how facts emerge via experimentation, without making the metaphysical leap to believing that those experiments and those facts are leading us to some fuller realization.

This doesn't mean that everything is relative, or that "all that is solid melts into air." Notice how that famous aphorism skips over the middle term between solid and gas, liquid. Look again at *Coastline Measure*—it's not about the evaporation of solidity; it's an image of solidity *and* liquidity *and* muddling through in their midst, all at once.

This doesn't mean that it would be easy to make a world without, say, electron spin. Seeing electron spin as the complicated sign-force of realitty that Feynman visualizes, gives it no less stability than seeing it as a simple brute existence in reality. But it does allow us to see that as mathematics slowly changes, as instruments become different, as the ensemble of physical theories in which "spin" is an important term continues to evolve, a new fact is someday going to emerge to replace electron spin, just as quantum-mechanical electrons have replaced the miniature billiard balls of classical physics. It means that realitty will change, because that's what realitty does best.

Moreover, shifting the focus of our thought-style from reality to realitty might better allow people to ask such questions as: If we're told that a definite low level of benzene in our drinking water in fact poses no risk to health, what realitty does such a statement depend on, or belong to? In what realitty is a statement like "intelligence is genetic" a factual, rational one—and are there any other realitties that make as much if not more sense?

It's not a perfect invention, this word-concept. It doesn't do everything we want it to do, and it will no doubt do too many disturbing things for others. The coastline of the sciences is a tumultuous, uncertain place. Does a rocky realitty indeed harbor a kind of truth, serve as a reliable fact? To paraphrase John Wheeler again, there's only one way to tell: start it up, and see if it doesn't go.

Our stories about famous scientists and their scientific achievements are better than the conventional accounts because ours are grounded in a complex realitty instead of paltry reality. The sciences themselves are better when they are situated in realitty. Realitty is neither discovered nor constructed. It is pursued and performed; it pursues the scientist. Realitty is never perfect nor direct, but always indirect—which is pretty damn good. Realitty is completely Other, inhuman, overwhelming, inexhaustible Nature; it is nothing without us. Realitty resists utterly; it yields supplely. Realitty is the beginning and end of experimenting.

Chapter 2

Articulating

Sacramental Swan

Gregory Bateson was . . . what *was* Gregory Bateson? An anthropologist studying human culture? An ethologist observing animal behavior? A philosopher playing with ideas? A theoretical biologist enmeshed in the social and intellectual feedback loops of cybernetics?[1]

It's hard to say exactly what Gregory Bateson was, which is why we begin this chapter with him. The questions we'll be staying close to in this chapter have to do with the problem of *saying things exactly*. If it's true, as Max Planck argued, that we can never have *direct* knowledge of the "real outer world," does that mean that all of our indirect knowledge is, in some sense, inexact as well? Can the sciences—at least some of which are referred to, sometimes, as "the exact sciences"—ever say something exactly? If they can, what language do they use? Are the answers they give in response to "really?" questions spoken in mathematics? Symbolic logic? The same language that we use for novels, poetry, road signs, and conversation? Or some combination of all of these? And a less obvious but equally important question: when the sciences make these kinds of speeches about what's really real, on what platform do they stand?

When Richard Feynman tried to articulate how he visualized the world when he was doing physics, and in particular, how he "saw" electron spin, we could say that he was speaking metaphorically. Should all statements about realitty, then, be taken as metaphors? That is, is the real really a metaphor? That would be a very unmetaphorical way of asking the question and would demand something like an equally unmetaphorical yes or no answer. Maybe there's another, less exact and less direct way to pursue the questions of exactness, language, and the sciences.

Consider Bateson trying to explain to his daughter what it means to speak in metaphors, in terms of something being sort of something else. They start their dialogue—in fact, "sort of" a dialogue, since Bateson wrote it himself, for his book

Steps to an Ecology of Mind—by looking at how a ballet dancer can sometimes be a sort of swan, and then move to the question of transubstantiation:

> *Father:* And then there is that other sort of relationship which is emphatically *not* "sort of." Many men have gone to the stake for the proposition that the bread and wine are *not* "sort of" the body and blood. . . . If we could say clearly what is meant by the proposition "the bread and wine is *not* 'sort of' the body and blood," then we should know more about what we mean when we say either that the swan is "sort of" human or that the ballet is a sacrament.
>
> *Daughter:* Well, how do you tell the difference?
>
> *F:* Which difference?
>
> *D:* Between a sacrament and a metaphor.
>
> *F:* . . . Well—I think it's sort of a secret.
>
> *D:* Do you mean you won't tell me?
>
> *F:* No—it's not that sort of secret. It's not something that one must not tell. It's something that one *cannot* tell.
>
> *D:* What do you mean? Why not?
>
> *F:* Let us suppose I asked the dancer, "Miss X, tell me, that dance which you perform—is it for you a sacrament or a mere metaphor?" And let us imagine that I can make this question intelligible. She will perhaps put me off by saying, "You saw it—it is for you to decide, if you want to, whether or not it is sacramental for you." Or she might say, "Sometimes it is and sometimes it isn't." Or "How was I, last night?" But in any case she can have no direct control over the matter.
>
> . . . I'll start again. The swan figure is not a real swan but a pretend swan. It is also a pretend-not human being. It is also "really" a young lady wearing a white dress. And a real swan would resemble a young lady in certain ways.
>
> *D:* But which of these is sacramental?
>
> *F:* Oh Lord, here we go again. I can only say this: that it is not one of these statements but their combination which constitutes a sacrament. The "pretend" and the "pretend-not" and the "really" somehow get fused together into a single meaning.
>
> *D:* But we ought to keep them separate.
>
> *F:* Yes. That is what the logicians and the scientists try to do. But they do not create ballets that way—nor sacraments.[2]

A good muddling scientist, when asked if she is saying something "really" true about nature, or just something "sort of" true, will know that she can't say. (She also knows that she has to say *something*, after she rolls her eyes, invokes the deity, and prepares herself to start over again—and then again. It's what Bateson would have called a double bind: knowing that she can't say what is asked of her, and knowing that she must say it.) She can't say it's *either* real *or* metaphorical because that kind of distinction is a secret, sort of. It's not that the scientist knows the difference between the real and the metaphorical, but won't say; nor is it the case that the scientist *doesn't* know, and so can't say the distinction. It's more like she knows, but can't say because she doesn't know that she knows.

Or, to put it in terms taken from Ludwig Wittgenstein, who thought long and hard about what it means to try to say something exactly: The statement that would speak the distinction between the real and the metaphoric cannot be part of the system of real and metaphoric statements; it is outside that system and hence unspeakable. You can try to make such distinctions and, oddly enough, it's important to try. But at the same time, you do not make art or sacraments that way. And you don't make the sciences that way, either.

Transparency

We invented the word realit*y* to stay close to the question "real or constructed?" rather than to resolve it—to keep our eyes on the collusions between things and thoughts, thoughts and experiments, and facts and theories. Now we move on to do some experimenting with this latter term, theory.

We know now that the ideal picture of building up (by either logical deduction or induction) shining, illuminating theory out of dull, dirty, brute facts just doesn't capture the messiness or richness of "the scientific method." Even the most curmudgeonly philosophers of science, such as Sir Karl Popper, know that theorizing is never simply a matter of induction from empirical facts, but is a more elusive and less perfectly rational process that might even involve the imagination.

Can we say what theory is exactly, then? Starting with the dictionary meanings, we quickly run into trouble, trying to reconcile various definitions ranging from "a coherent group of general propositions" to "guess or conjecture." Wherever it fits in this spectrum, it looks as though theory is located in another world, and applied to this one: *Sounds good in theory, but will it work in practice?* A theory is "abstract," mind-stuff, not of this material world. Theory is just a set of statements, diagrams and equations in a manuscript, words on a page representing ideas in the air.

The word's roots in the Greek *theorein*, meaning "to view," are no less riddled with different interpretive possibilities. But that older definition at least brings into view the idea that our habitual ways of thinking about "theory" are tied up with direct visual perception—images of illumination, enlightenment, eyes, and less directly, mind. It's hard to even think of the words "theory" and "theorizing" without getting the image of someone's head with a light bulb over it. Theory is part of an ideal realm that is as seemingly insubstantial as light itself.

Implicit within these conceptual habits of theory as immaterial light is a family of ideas tied to the notion of transparency. Theories in the sciences are like lenses, which enable us to focus on the real world or on truth, without distortion. Like light, the real and the true pass directly to our eyes via scientific theories. The privileged access to the real world that we grant to the sciences is presumed to be, must be, unmediated. That's exactly why we privilege them: the sciences give us direct access to the real, with none of the coloration, diffraction or other distorting effects of technologies, subjectivities, or cultural or other interpretive factors. No shadow should fall between the eye and the thing, between the idea and the actuality.

But why? What is the danger of mediation that it evokes such horror? Because it starts with the idealization of transparency, mediation in the conventional view can only be conceived negatively: to be mediated, to be in the middle is to be fallen, to deny or to have been denied the pure illumination of objects. (The subtle force of religious culture is at work here in this part of the sciences.) When the positive ideal is the absolute transparency of the "scientific perspective," a mediated and limited perspective can only be about distortion, the intrusion of error, the presence of pollution, the insertion of ideology or political force between us and it. This middle space must be excluded because only bad things can happen there; the middle can only be impure and contaminated. Ideally, a scientific instrument like a telescope or particle detector, or a scientific theory like evolution by natural selection, should fade away and allow things or facts to speak for themselves.

Let's be clear about the productive qualities of such an ideal. It always has prompted scientists to do their best work, making them obsessed with the proper functioning of their instruments. A good scientist is dedicated to getting instruments to "work properly" and produce "good data." That is the exhilarating, frustrating, demanding, and admirable work that goes on every day in hundreds of laboratories, observatories, and field sites around the globe. Millikan would have never defined the charge of the electron if he had not painstakingly tinkered with his experimental apparatus, checked and rechecked observations made with it, patiently gone over calculations. Spare and unreliable as Copernicus's observational data were, they nevertheless demanded that he refine his mathematical and geometrical tools and assumptions. Similarly, chromatic aberration and other effects of poorly crafted lenses were problems that plagued observation of cosmological and microscopical reality for hundreds of years—and that we learned to solve. You didn't have to be an expert in optics to see the difference in those before-and-after photographs from the repair of the Hubble Space Telescope as they made their triumphal appearance in the media. Somehow we know the difference between a bad photo of the edge of the known universe, and a good one.

But why do we think that, tens of millions of dollars and a space shuttle mission later, an enormous array of photodetectors, computer software and hardware, and thousands of other hard and soft components suddenly turned transparent and allowed us to see exactly, directly what the edge of the universe *really* looks like? Why do we think that we were seeing directly, for the first time, a part of reality that we had never seen before? Could we instead be satisfied with knowing that we had coaxed into our world a part of realitty that hadn't been here before? And just what, for heaven's sake, could that mean?

Articulating "Articulating"

To stay close to these questions, we need a few more alternative words. *Articulate*, *articulation*, and *articulating* might be made to carry a different set of associations from those of "theory" and "theorizing." Articulate has a number of meanings, as both adjective and verb, and the ensemble of these meanings gives us a better set of working concepts for reformulating how we use theory, and how we think about what it is.

First are those meanings that revolve around language. To be articulate is to be "endowed with the power of speech," or to be capable of "expressing oneself easily in clear and effective language." To articulate is to give words to, to try to express, describe, or invent something that wasn't previously a part of language or thought. If theory is the mental capture and representation of an illuminated world "out there," then an articulation is something one speaks rather than sees, that is expressed rather than mirrored. An articulation is something put as adequately as possible into words, rather than something seen in its true nature.

Then there are the anatomical or structural senses of "articulate"—disparate articles jointed into a larger, segmented structure. Imagine a lobster, or a Rube Goldberg contraption. It involves hinges where disparate elements turn on each other, finding new leverage points, creating new angles and forces. If theory is a mental construct, then an articulation is corporeal, an organic or mechanical (or some hybrid, in-between form of these) structure in which different parts meet and rub against each other. There are tensions and frictions, squeaky wheels, generated heat, noisy crashes and slips—yet there is an overall ability to accomplish or produce something important to us.

To think in terms of the articulations of science rather than scientific theories, of articulating rather than theorizing, would restore questions about the importance of language to the domains of the sciences, and questions about our ability to say things exactly. Thinking in terms of articulations would also help us pursue sciences in which theory is less a set of ideas that somehow mirror the world, and more an assemblage or contraption of different parts, humanly designed within the world to do certain kinds of work.

Theories in the sciences are lobsterlike entities made out of disparate, heterogenous elements, including numbers, relationships, machined metal, purified enzymes, hazy concepts, precise and imprecise metaphors, and many other things and nonthings. The sciences are "science" not because they reflect or transparently transmit the world, but because their many articulated connections enable them to crawl around, clutching and poking, within that world

These are rather abstract articulations of what articulation involves. The rest of this chapter fleshes out this starting structure, adding example upon example to show why it is better to think of the sciences as involving mediated articulations, rather than immediate, transparent theories. If we can learn to keep the terms of mediation in view and in question, to track down all the things that are being articulated, even down to the very tail of the lobster, then we will be better equipped for the kinds of critical inquiry into the sciences that democratic society now sorely needs.

Pens, Swords, Tongues, and Politics

"Controversye Is a Civill Warr with the Pen which pulls out the sorde soone afterwards."
Earl of Newcastle to King Charles II

In the first chapter we saw how the new experimentalists of seventeenth-century England were creating a new discipline and form of scientific knowledge, which they called natural philosophy. Here we take up this topic again, extending our view to encompass the broader set of articulations within which experimentation was embedded. Why was experimentation persuasive? What was particularly valuable about it, and the kinds of knowledge which it yielded? Can experimenting be valuable now for the same reasons?

When Charles II was restored to the English throne in 1660, overt civil war might have ended, but dissension and controversial beliefs of all manner remained pervasive threats to civil order. Anything having to do with the beliefs or knowledge claims underlying politics, religion, law, or the natural world was a matter of vital social concern. The Uniformity Act against dissent in religion resulted in the formal expulsion from their posts in 1662 of hundreds of ministers from Presbyterian, Anabaptist, and other sects. Laws were enacted to enforce fealty oaths to the king, and against "sedition in print." The number of licensed printers was cut from sixty to twenty, and placed under the authority of the Archbishop of Canterbury and the Bishop of London. In the words of one licensing official, "[T]he spirit of hypocrisy, scandal, malice, error and illusion that achieved the late rebellion was reigning still."[3] The new "coffee-houses" were placed under surveillance as potentially dangerous meeting places of sectarian groups.

Thus when Charles II granted to the newly founded Royal Society the right of assembly, the right to correspond with foreign members, and the right to publish without censorship (not to mention things like the right to dissect the corpses of prisoners), it's understandable that he would want certain assurances in exchange. The Royal Society intended, in the words of Robert Hooke, pioneer microscopist and drafter of the Royal Society's charter, "to improve the knowledge of natural things, and all useful Arts, Manufactures, Mechanics, Practices, Engynes and Inventions by Experiment"—*and* it would do so without "meddling with Divinity, Metaphysics, Moralls, Politicks, Grammar, Rhetoric, or Logick."[4]

A divide was forming, with knowledge (arrived at through experiment) on one side and power (in interestingly diverse forms) on the other. (It's a divide to which we will return in the next chapter.) Yet even though such a division was indeed emerging, its cross-border traffic was not so much nonexistent as very consciously managed—stylized, even. Thomas Sprat's *History of the Royal Society* compared the new houses of experiment to a theater, where disputes could be safely staged: "There we behold an unusual sight to the *English Nation*, that men of disagreeing parties, and ways of life, have forgotten to hate, and have met in the unanimous advancement of the same *Works*. . . . [I]t gives us room to differe, without animosity; and permits us, to raise contrary imaginations upon it, without any danger of a *Civil War*."[5] Experimental knowledge was literally a type of enactment, designed to perform the kinds of acts on the stage of knowledge that should be emulated in the larger house of power, even while insisting that the two spheres were separate. Differing "contrary imaginations"—different theories, in other words, or different articulations—could be managed by unlinking or disarticulating them from the experimental production of "unanimous" facts. Check your politics at the door and witness the production of consensus inside; not only would you produce the best possible knowledge, but you would enact the social ideal for managing—indeed, eliminating—controversy.

Agreeing to exclude controversial discussions or analyses of "Divinity, Metaphysics, Moralls, and Politicks" is one kind of deal, perhaps even an understandable one in light of the violently contentious times. But what threat to order, civil or otherwise, was perceived in "Grammar, Rhetoric, or Logick"? What was it about language that led Sprat to remark in his history that "[t]he Truth will be obtain'd between [men]; which may be as much promoted by the *contention* of hands, and eyes; as it is commonly injur'd by those of *Tongues*"?[6]

Enter Thomas Hobbes.[7] Hobbes shared some of the experimentalists' values. He was convinced that the production of knowledge could be entrusted only to gentlemen, and that those gentlemen should avoid "contumelious language." Only cool heads and proper manners could prevail, for they were necessary for the exercise of judgment (the topic of Chapter 4):

> It cannot be expected that there should be much science of any kind in a man that wanteth judgment; nor judgment in a man that knoweth not the manners due to a public disputation in writing; wherein the scope of either party ought to be no other than the examination and manifestation of the truth.[8]

But unlike Robert Boyle and other members of the Royal Society, truth for Hobbes could not be manifested by experimenting. Hobbes argued that by avoiding issues and questions of language and logic, or reason, and focusing only on experimental facts supposedly unencumbered by metaphysics, Boyle and the Royal Society could never arrive at real knowledge (and therefore, real intellectual and social order). It wasn't a matter of getting rid of metaphysical language, contended Hobbes, or putting it off-stage, but of finding the *proper* metaphysical language. That rival interpretations of an experimental phenomenon such as "the vacuum" were possible proved the necessity of such a proper language to Hobbes, if something like true knowledge was ever going to be had.

The model of such a proper language for Hobbes was geometry. Why geometry? In part because it was, according to Hobbes, "the only science that it hath pleased God hitherto to bestow on mankind." But more important than its divine origin was the fact that geometry provided irrefutable results and unquestionable truth. "All men by nature reason alike, and well," Hobbes argued in his *Leviathan*, "when they have good principles." Geometry was a better staging of how social controversy was to be managed than experiments were because it showed that the only way to *reach* agreement was to *start* in agreement on fundamental principles, and reason from there. Moreover—and here's where Hobbes's argument takes an interesting twist—at the solid foundation of geometry were definitions of words. "Settling the signification . . . of words" and "plac[ing] them at the beginning of reckoning" was Hobbe's first methodological principle. From there, "right reason" took over, an entirely straightforward affair of "adding and subtracting, of the consequences of general names agreed upon for the *marking* and *signifying* of our thoughts." Experimental facts were to Hobbes "nothing else, but sense and memory." They couldn't serve as the basis of knowledge, let alone as the basis for resolving social disputes. In other words, Hobbes knew that experimenting had to be *articulated*, joined to language *via* language.

Of course, Hobbes had a rather authoritarian view of language, as of most everything else. But our goal here is not to side with either Hobbes or the experimentalists, but to carve out some working principles somewhere in between. Seventeenth-century England, and the rise of the experimental sciences within it, is a complicated and contradictory space, and does not lend itself to simple conclusions and good-guy, bad-guy lessons. On the one hand, we can admire how the experimentalists eschewed discussion of "true causes" in favor of probable explanations. As we emphasize in Chapters 1 and 3, there is something in this aesthetic of knowledge that is well worth preserving: its (potential) sense of modesty, finitude, and tolerance of difference. History has certainly shown that experimenting can indeed be a wonderful source of new things and new ideas, and can break up old habits of knowledge and power. However imperfect or uneven, experimental science is an immensely valuable antiauthoritarian resource in many situations. (And in other situations, not.)

On the other hand, we can hold our noses and take an important lesson from the fairly odious Hobbes: Language matters. Language is essential to reason, and can't be gotten rid of so easily with a few new machines. Somewhere along the line—no matter how long that line is—every experiment, every mathematical equation, every pure numerical value will have to find its way into words. But rather than share Hobbes's faith that finding the one proper language and reason will guarantee knowledge, we only insist that attention be paid to the varieties of language articulations in the sciences, and their effects. If the question for Hobbes was what constituted proper language, the question for us is how languages participate in making knowledges proper. That unavoidably awkward-sounding question sits solidly in the muddled middle between Boyle and Hobbes, and is one to which, in various ways, the remainder of this chapter stays close.

Copernicus Revisited

In the previous chapter we began a more complicated story of what Copernicus was trying to do with his book, De revolutionibus, and how he thought he could do it. Was he merely trying to "save the appearances," to simply provide a possible, working model that would explain the apparent movement of the planets? Did he see directly to the truth of empirical planetary motions, his eyes and mind unclouded by philosophical or religious doctrines? Did he see himself in opposition to the Church and religious dogma, instantiating a new order of truth?

What historian of science Robert Westman calls the Vulgar Triumphalist version of Copernicus no longer holds up among professional historians, although you can still find it in science textbooks, popular magazines and newspapers, and the popular writings of scientists today.

In our culture, Copernicus is identified as a scientist allied with the truth, a man who defied the priests and their traditional, false beliefs. This is what we learned and continue to teach our children in school. We'd like to offer a different story, a version of history both better grounded in the facts, and better able to help us understand the sciences in productive ways. In this story Copernicus stands firmly in between the opposed terms of scientist and priest, and science and tradition; if anything, this portrayal should make him more admirable, not less.

Copernicus lived his entire life as a church official, a canon in charge of such things as local military defenses and health care. This position entitled him to collect rent from the local peasants. As was usually the case during the late Renaissance, he obtained this office and the kind of upward mobility it afforded through family connections. He further upped his income when his uncle, the bishop of Varmia, appointed him to the position of scholaster at the Church of the Holy Cross in Breslau. Had Copernicus chosen to do so, he might have made a career move into the papal bureaucracy in Rome.

He chose a more middle path, however, pursuing astronomy and mathematics while maintaining his church offices and seeking further patronage from the larger church system. Here's where the history of science, religion, and humanism come together in an interesting and productive intersection.

Historical analyses of the University of Cracow have articulated the profound intellectual and cultural shifts taking place in the early sixteenth century—larger changes of which Copernicus was one small part. There, at one of the greatest of the medieval universities, Copernicus not only reaped the benefits of one of the best faculties in mathematics and astronomy but was also exposed to the new "critical humanistic attitude which was transforming older cultural and educational values."[9] The new humanists of the late fifteenth and sixteenth centuries—lovers of language and experts in Roman and Greek poetry, philosophy, and moral and civic literature—placed themselves in service (often as diplomatic emissaries) to the papal court. Copernicus, then, didn't come out of nowhere, the lone hero of science who single-handedly provided the material to overthrow a repressive tradition. He was one of many who attended the universities at Cracow and elsewhere, and who would collectively transform the entire cultural and intellectual climate. And Copernicus obtained a whole bag full of tools and skills to help in this transformation—"science" was just one part of it.

Copernicus also had a network. Pope Clement VII's secretary, Johann Albrecht Widmanstetter, was one of these humanists, who explained Copernicus's theories to the pope and his associates. In exchange for these services, Clement gave Widmanstetter a gift as well—not a religious treatise, but a Greek manuscript of scientific essays. This gift illustrates the need to resist easy oppositions between "church" and "science" at this time. Another cardinal whom Widmanstetter served, Nicholas Schönberg, wrote to Copernicus in 1536 and asked him to send a copy of his manuscript to Rome. In a strategic move, Copernicus inserted this letter into De revolutionibus between the title page and his preface to the pope, in effect allowing Cardinal Schönberg to announce Copernicus's "new account of the World": "In it you teach that the earth moves," wrote Schönberg in his letter, "that the sun occupies the lowest, and thus the middle, place in the universe."[10] Contrary to what the "preface" of his editor Osiander might suggest, then, it seems that Copernicus did indeed want to say that the earth *really* moves and the sun *really* was at the "middle" of the universe—although he let the cardinal say it for him first. (At the time philosophers and other professional classifiers considered astronomy itself to be a "middle science," sandwiched between physics and mathematics, and thus somewhat less pure and lower in status than either of those pursuits.) Far from acting the defiant hero against the church, Copernicus was angling for papal acceptance and patronage.

But only somewhat. The pope at the time, Paul III, came from a wealthy noble family, had gone to school at the University of Pisa, and was known as a poet, a widely read scholar, and an aficionado of astronomy. (He also helped initiate the Roman Inquisition.) Copernicus could have easily placed himself in service to him; one astronomer who made favorable astrological predictions for Paul III (remember, astronomy and astrology were entangled at this time) was amply rewarded. Perhaps because he wanted to associate himself more closely with the church bureaucracy's humanist contingent, which didn't think highly of astrology, Copernicus did not follow this example.

Instead, his preface to De revolutionibus was an impressive concatenation of the sophisticated rhetorical devices (irony, confession, antithesis) and tropes (knowledge as a solitary voyage, an aesthetic of cosmic belongingness) then offered by humanist literature. The preface was meant to *persuade*, to subtly convince the pope as well as other readers of the veracity of a difficult and controversial argument—an argument that did not, and could not, proceed on the basis of evidence or reason alone. Copernicus embedded references to Horace in his text, a favorite author of the pope and other humanist scholars. He made subtle allusions in favor of the pope and against Martin Luther.

Even this more complicated version of Copernicus and his strategies is somewhat simplified. We recommend a reading of the full article by Robert Westman from which we've drawn this material. Westman concludes that Copernicus

> aimed to solicit reform sentiment from among those in the church who
> . . . valued the mixed mathematical disciplines but saw them as needing
> renewal through a return to a purified ancient tradition. . . . His reformist rhetoric was not stridently polemical; it was gently Horatian and

Erasmian: an end to controversy among astronomers; an internal cadre of humanist mathematicians to reform church teaching on the heavens by providing true principles from which planetary order and calendrical accuracy could be restored, the entire enterprise to be legitimated by papal authority and by appeal to a broad range of ancient, pagan sources. The approach evokes Erasmus's broad reconciliation of Christian and pagan letters in a *philosophia Christi*—a life of lay piety modeled on the true life of Christ. . . .

In *De revolutionibus*, Copernicus sought to bring the individual parts of the universe into concordance with a sun that he described to his ecclesiastical audience in the most classical, pagan images, not as the generative or emanative force of the Neoplatonists but rather as a properly placed lamp or lantern, an eye, a mind, an enthroned king, a visible god. His choice of language and imagery pointed the church away from mediative, astrological influences and instead returned it *ad fontes*, to an ancient truth: the primitive order of the creation.[11]

Copernicus *articulated*, in other words. He didn't simply theorize a new cosmology, he joined together old observations, traditional notions of order and symmetry, new humanist rhetoric and new physical-mathematical concepts, the pious life of a believer, and appeals to papal patronage and protections. It is of course still possible to say, as science purists would, that these are "outside" forces, incidental to the "real" science and mathematics that make up the bulk of *De revolutionibus*. But incidental to whom? They certainly weren't incidental to Copernicus, and we don't think they should be incidental to us. The Copernican achievement was a package deal, and even if his rhetorical strategies didn't mess up or bias the physical-mathematical system, neither are they superfluous. We need to understand rhetoric, the historical context, and the cultural and political scene to make better sense of the way the whole system of "truth" operates: what gives it its force, what makes it persuasive, and what makes it work.

Like scientists, historians are always returning to bump their heads against existing materials over and over, and each time they bring with them new interpretive skills, new data, new perspectives. If it was once difficult to craft a historical Copernicus who was anything other than the heroic, oppositional, pure scientist, we're now able to cast him differently, as expert muddler: an adept mixer of math and rhetoric, church reform and disciplinary reform, new truths and old, reform and tradition, change and conservatism, the radically new and the absolutely primitive. Like all scientists (and historians), Copernicus had to articulate a kind of sweeping narrative within which his science made sense. It was not the narrative of "vulgar triumphalism" so popular today among science purists, but of subtle, binding connections between religious, cultural, and scientific ideals which, taken together, would stabilize and authorize a social order that was both old and new.

Clarity as Illusion: It's Reigning DNA

Copernicus articulated a broad, overarching narrative as part of his work, but this is not the only way in which language is vital to the sciences. Language also works within the sciences at microlevels. Here we take another approach to the questions about saying things exactly and plainly—that is, without metaphor—that we started to explore with Bateson's dialogue.

The direct truths of the sciences, revealed through seemingly transparent technologies and methods, are supposed to be literal and self-evident, as plain as the nose on your face. This is why today, for scientists like Richard Dawkins who write popular books, the word "plain" has a place of particular honor and importance. "My own books have been both popularizations of material already familiar to scientists and original contributions to the field which have changed the way scientists think," Dawkins has told an interviewer, "albeit they haven't appeared in scientific journals or been languaged up with incomprehensible jargon. They've been written in terms that any intelligent person can understand."[12] The implication here is that science doesn't have to be hard or abstruse; not only can it be translated into plain language—or maybe "languaged down" would be more comprehensible—but translated such that the terms preserve their original, unequivocal meaning for *any* (intelligent) person.

But plain language says both more and less than what it seems to say, which is why it merits skepticism and close scrutiny. Asked by the same interviewer about Dawkins's work, computer scientist W. Daniel Hillis commented: "My only complaint about Dawkins is that he explains his ideas too clearly. People who read his books often walk away with an illusion of things being much simpler than they actually are. . . ."[13] Clarity, then, can be a form of illusion, and often comes at the price of necessary complexity. One of the complex things that Dawkins most frequently covers up—indeed, it is a covering up that amounts to perhaps his most significant contribution to the sciences—is the organism.

In his book *The Blind Watchmaker*, Dawkins spins an image that comes close to summing up his scientific viewpoint:

> It is raining DNA outside. . . . Up and down the canal, as far as my binoculars can reach, the water is white with floating cottony flecks. . . . The cotton wool is made mostly of cellulose, and it dwarfs the tiny capsule that contains the DNA, the genetic information. . . . It is the DNA that matters. The whole performance, cotton wool, catkins, tree and all, is in

aid of one thing and one thing only, the spreading of DNA. This is not a metaphor, it is the plain truth. It couldn't be any plainer if it were raining floppy disks.[14]

This is quite the vision, a plain vision (assisted by presumably transparent binoculars) that sees behind the illusions where less well-trained eyes might linger. But literary theorist Richard Doyle sees something else going on in this passage. Doyle analyzes the ways in which language works (and doesn't work) within the life sciences today, and contends that "this vision hides as much as it reveals . . . the 'program' or floppy disk of DNA is not itself sufficient for life. The 'fluff' that Dawkins disparages in the name of 'plain truth' is more than a mere husk or tool; in its movement and 'performance,' it literally makes life possible. In a sense, it is nonsensical, or at least certainly not 'the plain truth,' to speak of the spread of DNA without remembering the spread of organisms."[15] One might just as sensibly say, "It is the 'fluff' that matters. All DNA coding and transcription is in aid of one thing and one thing only, the spreading of 'fluff.'"

So Dawkins says less than he should, but also more. In the name of plain speech and transparent truth, Dawkins musters a whole series of metaphors. He "systematically deploys metaphors while refusing them," as Doyle puts it. That contradictory impulse results from the fact that language within the sciences, if it is to be at all productive of thought, is inescapably metaphorical.

In this and other examples from his book *On Beyond Living*, Doyle gives us another language with which to articulate how the life sciences work, via both the "hardware" of experimental technologies *and* the "software" of language. Drawing on such computer metaphors is indicative of the strategies and necessities of both Doyle's project and the projects of the life sciences today: symbiotic, hybridizing, recombinant endeavors. Imagine a software program without the machinery of a computer—it's not very interesting or effective. Now imagine a DNA program without cellular machinery—it's not very interesting or effective. And now imagine the sciences without language and its metaphors—ditto. Finally, imagine a book about the life sciences that didn't make use of metaphors and other linguistic tools from other disciplines and social domains.

Doyle's book is not "plain speech," but a precise and demanding discourse appropriate to the complexity of the object it is analyzing: the life sciences and, indeed, "life" itself. It respects the fruitful and unavoidable contradictions inherent within the languages, practices, and theories of life scientists. Doyle is not exposing molecular biologists or sociobiologists as frauds; he is not "anti-life sciences." He is bringing his expertise in language to bear on the problem of what effects language itself has on and within the life sciences, particularly in an era in which "life" itself is figured in terms of a kind of language, from genetic codes to neurochemical messages. It's a pursuit he shares with a number of scientists as well, such as the biophysicist Henri Atlan:

> [T]he notion of the genetic program, originally a metaphor proposed by biologists to target new problems and define new research directions more than as an answer to the eternal questions about life, has become the constant refrain and crux of reflections about the innate and acquired, which are so many false problems derived from the fact that its metaphorical character has been so quickly forgotten; the "dogma" of molecular biology, so designated by molecular biologists in an attempt to be provocative and facetious, has indeed effectively become a dogma, in the vulgate of this discipline. The zenith of this line may have been reached by the theory of "selfish genes," abusively applied to extend sociobiology to political sociology. . . . What was only a joke. . . . became a serious description of how life is supposed to be in reality, beyond the appearance of living beings, their behavior, and their functions, as we experience these. In all these cases, scientific discourses were taken as new dogmas to be handed down religiously after being dissected out from the context of the works and discoveries that motivated them.[16]

While we wouldn't say, with Dawkins, that "it is raining DNA," there's no question that in scientific culture of the life sciences today, and in our culture more broadly, DNA reigns. We would say, with Atlan and Doyle, that understanding what metaphors do for the sciences can help demystify some of the power the gene currently holds in the public imagination. Such an understanding can also help scientists pursue better, less dogmatic sciences—and perhaps help Dawkins write better, less dogmatic popularizations.

We'll take a deeper plunge into the complexities of contemporary life sciences in a later chapter. There we'll see how the "discourse of gene action" was articulated within the research context that Atlan alluded to above, connecting up with new tools and techniques to transform many areas within biology. We'll also see how the "dogma of DNA" might be mutating into something less dogmatic, under the developmental pressures created by that very transformation. This brief look at life here has only scratched the surface of the ways in which metaphor and other rhetorical features of language are inseparable from the practices of the sciences—indeed, are inseparable from the power and beauty of science. Metaphors are not simply good or bad, productive or unproductive, but both simultaneously: they limit and unleash, constrain and generate. They flow not only from science to the larger culture, but from the larger culture to the world of science, and are equally vital to each sphere.

Metaphors can be attended to, questioned, and rearticulated, but never eliminated. Perhaps ironically, the purity of supposedly transparent, plain, unmediated truth leads to a kind of vulgarity, in the same sense in which we saw the "vulgar triumphalist" view of Copernicus at work. To lose sight of metaphors, to forget about the ways in which they both enable and constrain the sciences, is to fail to appreciate the real subtleties and complexities that pervade organisms, the sciences, history, and language alike.

Gender Bending

In the last twenty years or so, scholars developing a variety of "feminist analyses" of the sciences have traced the many ways in which gender metaphors have been articulated within and about the sciences. The scare-quotes are there to remind you that saying *exactly* what constitutes a good "feminist analysis" of the sciences is a precarious, contentious, and, in the end, a somewhat quixotic endeavor. Within this diversity of feminist analyses of the sciences there are many intellectual and political insights, as well as inevitable limitations and blind spots. The best of them, however, share a commitment to pursuing sciences that offer "a more adequate, richer, better account of a world, in order to live in it well and in critical, reflexive relation to our own as well as others' practices of domination and the unequal parts of privilege and oppression that make up all positions," in the words of one practitioner, Donna Haraway; the problem is "how to have *simultaneously* an account of radical historical contingency for all knowledge claims and knowing subjects, a critical practice for recognizing our own 'semiotic technologies' for making meanings, and a no-nonsense commitment to faithful accounts of a 'real' world, one that can be partially shared and friendly to earthwide projects of finite freedom, adequate material abundance, modest meaning in suffering, and limited happiness."[17] It's the problematic project we're calling "pursuing sciences" here, the challenge of muddling through realitty.

Grand articulations involving rigid categories aren't very helpful to such endeavors. To argue that the sciences are not only a male enterprise (i.e. socially dominated by male practitioners), but a masculine one—the intellectual embodiment of a knowing, masculine subject which dominates and tortures a feminine nature, with disastrous ecological and social consequences—is not going to get us very far. Nor will reversing the privileged terms: Nature is a Woman, and because women are more organically connected to it, they constitute better "knowing subjects" employing a "feminine epistemology."

What will be more productive in the long run are those many feminist analyses that detail the historical, social, semiotic, and technical specificities that go into and out of particular episodes and endeavors in the sciences. One such area has been primatology, a field that provides powerful understandings, or articulations, about what it means to be human through the study of other members of the primate order.

MONKEYS IN THE MIDDLE

One of the most thorough, thoughtful, densely articulated, and popular books has been Donna Haraway's *Primate Visions*, a kaleidoscopic treatment of the kaleidoscopic production of this branch of science. In a zone of myriad intersections, it seems Haraway has explored them all: where the history of primatology has had to cross with racism, colonial projects, space exploration, popular culture, changing professional and disciplinary standards, shifting developments in biology, and many others. This book, and the rest of Haraway's work, is already having profound effects on the teaching and thinking of sciences today, and perhaps of all the books we refer to here most deserves to be called essential reading.

Haraway's work is a real contribution to the field of primatology, not despite, but precisely because of the fact that she incorporates historical, feminist, cultural, and political analyses into her writing and research. This would be true even if she didn't have a Ph.D. in biology from Yale. But the work of another, more conventionally credentialed practicing primatologist, Sarah Blaffer Hrdy, serves as a better example of how scientists themselves have a much more expansive view of what the sciences are than is traditionally thought or claimed.

Hrdy has identified herself as both a feminist and a sociobiologist, two words that are not generally found together. Sociobiology is most frequently associated with Harvard biologist E.O. Wilson, and readers might be familiar with the debates on sociobiology that began in the 1970s, debates both within the scientific community and around it. "Left-leaning" biologists like Wilson's Harvard colleagues Stephen Jay Gould and Richard Lewontin publicly criticized sociobiology, arguing that its emphasis on reproductive success (focused at the genetic level), its extrapolations from the behaviors of other primates (and other species altogether) to the behaviors of humans, and other features made its status as a science at best problematic, at worst a kind of pseudoscience. Moreover, such critics argued that sociobiology was shot through with "conservative" values: a valorization of competition, a deterministic (and pessimistic) view of human nature, a tacit legitimation of violence and male sexual domination. There is much merit to some of these arguments about sociobiology; a good guide to the articulations made among evolutionary theory, ethology, social theory, and political science is Philip Kitcher's book *Vaulting Ambition*. But there is also something to be gained by leaving these arguments aside, and zooming in to a tighter focus, where we discover Hrdy rearticulating the supposedly essential connections between "conservative values" and sociobiological theory.

Charles Darwin (whom we look at in more detail in the next chapter) had described in his 1871 book *The Descent of Man and Selection in Relation to Sex* what Hrdy calls the "partially true assumptions" that would shape this particular field of inquiry for the next century. "The males are almost always the wooers," Darwin wrote, and the females of the species were mainly restricted in their activities to choosing the best suitor of the bunch. Darwin didn't see himself as making assumptions, but as reporting the transparent facts: "It is shown by various facts, given hereafter, and by the results fairly attributable to sexual selection, that the female, though comparatively passive, generally exerts some choice and accepts one male in preference to the others."[18]

Is that a statement of fact? Is it a theory? It could be named a number of things, but in calling it an articulation, we suspend those questions and adopt a kind of ethnographic perspective on it. It was written by an eminent scientist, an expert (if not *the* expert at the time) in evolutionary theory, according to the rules for what would count as fact or theory in that context. Whether or not it was a solid fact or a good theory at the time, or whether we would judge it so now, this articulation was *powerful*. Hrdy renames this articulation "the myth of the coy female," which sets up a myth/science distinction that we want to suspend for now. But we can still follow Hrdy as she traces the descent of this myth (in her terms) or powerful articulation (in ours) through subsequent articulations of evolutionary theory.

The 1930s and 1940s saw the growth of what was later called the "evolutionary synthesis," in which Darwinian thought was combined with, reinterpreted in light of, and reworked with new theories from genetics, new statistical methods of analysis, and much new empirical data. A key paper in the field was published in 1948 by the plant geneticist Angus John Bateman, who in this case was working not with plants but with the fruit fly *Drosophila,* the research organism most important to the development of (animal) genetics.[19] Conducting a series of sixty-four experiments with these flies, Bateman found that "successful" males could produce three times the offspring that a "successful" female could. He concluded that male fruit flies could gain (in evolutionary terms, by simply having more offspring; in more sociobiological terms, by reproducing their genes) by mating again and again, while the female fruit flies had little to gain from multiple mating. (The language of "gain" in these kinds of articulations signals a muddling between evolutionary and economic theories, which will come up again in the next chapter.) Bateman's extrapolation from flies to all of nature made his 1948 publication the second most widely cited paper in the literature of sociobiology. "There is nearly always a combination of an undiscriminating eagerness in the males," he wrote, "and a discriminating passivity in the females. Even in a derived monogamous species (e.g. man) this sex difference might be expected to persist as a rule."[20] It didn't matter whether one was talking about female or male flies, female or male apes, or female or male humans; to a sociobiologist, they were all equivalent, since they all had genes.

This is how the concept of "coyness," along with that of the "nurturing female," came to operate so powerfully within sociobiology. These articulations continued, for all practical purposes, to work as a fact or solid, true theory. "At the root of this generalization concerning the sexually discriminating female (apart from Victorian ideology at large)," Hrdy writes, "is the fact of anisogamy (gametes of unequal size) and the perceived need for a female to protect her already substantial investment in each maternal gamete; she is under selective pressure to select the best available male to fertilize it."

(The wry reference to "Victorian ideology" places it on a different level from that of the "fact of anisogamy." It would be no less legitimate to say that these are both articulations, both are constitutive of the "coy female," and both operate as powerful "sign-forces" out of which scientific theories [and realitty] are built. If anisogamy seems more solidly factual, remember that these kinds of articulations about coyness still get picked up regularly in popular venues from *Playboy* to *Newsweek*. Victorian ideology is as much a signal that resists as the fact of anisogamy.)

These articulations dressed as theories or facts of female sexuality began to fall apart under a number of pressures. One of these was empirical: Behavioral studies of baboons and chimpanzees began by the 1970s to yield different stories—stories about female primates exhibiting what Hrdy and other researchers call polyandrous (rather than monogamous) behavior and relationships. But still, it wasn't until 1980 that the subject began to be scrutinized carefully. That is, it wasn't until 1980 that primatologists began to rearticulate their own articulations, and pay attention to them *as* articulations.

Hrdy is careful to qualify her thoughts about why this was so: "In my opinion, no conscious effort was ever made to leave out the female sides to stories. The Bateman paradigm was very useful, indeed theoretically quite powerful, in explaining such phenomena as male promiscuity. But although the theory was useful in explaining male behavior, by definition . . . it excluded much within-sex reproductive competition among females, which was not over fertilizations per se but which also did not fall neatly into the realm of the survival-related phenomena normally considered as due to natural selection. . . . The realization that male-male competition and female choice explains only a small part of the evolution of breeding systems has led to much new work. We now have, for example, no fewer than six different models to explain how females might benefit from mating with different males."[21] Articulations proliferated, in other words, as researchers created more complex sets of articulated elements: observations, statistics, cultural baggage, literary forms, and so on.

"Improved methodologies and longer studies would not by themselves have led us to revise the myth of the coy female," Hrdy suggests, although these are important factors that can't be dismissed. In some cases, those conventional "scientific methods" played a critical role in revising primatological and evolutionary theory. But in the case of the coy female they were less important, "simply because the relevant information about 'female promiscuity' was already in hand long before researchers began to ask why females might be mating with more than one male." Instead, Hrdy thinks that "something motivational changed" and she is still exploring possible explanations for that hard-to-define something.

Chief among these was the sheer increase in the number of women researchers working in primatology. Is it that women were better observers? Maybe. Is it that women researchers have a more pronounced capacity for empathy with female subjects? Maybe, but that would run the risk of naturalizing "feminine" and "masculine" capacities in ways that might come back to haunt. After exploring some of these possibilities, offering some arguments and references to the literature on both sides, Hrdy defers to another kind of expertise: "I leave the general answers to such questions to social historians, who are more qualified than I to deal with them." Recognizing both the importance of the questions, and the ability of another group of scholars to bring their methods and theories to their further exploration, if not their answering, is refreshing and admirable.

Hrdy further demonstrates her eminence as a scientist by reflecting on her own training and experience, tracing her career and research interests from graduate school at Harvard where she began research in infanticidal behavior by males, to her focus on female reproductive strategies a decade later. The transformation had multiple causes, from new encounters with different species, new theoretical perspectives, and a "dawning awareness of male-female power relationships in my own life, though 'dawning' perhaps overstates the case." Field biologists and primatologists, she reflects, are often profoundly "politically unaware," leading isolated lives, hesitant to join groups or movements. At the same time, she thought many feminist writings reflected poor scholarship. Nevertheless, "there were two (possibly more) interconnected processes" involved in her professional transformation: "an identification with other females among monkeys taking place at roughly the same time as a change in my definition of women and my ability to identify and articulate the problems women confront." The admission of that intersection between professional and personal, scientific and social (and possibly more), "raises special problems for primatologists. My discipline has the choice of either dismissing me as a particularly subjective member of the tribe or else acknowledging that the tribe has some problems with objectivity. It is almost a cliche to mention now how male-biased the early animal behavior studies were. But, in the course of the last decade of revision, are we simply substituting a new set of biases for the old ones?"

Hrdy remains committed to certain scientific ideals, such as objectivity, even as she puts them under question and pursues their frayed edges. She suggests that not only primatology, but the sciences more generally are "inefficient, biased, frustrating, replete with false starts and red herrings, but nevertheless responsive to criticism and self-correcting, and hence better than any of the other more unabashedly ideological programs currently being advocated."[22] In other words, simple commitment to a political ideal is not the best way to pursue primatology; putting those commitments within a more multiple, more open, more questioning and critical system of inquiry and engagement makes for a much better pursuit. "Feminist perspectives," in Hrdy's view, "can lead to more balanced observations and to the construction of more insightful, comprehensive models."[23]

Hrdy's is a far more nuanced view of the "self-correcting" nature of the sciences than the one we usually encounter. Correction and criticism can come not solely from "the facts" of the material word, and not just from "new theory," but from careful attention to the ever-present work of metaphors, social beliefs, personal and professional positioning, *and* more encounters with a seemingly infinite world of differences and sign-forces. Correction and criticism in pursuing sciences comes, in short, from reality.

Feminist perspectives can show, in other words, how primatology is neither pure science in the traditional sense, nor is it merely a reflection of social interests or cultural values. Primatology is always in the middle, in between these ossified categories, which is precisely what makes it one of the sciences. Indeed, being in between in this fashion makes it not an "it," but a "them:" primatologies. An emphasis on the plural allows us to see the differences between, say, "Western" and "Japanese" primatologies (inexact terms which shouldn't be taken too seriously) as just that: differences, requiring respectful analysis and even cultivation, rather than dismissive eliminations.

Consider the work of Pamela Asquith, an anthropologist who has spent more than fifteen years studying primatologies and the primatologists who produce them, especially in Japan. Asquith accounts for the differences between Japanese and Western primatologies and primatologists in terms of differences in both cultural beliefs and practical constraints. The observations and scientific reports of Japanese primatologists were considered "too anthropomorphic to be taken seriously by most Western researchers;" they seemed "atheoretical, uncommitted, or even illogical." Yet Asquith shows how Japanese primatologies were just as rigorous as their Western versions, and indeed were capable of articulating insights that were at first resisted by Western primatologists only to be accepted as true accounts decades later. Japanese primatologists, for example, had more complex and nuanced notions of authority, kinship, and individuality, articulated via a different set of cultural values and social practices. These articulations produced much richer accounts of primate behaviors than did the Western accounts, limited as they were to concepts of authority grounded almost exclusively in terms of physical size and strength.

The Japanese primatologist Yukimaru Sugiyama offered the following comment on Asquith's initial research proposal: "The biggest problem which we can't write in our scientific articles in English is the basic way of thinking by the language barrier." That noisy translation can serve as a quite appropriate and even exact articulation of the situation: "Thinking by the language barrier" suggests something akin to "staying close to questions without resolving them." Rather than try to get over or breach language barriers, perhaps we can think *by* them: standing near them, in their shadow, tapping at their very solid differences. (And there's always the chance that we can get by those language barriers, over them, even if only temporarily and partially.) Even what might be a bad translation from Japanese to English can become a place to start inquiry, a place for thinking by confronting and respecting the barriers of difference.

Asquith points to some of these differences. "Scientific reports written in Japanese do not typically state conclusions. Instead, they try to describe one fact from various points of view. These points may be connected by imagery to other points, rather than by strict logic. Why do Japanese scientists not state a firm conclusion? It would, they say, close their world. A typical example is found in biologist Motokawa's observation that a conclusion means that 'this statement is the truth.' Such a statement, he maintains, is definitely false because our words can never be absolutely true."[24] Truer words, as the saying goes, were never spoken.

A CULTURE OF NO CULTURE?

These differences of language and culture might operate not just in primatology, but even in the supposedly "purer" and "harder" disciplines of physics. Sharon Traweek is a cultural anthropologist who has studied high-energy physics and physicists, and what she wryly calls its "culture of no culture." Her book *Beamtimes and Lifetimes: The World of High Energy Physicists* describes in great detail the formal and informal educational mechanisms, the daily practices of dress and speech, and the more professional practices and ways of speaking that together produce a world which seems to be the most above-it-all of all the sciences, the "culture of no culture." Her later writing extends the comparative work she began there between U.S. and Japanese high-energy physics. Listening "by the barrier" here has led her to ask questions about what we regard as most "natural," and what room there is for difference and pluralism:

> Just how did we get to believing in those peculiar singular generics: science, man, woman, state, justice, evil, god, love, truth, beauty, logic? Why is it, in our time, in our country, in our academies, considered so

very blasphemous to add an *s* to those words? Why is it so horrifying to suggest that we might think more interestingly, and perhaps more carefully, if we stopped, just for a while, using any singular generics? Just saying this in a seminar once led to a philosopher announcing that in the future he would refuse to be in the same room with me.[25]

In helping some of the Japanese physicists with whom she works revise drafts of scientific papers they had written in English, Traweek realized that a key translation problem was that the Japanese physicists were unaccustomed to the demands of definite articles such as *a*, *an*, and *the* on which the English language and culture relied. "I knew that the Japanese had done perfectly interesting physics for a century without recourse to the singular generic, the indefinite article, and the definite article. Obviously, it was not necessary for science."

But it is nevertheless a persistent cultural habit that works on many levels. Karen Barad is a physicist who has written an exemplary article about her equally exemplary "Feminist approach to teaching quantum physics." Like Traweek, she too wonders about the habit of singularity and the effects it has on how physics is taught and, ultimately, done. One of Barad's questions: "How is it that many physicists believe that there is *a* Copenhagen interpretation, based on the understandings of Bohr and Heisenberg, when these two physicists had profoundly different philosophical perspectives and therefore disagreed about many of the essential issues?" By looking at the differences hidden by the definite article and singular phrase of "the Copenhagen interpretation," Barad gets her students to think critically about "the role of descriptive concepts in scientific knowledge construction."[26]

Barad's description of how quantum theory is usually taught at the college level is right on target. We know this firsthand: One of us (Fortun) veered from a three-year path as a physics undergraduate major and into the history and philosophy of science, when the *really* interesting questions were deemed out of bounds in precisely the manner Barad describes; the other (Bernstein) has developed and teaches a course called "Quantum Mechanics for the Myriad" which doesn't cover up the conceptual conundrums in quantum theory, but instead uses them to spark creative and critical thinking in physics. And it's that kind of critical inquiry, on which the sciences so depend, which is precisely what the conventional way of teaching quantum theory stifles. As Barad points out, when the curious college students starts to ask probing questions about how it is, for example, that Feynman's electron doesn't really have spin until it is measured, the usual response from the physics professor is what Barad calls "mystifying": "That's just the way it is. This is what quantum mechanics predicts, and this is what we find in the lab. . . . We don't care about whether falling trees make noises in forests when no one is there to hear them. Using this theory we can calculate he outcomes and

successfully predict an incredible range of phenomena that can be verified experimentally." "This learned ignorance," Barad writes, "is part of the culture of extreme
objectivity in physics," or Traweek's "culture of no culture."

Barad returns to the work of Niels Bohr to reinterpret and rearticulate Werner
Heisenberg's famous "uncertainty principle." In this area, painstaking attention to
subtle questions of language is absolutely essential, for both the doing of physics
and for popular understanding of it. As it is usually taught, and as most popular
accounts discuss it, Heisenberg's uncertainty principle is about just that: uncertainty. But, says Barad, "the word *uncertainty* has the connotation that an object
possesses definite properties but we're just not certain what they are." In the case
of Heisenberg's principle, the usual view is that a particle such as an electron or a
photon *really* does have a certain momentum or a definite position, but because
this microworld is so finely grained and our measuring instruments are so comparatively clumsy, we just can't be certain what that momentum or position *really*
is. It's Planck's "real outer world" that we can never directly know. Although a
seemingly radical reinterpretation of the physical world, because "we" always "disturb" "it" by our interactions (in the subject-verb-object articulations that are so
hard to avoid), the term "uncertainty principle" actually preserves—while complicating—the same old dichotomies.

Well, as we'll see in some more detail in Chapter 8, we can now say that this articulation of the Heisenberg indeterminacy principle—and the articulation of the
real outer world that goes along with it—is, in realitty, wrong. Barad shows how
Bohr's articulations of these questions, concepts, and experimental results differed
from Heisenberg's. Rather than speak of an inaccessible reality that physics tries to
get at directly (and then fails to do so), Bohr spoke of "the phenomena." All quantum properties, such as electron spin, are properties of phenomena rather than
"things out there." You could say that phenomena are formed as the "interaction"
between "object" and "instrument," as long as you remember that you're really not
doing very sensible physics when you talk about an "object" to which you have no
access. Phenomena are "in between," and the in-between is all we have in quantum physics, or at least all that it makes sense to talk about. "That is," Barad explains, "within a given context, classical descriptive concepts can be used to
describe phenomena, our intraactions within nature." Barad deploys the term "intraaction" "to emphasize the lack of a natural object-instrument distinction, in
contrast to *interaction*, which implies that there are two separate entities; that is,
the latter reinscribes the contested dichotomy."

As Bohr, sounding very much the pragmatist and even possibly *feminist*, put it:

> The extension of physical experience in our days has . . . necessitated a
> radical revision of the foundation for the unambiguous use of elementary

concepts, and has changed our attitude to the aim of physical science. Indeed, from our present standpoint, physics is to be regarded not so much as the study of something a priori given, but rather as the development of methods for ordering and surveying human experience.

We meet here in a new light the old truth that in our description of nature the purpose is not to disclose the real essence of [physical objects] but only to track down, so far as it is possible, relations between the manifold aspects of our existence.

"Feminist" readings of the sciences are about much more than simply gender. Many of the working perspectives and concepts we use throughout this book, and in our practical and intellectual work generally, draw on feminist theories. They can be particularly good at analyzing what gets overlooked or deliberately excluded from traditional perspectives: the daily work, the often uneven distribution of that work, the ways in which innocent metaphors can have not so innocent effects.

Indeed, the term that Haraway has developed to articulate her concept of what the sciences are and should become, "situated knowledges," has many affinities with our own views. Haraway writes about how she often begins a new class "with the serious logical joke that, especially for the complex category and even more complex people called 'women,' A and not-A are likely to be simultaneously true. This correct exaggeration insist that even the simplest matters in feminist analysis require contradictory moments and a wariness of their resolution, dialectically or otherwise. 'Situated knowledges' is a shorthand term for this insistence. Situated knowledges build in accountability."[27] Learning to think of the sciences as "situated knowledges," and then situating them differently, means to never claim them as direct, revealed, transparent truth, but as always depending on particular accounts (in the many senses of that term—fiscal, narrative, and legal). Situated knowledges are knowledges of realitty.

In our recounting of Hrdy's analysis of the "myth of the coy female," the primary sense of "articulating" on which we focused was in the form of statements about sexual behavior, sexual difference, and "gain." In the not too distant background, though, we saw how those statements were also hooked up, or articulated, with field observations, genetic experiments, statistics, and other elements. This was primatology as lobster: it moves and grasps because it is not simply political ideology, not simply myth, nor simply observations, experiments, and pure theories; primatology (and physics) works because it is all of these and many other segments linked in an articulated structure. Our next examples, drawn from the worlds of biomedical research, extend this latter sense of "articulating." And we'll see again how good scientists know how to pay attention to all the elements being articulated, what their limits are, and what statements are responsible ones.

This Is Your Brain on PET

PET (positron emission tomography) scans are a relatively new biomedical imaging technology in which a radioactive substance is tracked as it courses through various parts of the brain. PET technology produces both images and data, and this is an important distinction to bear in mind. You've probably seen the images, which are quite impressive and beautiful: multicolored cross sections of the brain, often presented in paired sets over the captions "normal" and "abnormal." The data that underlie these images are comprised of thousands of measurements of certain chemicals in the brain, organized by computer as they fluctuate. But both images and data are geared to making statements about brain functions and processes. This is what distinguishes PET from technologies like x-rays and MRIs (magnetic resonance imaging), which are designed to give us information about physiological structure or morphology and therefore yield static pictures. PET images, on the other hand, represent changes over time—for example, how quickly different regions of the brain are metabolizing a chemical source of energy.

PET evidence is being used in the courtroom, in legal arguments over mental competence. It also appears in such popular venues as *Newsweek* where the flashy color images give the public an impression of clear, scientifically established differences between "schizophrenic" and "normal" people, between less intelligent and more intelligent people, or between men and women generally. These images and the differences that they are said to represent bring to mind the term "transparency" as outlined earlier in this chapter; that is, that there *really* are biological differences between, say, the certifiably mentally ill and the rest of us, and this technology allows us to see those differences clearly.

Some scientists associated with PET certainly hold something like this view. But other scientists (including some who helped invent PET technology) are more aware of, and more careful to point out, the mediated nature of both PET images and PET data. Joseph Dumit is an anthropologist who has been interviewing the scientists who invented PET, who now use it in research, and who have become both expert witnesses and media figures. Dumit's work gives some glimpses into how these scientists are often well aware of the nontransparent qualities of their inventions and the knowledges they produce.

Michel Ter-Pogossian is a physicist and is the director of PET at Washington University in St. Louis. As a scientist who played an important role in the development of PET, his discussion of the mediated nature of the images deserves serious attention:

. . . We can have a color scale and each different color represents a five-percent increment. But often colors place boundaries where they don't exist, you can jump from one color to another. Otherwise, people like them, and they are just more attractive to show in color. The dynamic range is increased, though in an artificial way.

As a matter of fact, it can mask things, or rather, put in boundaries where there aren't any. All at once you go from green to yellow and you see a boundary of some sort and it is just a small increment one way or another. Still, in general we try to show [them] in color because they are attractive.

Dumit: . . . [O]ne of the things that I am interested in is the color pictures in terms of the different things that they can signify. In one case they signify that there is a lot of activity going on here.

Ter-Pogossian: Well, yes, they signify whatever you want them to signify. This is the pitfall, of course. You can emphasize, for example, a given phenomenon, very artificially so, if you want to do it with color. It is dangerous, too. You have to be very careful when you are using it. . . . Parenthetically, the pictures that are particularly attractive that you have seen in general are doctored, in the sense of making them more attractive than they are at first.[28]

The fact that PET is a mediating rather than a transparent technology does not invalidate it. It does mean, though, that a scientist is required to be careful, aware of the built-in dangers and pitfalls. Part of that care involves being attentive to the mediating properties of the technology itself, and of course trying to improve it. Being careful also means ensuring that the public knows how these images are constructed to meet aesthetic as well as scientific requirements. This is especially important for the law, as Ter-Pogossian goes on to explain:

[T]he courts should use a certain degree of skepticism. . . . It is not enough to have one individual tell them that it is great. Why not go to three other individuals? I can find twenty individuals who will stand up and say that this is questionable. And they can show images that indeed deny that, in other words, show images of schizophrenics that are absolutely normal. It is not clear to me that one can analyze these images to the point of making them useful to courts. But as you say, the courts are quite willing to take a test, I mean, anything that shows color pictures on a screen—click—all at once, this is it. . . .

I'm not a judge. But showing PET images now to a jury without extensive preparation doesn't make any sense whatsoever. I mean, if whoever shows these pictures was given a stack of twenty pictures of perfectly normal subjects and twenty pictures of schizophrenics, and then you shuffle the pictures, I doubt that he or she would be able to stack them, to unscramble them. There are some areas, which seem to be associated sometimes with schizophrenia, but it is a minefield.

Another PET expert, psychologist Richard Haier, says similar things but with a slightly different twist:

> . . . [Y]ou could clearly say, "This scan shows hypofrontality, which we believe is consistent with the diagnosis of schizophrenia." So you know you have the usual parade of psychiatrists coming in saying this person is schizophrenic. Now you have PET. Well, the PET scan is consistent with schizophrenia, but hypofrontality by itself is not diagnostic of schizophrenia. Other people have it, too. Let me put it this way: Some juries look at a PET scan that has clear brain damage and decide this is a mitigating circumstance. Other juries will look at a similar PET scan that is obviously brain damaged and decide, "Tough. The facts of this case are so horrific that we don't really care that this guy has brain damage. He gets the death penalty." So the PET turns out to be another piece of evidence for the jury to consider, and you can't judge how the juries are going to take it.
>
> [O]ne judge, in pretrial, decided that he would admit PET scanning—no judge has decided that he wouldn't—but this judge decided that he wouldn't allow the jury to see any of the pictures. He would just allow the testimony about what was in the pictures because he felt that the pictures in themselves were prejudicial. This strikes me as absolutely true. This seems to me to be a very wise decision. Because those pictures are very compelling, and what I told the superior court justices is that if you want to manipulate PET, it was very hard to fake it by saying, "What can I think of now to activate my left anterior thalamus?" But as an operator, I can choose the colors on the scale and I can choose the interval on the scale, and I can make a lot of areas black. And that would look very dramatic.

What we have in PET are articulations within articulations. First, the image itself, the "picture of brain processes," is a computerized articulation of colors of varying and variable intensities. That articulation is made on the basis not only of measured levels of annihilation reactions indicating chemical activity, but on interpretations of statistical significance, the PET operator's level of knowledge and skill (this difference is more important than that difference, and should be emphasized), software capabilities, and even a certain aesthetic sensibility. What the "picture" *means* is another kind of articulation, where an image showing "hypofrontality" is linked up to a diagnosis of schizophrenia. (A diagnosis of schizophrenia is yet another tangle of articulations—involving professional training, changing diagnostic criteria, new pharmaceuticals, institutional pressures, national differences in diagnostic trends, and a great deal of human and familial suffering—that we can't open up here.) Or an "abnormal" image is compared against twenty other "normal" images; where a trained expert may be able to articulate the differences, a jury member may not. (It will be the prosecuting or defense attorney's job to either shore up the expert articulations, or attempt to tear them apart in cross-examination.) And then these articulated images and their articulated meanings are assembled within the public or legal spheres: dramatic images and clear meanings for the readers of *Newsweek*, expert testimony for the courts—or more precisely, for one jury and another, each articulating that testimony with its own differing concerns.

All right, it might be objected, we have to be very careful about presenting mediated PET images, and not making them prejudicial. Perhaps the images should be banned from the courtroom, but the data that underlie those images—they're raw and unmediated, and a truer, more transparent source of theory. But that's not the case, and we again can take that lesson not from the cultural analyst Dumit (as science purists might expect), but from the PET expert himself, Ter-Pogossian:

> *Dumit:* I have been looking at some of the mass-media descriptions of what PET can show. It can show efficient brains versus inefficient brains.
>
> *Ter-Pogossian:* Lord, really? I mean, efficient? How can you distinguish that with PET, efficient brains from nonefficient brains?
>
> *Dumit:* It was a theory that you use PET to measure intelligence.
>
> *Ter-Pogossian:* How did you measure intelligence with PET? What pharmaceutical do you use and what model do you use?
>
> *Dumit:* This was a DFG. . . . The theory was that with more intelligent people and less intelligent people, or people who have trained in something well versus those who haven't trained in it, like a video game or something like that. And they thought that the more intelligent, the more that you would use, and it turned out to be the opposite, the more intelligent, the less. So they thought that this might be inefficiency that causes it one way or another.
>
> *Ter-Pogossian:* Oh, I see. In order to make that kind of study valid you would need a very, very tight series; it has to be judged double

blind. The numbers that you need are enormous, the criterion that you need. No, it is doable, it would be an interesting study, but reaching any kind of conclusion on the basis of small numbers that we have now is not very responsible. Anyway, people like to do these things. . . .

But one has to be careful, it doesn't mean that the tool is not there, I disagree with that. . . . But first we have to understand what DFG does also, I mean, this is energy, metabolism. The fact that the brain uses more or less DFG really means nothing else than that. That it uses more or less DFG. You can't extrapolate. You can't even extrapolate to normal glucose. DFG is not glucose. It is 2-deoxyfluoroglucose, and its behavior mimics glucose only to a point.

Ter-Pogossian knows that there is no simple, immediate linkage between the concepts of efficiency and intelligence, and the swarm of data generated by PET technology and science. There is no reflection or mirroring; these different elements have to be articulated—and the denser the articulation, the more elements there are and the stronger the links between them can be made, the better off you are. Maintaining a constant awareness of the mediated structure of your observations and articulations is both a means to improving those observations and theories, and an important way of being careful, and scientifically and socially responsible.

From that perspective, you don't reject out of hand things like PET scans, but you do try to be very precise about how they are produced, what their limitations are, where the logical leaps are located, and, most importantly, how these refractory mediating effects should be presented in public arenas such as the courtroom or weekly news magazine. The makers of PET technology know and admit that they're "producing" images, although they sometimes lapse into naive realism. It's imperative that we become more literate about how such images are produced, how they can vary, what uncertainties are involved, what effects they produce in certain situations. We have to be literate not about received, unquestionable theories and findings, but about articulations.

There is a paradox here worth noting, if only as a good example of how to employ the deconstructive method (although such a thing only exists to the same extent that the scientific method does). We started with one of those pairs of oppositions on which our thinking relies, "transparent" and "mediated." We noted how transparency has been accorded an ideal status, against which the qualities of mediation are then defined, making mediation a negative term. Rather than take the simple move of reversing the privilege—which would make the fine slogan, "Down With Transparency; Everything's Mediated!"—we have only "traced the limit" of how these terms operate within a system of language and thought. That system, which we are more or less stuck with, depends on paired, binary terms, one of which always gets the short end of the stick. But the terms themselves don't *permit* a reversal: To say "here's how everything is mediated" depends, oddly, on the concept of transparency—the ability to fully perceive and delineate what constitutes "mediation." What the deconstructive method leaves you with is not the overthrowing of one term by another, but the same pair of terms. What's the difference, then?

That's a very good question. For now, let's just say that transparency is no longer so pure or easy to come by, and what was once a stable and safe opposition now shakes and quivers.

The Peircean Triad: Who's on Third?

We introduced our invented word "realitty" in the previous chapter. If reality is made out of atoms, to put it crudely, then what is realitty made out of? What's the fundamental structure of articulated worlds, or of the articulated, situated knowledges which map out those worlds?

Some of the most interesting and significant work within philosophy in this century has centered around questions of language. We cannot give an exhaustive treatment of this body of scholarship here. We draw on it implicitly throughout this book (as we did just above with our brief example of the deconstructive method), and explicitly from time to time. Rigorous attention to the structures, workings, and play of language has begun to profoundly affect the way we pursue not only academic work in philosophy, linguistics, literary criticism, anthropology, history, feminist theory, and other disciplines, but also how we pursue the visual arts, dance, architecture, and, of course, literature.

Because this intellectual work stays close to foundational questions about the sciences and traditional notions of truth, it has upset many scientists and other commentators. They either dismiss it as nonsense, irrational, or gibberish, or fulminate against its insidious, degrading effects. We hope that our brief, superficial account of some these developments is enough to convey our belief that it is vital to the kind of revitalization of the sciences we are advocating here.

At least since Plato, philosophers (and their scientist kin) in "the West" have been anxious to separate the literal from the figural. Other conceptual oppositions can be lined up underneath this one: prose/poetry, speech/writing, science/literature, objective/interpretive, truth/fiction. Truth had to be a literal truth: plain, transparent, not subject to different interpretations. To fulfill its own dream, philosophy had to get rid of the polluting aspects of figurative language—irony, metaphor, metonymy, synecdoche, hyperbole, catachresis, and so on. Like the heliotropic sunflower which subtly but surely turns toward the passing sun, these literary tropes turned thinking in almost imperceptible ways. So to keep our thoughts from being turned and swayed from the truth, against our intentions, most philosophers (and scientists) have taken it as their task to strip language and thought down to their literal core, where meanings were pure and unequivocal.

At the heart of this scheme (to use an inexact bodily metaphor) was the relationship between the name and the thing, between the signifier and the signified. There was supposed to be a tight, unshakable, direct, and *transparent* bond between signifier and signified, between name and thing, between scientific or philosophical knowledge and reality. The higher, purer signifier took its meaning directly from the lower, baser signified thing:

This Platonic logic has never gone unchallenged or unquestioned, but it has remained dominant throughout the history of Western culture. In the twentieth century it has come under increasing question from a variety of disciplines. The rigorous structural linguistics of Ferdinand de Saussure revealed that languages simply didn't work in that Platonic way. The relationship between signifier and signified, while not exactly severed, was problematic and complex. (Something which at least some scientists have indeed recognized from time to time, as suggested by Michael Faraday's remark that names were one thing and science another. In a similar vein, we can also recall reading somewhere that Richard Feynman's father taught him to recognize birds only on the basis of their songs, colors, shapes, and behaviors, and to ignore their names.) Investigations into languages and their articulations developed further in the work of anthropologists and linguists (such as Emile Benveniste, Benjamin Whorf and Edward Sapir, Louis Hjelmslev, and Claude Lévi-Strauss) as well as people most easily termed philosophers (most famously, Ludwig Wittgenstein and Jacques Derrida). One of their lessons has been that the meaning of individual words and terms (such as "bat," "cat," or "catecholamine neuroreceptor") is constituted not from the things to which they refer, but from their linkages to other signifiers in an entire system of signifiers, and the differences within this system. All meaning and therefore knowledge is constituted out of chains of signifiers:

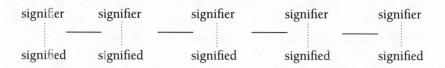

Great work has been done in this tradition, but another system of semiotic analysis is better suited to our purposes here—and our penchant for threes instead of twos, and middles instead of opposites. This is the "semeiotic" [sic] system of analysis developed by the American pragmatist Charles Sanders Peirce. Among the muddle of Peirce's various identities—philosopher, linguist, logician—is his work as a scientist, a toiler with pendulums and other precise instruments for the U.S. Coastal and Geodetic Survey. Peirce thought extensively about nature, reason, and other big questions, and his thinking was greatly enhanced by his working encounters in the sciences.[29]

As distinct from the dyadic structure of de Saussure, Peirce put an unbreakable triad at the heart of his "semeiotic" theories. This triad, a sort of atom of meaning, consisted of sign, object and interpretant, or what Peirce sometimes called Firstness, Secondness, and Thirdness. It can be diagrammed as such:

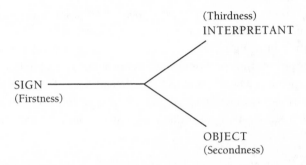

(Thirdness)
INTERPRETANT

SIGN
(Firstness)

OBJECT
(Secondness)

In Peirce's system, notes T.L. Short, "nothing is an object which is not signifiable; nothing is a sign which is not interpretable as signifying some object; and nothing is an interpretant that does not interpret something as signifying an object." So the terms Firstness, Secondness, and Thirdness are somewhat misleading, since they imply a causal or temporal order that isn't really there. The three elements collide simultaneously into an indissoluble collusion. Indeed, Peirce's overall schema can be thought of as the result of a collision between the kind of epistemological realism toward which science purists gravitate—where the real is "really real" (albeit inaccessible)—and the epistemological idealism that so disturbs them. Such a collision or collusion, as the anthropologist E. Valentine Daniel points out in his book *Fluid Signs*, "contains within it elements that are likely to disturb the idealist as well as the materialist. This is good: at best, such a disturbance should be enlightening, at worst, sobering."

Firstness, or the sign, is something we dream up. It comes from our imagination, our ideals, our minds. But of course things don't stop there, or as Peirce put it: "We find our opinions are constrained; there is something, therefore, which influences our thoughts, and is not created by them." That's what Peirce referred to as Secondness, which fits our conventional definitions of "the real." He phrased it somewhat more prosaically in another essay: "Let the universe be an evolution of Pure Reason if you will. Yet if, while you are walking in the street reflecting upon how everything is the pure distillate of Reason, a man carrying a heavy pole suddenly pokes you in the small of the back, that moment of brute fact is secondness."

Isabel Stearns puts the collusion between Firstness and Secondness quite nicely: "Firstness for Peirce has the status of a dream, a quality in itself unanchored to any laws, being a mere 'may-be,' fugitive and evanescent. Secondness on the other hand represents the encounter with hard fact, the undeniable shock of contact with the outer world. Secondness is the category of effort, of struggle and resistance." But again, things don't stop there, at two. These first two elements or categories are relatively familiar to us, and easily map onto our conventional schemes of mind/matter, subject/object, and so on. Thirdness is a bit more difficult, unusual, and subtle.

Stearns helps us out again: "[T]he category of thirdness is above all the category of mediation. . . . Thirdness is the factor of final causation which manifests itself as a 'gentle force' bringing together in a certain measure all that which without it must remain in arbitrary and unmediated opposition." As a kind of gentle, middling "interpretant" among brute fact and dreamy idea, Thirdness according to Daniel "is the locus of interpretation, that by which a sign is contextualized, that which makes signification part of a connected web and not an isolated entity." Daniel continues:

> Without an interpretant, the First and the Second, the *representamen* and the *object*, will forever remain unconnected, existing apart from any *meaningful* reality. . . . In other words, for a sign to function as a sign, there must be present in it all three correlative functions. Objects may exist in the universe as individual empiricisms or existent facts, but they do not become real until and unless they are represented by a sign, which representation is interpreted as such by an interpretant. This process of signification is as real in nature as it is in culture.

What can this mean for the sciences? Combine Niels Bohr's advice cited earlier in this chapter, that physicists should think in terms of encountering phenomena which are formed from the intraaction among their instruments and the badly named "real outer world," with Feynman's articulation of electron spin from the previous chapter. Electron spin is a phenomenon, and a phenomenon is a Peircean triad. It is an indissoluble collusion of unnameable, "brute" Second; and a First, those concepts, mathematical formulas and other signs "wrapped into and around, in a vague way"(in Feynman's words); and a gentle, interpreting Third that brings a final sense to the whole arrangement, in part by connecting it to other similar triads.

It's hard to grasp, we know, and something like electron spin is a relatively easy case, because it is in a sense so atomistic. What kind of Peircean triad is, say, "schizophrenia" or "sexual competition"? To pursue those, you have to remember that we've abstracted both the atomistic "electron spin" and the atomistic Peircean triad just to simplify their introduction. Neither one of them ever occurs in isolation. Just as no single signifier had any meaning apart from the "chain of signifiers" in which it had to be placed, so all Peircean triads find themselves linked up in a spreading, disseminating structure that looks something like this:

If we had to draw an image of realitty, it would have this kind of structure: a densely quilted network of Firsts, Seconds, and Thirds; concepts hooked up to things colluding with metaphors joined to other things cemented to ideas bolted to formulas muddled with cultural practices and so on and so on. . . . Most importantly, the network is always open-ended. As Val Daniel writes, "The extremities represented in the diagram are, in fact, heuristic fictions. For no polyadic relation has reachable end points. For this reason, every semeiotic study of however small a cultural aspect must, by definition, remain open and incomplete."

Daniel was writing as an anthropologist who had studied the conceptual and cultural systems of a Hindu village in Tamil Nadu, and writing primarily for anthropologists. But his words are as relevant to the sciences, and to scientists. Like any system of culture, the sciences remain open and incomplete, or, in Feynman's elegantly inelegant articulation, "half-assed."

Peirce would define reality as "what will be thought in the ultimate opinion of the community." This was a community of inquirers, and is very close in spirit to what Fleck alluded to with his term "thought-collective." (This suggests a number of questions we'll be raising throughout the rest of this section, and in Section II: What are the divisions and differences within the "community" of scientific inquirers? What happens when different communities reach more than one ultimate opinion? And what happens when the community of inquirers includes members of other communities—citizens affected by the sciences under question, members of other scholarly communities—in its knowledge-producing processes?) Their opinions are best thought of as probable. However disciplined those opinions may be, however carefully and judiciously the thought-style has been developed to inform them, however tightly they are articulated with instruments we have built to the most exacting specifications, those opinions will change. Reality isn't "out there." Realitty is between a community and its practices, between trained practitioners and their thought-styles, between disciplined experimenters and the instruments which they invented. Realitty is a kludge job.

Kludge Jobs

In engineering and in the sciences (particularly the computer sciences), "kludge" is a bad word—or a bad/good word, because it's something to be avoided, yet admired at the same time.

If you've spent any time in a lab, machine shop, or computer or electronics operation you know, ironically enough, *exactly* what kinds of processes and products are signified by this ugly duckling of a word. You've relied any number of times on kludging together tools, protocols, and even explanations. You've had to scavenge old machines for reusable parts, wedge a crumpled wad of foil in an open gap, jam a half-baked causal link between two irreconcilable results. You trust your skill, flick the switch, and hope for the best.

For those readers who haven't spent a lot of time in the lab, kludge (pronounced KLOOJ) is a word with contradictory connotations, whose final meaning can't be resolved. The primary connotations, according to the *American Heritage Dictionary*, are those of "a system . . . that is constituted of poorly matched elements or of elements originally intended for other applications," "a usually workable but makeshift system, modification, solution, or repair." Frankenstein (the monster) is a good example of a kludge job. A "kludge job" lacks a certain aesthetic, there's something grotesque about its mismatched parts—and yet there's something quite elegant and impressive about the whole lumbering entity. (Frankenstein [the doctor] was a kludge job, too: a little Oedipal legacy here, a bit of cocksureness there, some medical school, a chunk of suppressed literature grafted onto respectable science. And *Frankenstein* (the novel) is a fantastic kludge job itself: Mary Shelley stitched together some Romanticism, some science, some popular anxieties, some nascent feminist critique, and some personal experience into a literary machine that still runs beautifully today.) A kludge job may not always be beautiful, but it works—for a while.

The word may have been amalgamated from "klutz" and "nudge," but the problem with that theory is we would then pronounce it like "fudge." It may stem from the German *kluge*, meaning "intelligent, clever," so kludging would be clever and klutzy at the same time. The word almost certainly dates back no further than World War II. But true to its own form, its sources and full connotations are rather kludgey. As the *American Heritage Dictionary* speculates: "It seems most likely that the word *kludge* originally was formed during the course of a specific situation in which such a device was called for. The makers of the word, if still alive, are no doubt unaware that etymologists need information so they can stop trying to "kludge" an etymology together."

The qualities of kludge and kludging, then, run deep. They also range far and wide. When Claude Lévi-Strauss tried to articulate the differences between the "savage" and the "civilized" mind, he compared the latter to an engineer working out a rational system, and the former to a *bricoleur*, piecing together fragments ready-to-hand to make a working explanation. The ethnocentric biases of that fine distinction have been commented on by many, but we shouldn't forget that Lévi-Strauss's intent here was to validate and legitimate different "ways of knowing" in non-Western cultures, and for that important intellectual project he deserves much credit.

In addition, we should note that anthropologists like Lévi-Strauss and Marcel Mauss, who developed their thinking in France between the wars, were deeply influenced by the surrealist propensity for collage and pastiche—although this rhetorical construction so common to historians, "influenced by," is rather passive. Better to say that these ethnographers in France actively articulated the concerns and methods of their own discipline with some of the concerns and methods with which the visual and literary artists of the surrealist movement were experimenting. For the surrealists, collage and pastiche were a kind of highly aestheticized kludging of disparate words, borrowed images, art history referents, found objects, and "original" swaths of paint into a vibrant whole that was at once cohesive and dis-unified.

Who faces the most signals that resist, who has to operate and experiment within the most severe constraints? A Joseph Cornell arranging elements in a box to create a profound aesthetic effect, a Lévi-Strauss building a text around the similarities among and slight variations between hundreds of related myths and symbols, a Sarah Hrdy organizing a mountain of field observations along with evolutionary theory and changing social awareness, or a Richard Feynman entangling mathematical functions with physical concepts, visual or diagrammatic representations, and "things," to get a working theory-practice of electron spin? The answer would result from a rather pedantic and immodest argument, which could only have value, moreover, for creating an even more immodest hierarchy. Better to see each one as a different kind of articulation, and to describe and analyze them according to their specific differences, how those differences are linked together, and toward what ends they are turned.

The honeycomblike structure of connected Peircean triads can be read on a number of levels. It's a diagram of how language has to be articulated to make sense, or meaning: Words, phrases, and statements get hooked together, according to certain rules which are relatively stable but allow for change and slippage. A certain amount of ambiguity always remains, and meanings can be rearticulated by joining them up with new elements at the open ends of the honeycomb. In expert terms: The lines describe a *discursive field* in which we think, speak, write—and question.

The diagram can also be read as a schematic representation of the kludged elements of the sciences. On that level, meaningful articulations are themselves articulated with old and new experimental protocols and observational methodologies, technologies for manufacturing and imaging, and the social institutions (universities, corporations, courtrooms) in which these articulations are produced and interpreted. And of course, it's a diagram of reality itself.

(Keep in mind as well that this diagram is itself a heuristic fiction. The honeycomb structure is far too regular and rigid to serve as an adequate projection of a messy, irregular and ever-changing reality. But it's not a bad kludge job. It will work well enough for a limited time, and for a limited set of tasks. We'll have to do some more work on it in the next chapter, to try to make it run a little better, but even so, you shouldn't get too attached to it.)

Is reality fundamentally material? Yes. (And no.) Is reality fundamentally conceptual or semiotic? Yes, and no. Is reality fundamentally a matter of interpretation and culture? Yesandno. Consider some of our stories in more schematic form. In the case of the coy female in primatology, new observations and new experimental outcomes from other scientific disciplines (call them Secondness) combined over time with new fieldwork methodologies, new ideas, and new articulations (call them Firstness), and the more "gentle," subtle, and elusive changes in cultural attitudes and even "political ideologies" (call them Thirdness). Molecules coursing through the brain (Secondness) are only part of reality because they're already articulated with concepts of normality or abnormality, intelligence, or sexual difference (Firstness), and stylized by an interpreting community of inquirers in both laboratories and courtrooms (Thirdness).

The patterns of connection can be relatively stable and reliable, but do not have to be rigid. When we left the Hrdy story, there were six different ways to articulate how females might benefit from mating with different males. There were at least two "Copenhagen interpretations" from Barad's story of Bohr and Heisenberg. And the web in which Japanese primatologists conducted their science was different enough from that of their Western counterparts to allow for different, but no less robust, articulations about primate hierarchies and kinship.

Indeed, "robust" is a word that some of the scholars who study the practices and cultures of the sciences prefer to use instead of "objective." It's a substitution that preserves some of the better parts of the ideal of scientific objectivity—for example, reliability—while jettisoning the objectivity-ideal's absolutist tendencies and keeping a space open for pluralism, probabilism, and change. Learning to think about the sciences in terms of robust or dense articulations, and learning to *pursue* the sciences in those terms, can be no less productive, effective, and even "truthful" than doing so in terms of objective, transparent theories that have direct access to a world "out there." It has the added advantage of allowing more room for change, critical questioning, and the modesty that comes from recognizing ineradicable pluralism and probabilism.

That may not seem like a good bargain to some people: Weren't Darwin's articulations about male wooers and passive females robust, and *wrong*? Don't we want to be able to say that with all the force that transparent, direct, scientific truth affords? Despite all our careful rearticulations, doesn't a kind of relativism rush into the vacuum left by the abandonment of stringent objectivity?

Those are pressing, troubling questions. If they find some resolution in Chapter 4, it's only by way of staying close to what they ask us about the relationship between knowledge and power.

Chapter 3

Powering/Knowing

Reprogramming Galileo's Brain

The U.S. spacecraft *Galileo* arrived at the planet Jupiter in late 1995, released a probe that fell into the Jovian atmosphere and parachuted into its turbulent depths, and began sending back megabytes of information to eager scientists (and at least some eager members of the American public). It was a remarkable demonstration of scientific and technological know-how. But the event suggested a few other remarkable things about scientists as well, such as a persistent reliance on simplified beliefs about their ancestors and about their own practices—even in the midst of empirical evidence that should have cracked those beliefs wide open.

Let's begin with a passage from a *New York Times* interview with Torrence Johnson, a scientist on the *Galileo* project:

> It has taken *Galileo* half the space age to reach its destination. But that is almost no time at all compared with the 385 years separating the project and the Italian astronomer who first spotted the Jovian moons through his telescope. Even so, Johnson says, "[I]f Galileo himself walked in the door, once we'd overcome the Italian language barrier and so forth, I could easily brief him on what Galileo was doing. He'd be a much easier audience to brief on this mission than any lawmaker in Congress and the Senate."[1]

It's a fantastic belief: a temporal distance of four centuries, vastly different cultural and political circumstances, and today's planetary scientist thinks all he has to do is listen to a few Italian language tapes and Galileo and he would be getting along like brothers. The sciences are supposed to transcend time and space, even when the practicing scientists have long been returned to dust. That belief is all the more fantastic in light of the fact that Johnson and his team of scientists know all about failures to communicate, the breakdowns in connections, and the creative kludging which these errors and gaps evoke—which is what the second event is about.

When the main antenna of *Galileo* (the spacecraft) broke down on its way to Jupiter, the mission scientists managed "a change so profound and complicated that Torrence Johnson says they effectively 'reprogrammed its brains'." The craft had been damaged not so much by the launch or space journey, but by "loss of lubricant during cross-country trips between Cape Canaveral and Pasadena: the most perilous part of *Galileo*'s two-and-a-half billion mile journey may well have been a truck ride along Interstate 10." The most mundane aspect of this scientific megaproject imperiled its most heavenly. To prevent *Galileo* from being just another hurtling hunk of metal, the experts at the Jet Propulsion Laboratory muddled through magnificently: reevaluating, reprogramming, retooling, and revising their expectations downward. They wouldn't get everything they wanted, but they would get as much as they could, and that would be a lot.

It all amounted to what the *Times* called "a reconstitution of Galileo," and while it was referring only to what Johnson and his team did to the spacecraft, the phrase applies just as well to what Johnson did to Galileo himself. Like Tang mixed into water, the historically freeze-dried Galileo of seventeenth-century Padua was reconstituted even after centuries of existing in a powdered state, emerging miraculously intact and identical to the scientists of a southern California shaped by the aerospace industry. Johnson's resurrected Galileo has a message to transmit, too: Scientists speak the same language everywhere, every time, and are always at odds with the Powers That Be, who just don't understand. Science escapes time altogether, and is just as refreshing and thirst-quenching today as it was centuries ago.

That's not entirely untrue. We can indeed draw congruences between the world of Galileo-astronomer and the world of *Galileo*-spacecraft. The more nagging problem is that the Tang-Galileo promotes a kind of Tang-view of science: no kick, easy going down, a little tart but really quite *neutral*. And just as Tang was one of the spin-offs of the early U.S. space program, *power* is usually thought of as nothing but a spin-off of science; it's at best an incidental by-product of the search for truth, a nice bonus. Science happens in Pasadena, power happens in far-off Washington, D.C.; while some communication and negotiation may be an unfortunate necessity (especially around budget time), it would be disastrous to bring them fully together. Just ask Galileo.

Our preference, however, is to reconstitute Galileo into a more bracing drink. There's no question that we do need him, or at least the story of his work, to help us understand what the sciences have been, and the kinds of literacy skills we need to shape the sciences of the future. But that other story of Galileo will take a little brain reprogramming of our own, a mix-up of a much more mixed-up Galileo (and Darwin, while we're at it) with the help of social and cultural historians. These new concoctions have to be more bracing because they have to be more *em*bracing, and allow us to bring into our stories about the sciences that which is most often excluded: the sometimes destructive, often productive, but in any case inevitable effects of power.

Caution: Metaphors at Work

The concept of power is overwhelmingly large, complex, and therefore imprecise. It's a mistake even to write the word "power," as if the name were some kind of bucket that could hold all those complexities without spilling. Trying to write about the relationships between power and the sciences, or power and knowledge more generally, only compounds the error. The amount of words thrown by hundreds of scholars at these questions in the past ten years alone is truly humbling. It's a sign that the relationships between power and the sciences tenaciously resist any easy resolution. It's also a sign of the urgency of asking these questions. We desperately need new ways of thinking about what happens in the middle space between power and knowledge.

We can begin by recommending at least one thing *not* to do. Don't make any easy reductions or direct equations, such as the ever-popular "knowledge is power." That phrase (traceable to that other famous figure in the founding of the new experimental sciences in England, Sir Francis Bacon) is nearly meaningless. To mention just one example: Thanks to new Superfund regulations, citizens in a community might *know* exactly what chemicals are being released into the local stream by an industrial facility (if their local public library hasn't closed for budgetary reasons, and if it catalogues the Toxic Release Inventory). They might even know that one of those chemicals is associated with a higher rate of birth defects in laboratory animals. That doesn't mean that they know that their children's health problems can be traced to that chemical, or that they have the power to get an answer to that question, or that they have the power to demand that something be done. If we've learned anything about what the connections between our various forms of knowledge and the many operations of power have become over the span of this century, it's that they are far more multiple and indirect than simple, direct equations like "knowledge is power" would indicate. We have to try to be much more precise.

But the relationships between and among power and the sciences seem to invite imprecision. Our articulations of these questions often depends on a tangle of metaphors and vague concepts that are always sliding into and over each other. The sciences are supposed to be value-free, or value-neutral. Are "neutral" and "free" supposed to mean the same thing as "pure"? What exactly is a value, anyway? Are we talking about personal values, ethical values, moral values, social values? Are values the same as ideologies? Are ideologies the same as social interests? Are ideologies and social interests forms of power? How can something be both neutral and a source of great power? Does it only get powerful when it's "applied"? But if something is really "pure," doesn't that mean it could never turn into something "applied"? Is neutral the same as objective, or is neutrality just a kind of prerequisite for objectivity? If we say that the sciences are neutral, does that mean that they don't take sides in a conflict, like Switzerland doesn't (or like Switzerland *says* it doesn't)? Is saying that the sciences are and should be depersonalized, free of subjective bias, the same thing as saying they are and should be depoliticized, free of social control? Is all commercial research "impure" and "interested," and all academic or government research "pure" and "disinterested"? Can the sciences be both "interested" and "objective"? When people say that the sciences are value-laden, do they mean that the sciences are like a cart on which you then load a bunch of things called values? And does that mean you should just try to unload it?

That load of questions is too great a burden for one book, let alone one chapter, to bear. We're going to limit our discussion to two terms, tools and neutrality. We judge these two terms to be the most important to think about, revise, replace, and/or supplement. Rather than a one-for-one substitution, though, we're going to give you an ensemble of possible substitutions, to remind you and ourselves that the way we articulate things has enormous impact on how we can think about them, and what we do with them.

ON THE ROAD

Perhaps no metaphor of the sciences has been more persistent than the medieval one, the sciences as double-edged sword. The sword—like hammers, pens, screwdrivers, Uzis, Geiger counters, artificial hearts, and other basic "tools"—is something that is held in the hand. The hand is connected to the command center—brain, ethical subject, Pentagon, bioethics advisory committee—at some distant location. Impulses flow only one way, from the command center to the tool-wielding hand: swing this way (good) or swing that way (bad). No responsibility accrues to the hand, since it can't tell the command center which way to swing; it's just obeying orders. The tool itself is neutral, and the hand is just a transmitting conduit for a signal that originates from a great, safe distance.

These are powerful metaphors. They color much of our thinking about the sciences (and technologies as well); they help make a world in which the sciences are just a means to certain ends. Whether those ends are just, ethical, or responsible is not supposed to be the scientist's concern. Like the geographical distance between Pasadena and Washington, ends are clearly separated from means, and tools are independent of power.

But the sciences aren't like tools, they're like a transportation system. The sciences run on roads, rails, channels, and other systems of support that are collectively constructed with all kinds of social, economic, and political resources. Scientific inquiry is always carried by an infrastructure that includes not only the instruments, theories, and language tools described in the previous chapters, but larger institutions and their material and cultural resources. We wouldn't have lasers, computers, or the Internet, for example, without decades of massive military funding for such things. We wouldn't have recombinant-DNA technology, or know what we know about viruses and how they replicate, without similar levels of long-term support from the National Institutes of Health. The question of how this "inquiry infrastructure" matters, however, is not simply a question of government or industry support of some types of research rather than others—although this is certainly important. We also have to question how broader social factors shape what will count as nature, reason, or truth in particular social and historical circumstances.

An inquiry infrastructure isn't a tool that scientists pick up and use; it's a road that they help each other build so they can be transported to *someplace new*. It's a system that constantly has to be maintained with physical and mental labor (not neatly divided), and at great expense. The actual shape of the thing is a complex phenomenon that emerges from the intricate interactions of local and nonlocal politics, geography, construction materials, the demands of industry, infuriating detours, and the lure of scenic overlooks. And while new vistas of both the natural and social order may open up at unexpected turns in the road, while new worlds appear with each new off-ramp and modernizing paving project, you can't always go exactly where you like. We can always question the broader system of transportation and ask: How did we get *here*? What does the view look like from *over there*, and why isn't there any access road?

That's the crash-course version of inquiry infrastructure. The following stories about Galileo and Darwin will help map out these metaphorical systems in more detail, and both sharpen and soften some of the rougher edges. Later chapters in Section II will add even more analytic and historical specificity, particularly regarding the military support of physics and federal support of basic biomedical research.

Now, let's shift to neutral.

JUST CHARGE IT

The tool metaphor persists in our conventional understandings of the sciences because it's linked to a promise of neutrality. In terms of this promise, the sciences just *are*, they're beyond value judgment. They're a resource, a capacity to draw on, and it's only when you get into questions of how to use those tools, how and where to apply them, that questions of power, value, and ethics come up. You can't *not* want to know, and you can't place political or moral restrictions on knowing. It's not the tool itself that's the problem, but what you do with it. In perhaps their most extreme form, these examples go something like this: "You can use nuclear weapons to inflict pain, misery, and death on a few hundred thousand people in Japan—or you can use them to make a new canal across Central America. It's up to us to decide. Ergo, nuclear weapons are neutral."

These days, critiques of the "value-neutrality" of the sciences are a dime a dozen. To write at the end of the twentieth century that "the sciences aren't value-free" has become as cliché as the phrase "a dime a dozen." People have been making that argument for a hundred years or more—smart people, and good arguments. Yet this metaphor of the neutral tool remains popular and powerful. Why? The easy answer is because it's a convenient myth which serves the interests of both "power" and scientists, preserving the convenient relationship between scientists and the state which was emerging in the seventeenth century in our stories of Boyle and the Royal Society. The state appears to be acting disinterestedly, on the objective advice of neutral experts, and scientists can wash their hands of any dirty dealings and pursue whatever sparks their curiosity.

Our answer, as you might guess, is that there's no easy answer. We always end up falling short or contradicting ourselves whenever we talk about these huge abstractions like "the sciences," "tools," "neutrality," and "values"—and we don't as yet have very good strategies for living with those shortcomings and contradictions.

A good first step is to make the questions harder. For example: What are the various meanings that the word "neutral" can put into play, and what work do those meanings do?

Since we've started with the highway metaphor, let's shift to that familiar, strange, liberating, enslaving, productive, and destructive American sociotechnical development, the car. "Neutral" is that position in which a set of gears is disengaged so that power can't be transmitted from the engine to where the rubber meets the road (a metaphor often applied to the sciences and their experiments). The sciences are supposed to be a transmission in neutral. (Never mind how you're supposed to get anywhere.)

That meaning of neutrality suggests an opposite, that the sciences are driven. Many people argue that it's the engines of values, social and economic interests, or political power and ideology, that drive the sciences in one direction or another. It almost goes without saying that we regard our task as steering between these opposites. And for this we need some more meanings of "neutral."

In physics, to be neutral is to be without charge. A neutron appears in cloud-chamber photographs as a straight line. It doesn't swerve, it curves neither left nor right. It is unswayed by the immense and powerful electromagnetic fields around it—and that's the key point: it has no invisible forces acting on it. It doesn't have anything to do with the surrounding field. The neutron is a sovereign particle—it changes course only if it hits something solid and real—a carbon nucleus, say.

That's a pretty good metaphor for how we think about the sciences convention-ally. They move in a straight line, and only change their course when they collide with reality. Any other deviation must be an anomaly.

But what if we think of the sciences not as neutral, but as charged particles? Then we could see them spiral and curve, we could track how they respond to the invisible—yet with the proper technology, detectable and even manipulable—forces around them. We would conceptualize them as exquisitely responsive to the field in which they move—generating changes in that field as they move.

To think of the sciences as charged rather than neutral would take advantage of other meanings of that term. To be charged is to have a responsibility or duty im-posed. To be charged is to have incurred a debt, and to be financially liable (al-though you can "charge" that debt and postpone it—for a while). To be charged is to be excited, thrilled, alive with tension; to be charged is to be saturated, loaded, palpably permeated with a certain mood or tone. And "charged" is what happens to the members of a jury when a judge instructs them about the laws under ques-tion, how they should be applied, and how evidence should be weighed.

Does this metaphorical shift get us thinking about or acting any better toward the sciences than the tired old claim that the sciences are not value-free? Is it any less vague, or any more generative? For now, the metaphors of "charged sciences" and "inquiry infrastructures" are good kludge jobs for dealing with these vast, vague, and cumbersome concepts of values, power, and knowledge. We'll need to exercise some caution, though, to prevent "infrastructure" from becoming rigid and keep "charges" from escalating into a kind of strong electromagnetic force holding everything tightly in its power.

Perhaps after working through some examples grounded in historical specifici-ties, we'll be in a different place, a place of *contingent affinities within an assemblage*. (It's not as scary a place as those italics might make it appear.)

Trying a More Complicated Galileo

We'll look at some of the issues raised in Peter Huber's book *Galileo's Revenge: Junk Science in the Courtroom*[2] in Chapter 4. But since it offers another of the more strik-ing examples of how *not* to reconstitute Galileo, we introduce it briefly here.

According to Huber, junk science looks like science, but really isn't. It's just a mess of empty articulations, substanceless except for the political or social agenda it serves. Junk science is so horrifying to Huber because its proponents often claim to be "a new Galileo, a lonely, misunderstood genius who can see wonders that others neither discern nor understand." Hence the "revenge": Thanks to the example of Galileo, the most far-out truth claims can now be packaged and sold as the unappreciated work of farsighted, antiestablishment free thinkers.[2]

Huber is wrong about many things, but perhaps nowhere is he more blatantly off track—and just factually *wrong*—than in this reconstitution of Galileo. Recent historical work shows Galileo to be far more interesting than just a man of facts and experimentation, and with a relationship to the Pope that was far more complicated than the usual heroic opposition between the Man of Reason and the Tyrant of Dogma that is so much a part of our modern mythology (a mythology continually reinforced by figures such as Huber). Mario Biagioli's book *Galileo Courtier* portrays the man and his time with the kind of complexity we need now to think about the sciences and about scientists. The book shows that Galileo was not simply misunderstood by the powerful Pope Urban VIII; they in fact shared a very interesting and subtle set of understandings. Galileo and Urban had a complicated history, and not just a simple, brief trial. And far from being lonely, Galileo was intensely social. That's precisely what made him a genius—the one point on which we might agree with Huber (it just happens to be the least possibly interesting thing to say about Galileo).

Because these re-visionings of major scientists tend to provoke anxiety and misunderstanding, we want to articulate a few things quite carefully, and try making the more predictable and reactionary misreadings less likely. Thinking about Galileo in the context of seventeenth-century Italian court culture doesn't rob him or his work of "genius." Saying that Galileo *made* his science and made it popular (and made a new cultural and professional identity for himself in the process) isn't saying that he *made it up*. The questions are about what kind of infrastructures carried Galileo's projects and allowed him to move, think, and create, what kinds of charges his new knowledge held, and what material things, articulated ideas, personal skills, social institutions, and cultural patterns made it possible for him to make a new science and a new "world system"—a new realitty.

Galileo made many new observations about the Earth's moon, the phases of Venus, the sunspots, the moons of Jupiter—all of them remarkable achievements, given the novelty of these efforts, the need to invent and improve simultaneously the observational technology (the telescope), and the patience and scrupulous attention to precise detail that was necessary. Galileo (and a number of contemporaries) introduced a new language and practice into what was then called natural philosophy—the language and practice of mathematics—and they have been inextricably and productively intertwined ever since. Opening up the question of how Galileo and what we might call his "knowledge practices" and "knowledge claims" *meshed with* what Biagioli calls "the culture of absolutism" that prevailed in princely and papal courts in this period—and in the end, how the gears slipped, unmeshed, and produced a horrible grinding—doesn't denigrate that knowledge, but only complicates how we need to think about it.

To start complicating Galileo, then, let's begin with the simplest and most common statement: Galileo was a committed Copernican simply because Copernicus's system was, as we (think we) know now, true. What could be more natural than for a genius to believe the right thing?

But what did Galileo know, and when did he know it? Among the difficulties the historian immediately confronts is that of weighing public, written statements against private letters to colleagues like Kepler, and both of those against judgments about what Galileo *really* believed but didn't necessarily say. Like the case of Copernicus himself, what appears in Galileo's books and what was happening in his "true," interior belief system might be two (or more!) different things. Nevertheless, we can say that at the time of the publication of his book *Siderius nuncius* in 1610, which described his observations with his recently invented telescope of what we now call the moons of Jupiter but which at the time were called the Medicean stars (not a trivial difference, as we'll shortly see), Galileo sympathized with the Copernican system, but wasn't particularly committed to it. And the Italian elite who were Galileo's audience could accept his observations and theories as spectacular marvels and exciting discourse without thinking they were really true. The Jesuits, for example, whom Galileo would come to see as his most important rivals in his bids for scientific and cultural authority, could safely articulate Galileo's observations of the phases of Venus within Tycho's system (where the planets went around the sun, and that whole ensemble wheeled around the earth).

This is one reason why it's better to think about the sciences in terms of articulations. "The connection between Galileo's discoveries and Copernicus was not an automatic one," as Biagioli points out. "Depending on one's beliefs, socioprofessional identity, and patronage outlook, the Copernican dimensions of the discoveries could be legitimately emphasized or effaced."[3] The observations not only had to be articulated to a larger world view or system, such as the Copernican or the Tychonic, but also connected up to a system of financial and cultural support.

Here Galileo was lucky and savvy, as well as smart. Around the time he was making his first observations of the four points of light changing positions around Jupiter, the Medicean Prince Cosimo, whom Galileo had been tutoring for a number of years, ascended to the throne and became the Grand Duke of Tuscany. Dedicating his book and the moons to the Medici, Galileo made himself one of the most well-paid figures at court (a fact which, by the cultural standards of the day, guaranteed rather than undermined his autonomy) and the new "mathematical philosopher" most to be reckoned with. In return, the Medici got the good publicity and public sentiment that kept their power flowing, a smart figure who knew how to argue elegantly in their court, and four heavenly objects. Everyone was happy.

This mutually satisfactory sociocultural arrangement, Biagioli suggests, had its effects in the epistemological register as well—that is, on the Nixonian question of what Galileo knew and when, as well as what other people knew about both Galileo and the cosmos. The Medici ambassadors helped spread Galileo's discoveries and ideas, as well as his telescope and his name, across Europe. Galileo got a new cultural authority which earned him both respect and, inevitably, challenges. A complex feedback developed, rendering him "more powerful and more vulnerable at the same time":

> [B]y operating under the pressure put on him by his new socioprofessional identity, Galileo produced *more* discoveries. . . . Not only did these new discoveries and disputes confirm the reliability of the telescope and further upset his opponents, pushing them to counterattack Galileo on his Copernican agenda, they also gave him more resources to meet the attacks. In the process, his initial conceptual sympathies for Copernican astronomy were strengthened by his further patronage-driven discoveries. . . . By defending his newly acquired honor, Galileo became a full-fledged Copernican.
>
> [B]y 1610, dropping Copernicanism was not a real option for Galileo. Not only was he sympathetic to Copernican astronomy, but, because of his newly obtained court position, he was expected to maintain a high profile by producing new philosophical claims and engaging in controversial debates. By giving up Copernicanism, Galileo would have been undaring—he would have become "normal." More important, a *realist* reading of Copernicus allowed Galileo to live up to the title he desired so much: that of philosopher.[4]

In other words, to whatever extent Galileo believed that the earth really went around the sun, understanding this belief only in terms of "neutral knowledge" would be a mistake of simplification. At the very least, this neutral knowledge carried a charge: In Italian court culture, realism was a daring style, the mark of a new philosopher.

But before you start thinking things are straightforward, you should know that there were instability and contradictory forces in this system. If the culture allowed and even encouraged realist, controversial claims that carried a charge and created electrifying personalities, it also valued and supported another kind of charged knowledge at the same time. This was knowledge based on a virtuoso's play with a number of theoretical explanations, without being captivated by any one. And Galileo may have found himself in the uncomfortable middle, trying to play two charged games at once.

THE FABLE OF SOUND

Why did Pope Urban so love Galileo's book on comets, *The Assayer*, that he was having it read to him during his meals? One passage of which the Pope was particularly fond concerned the "fable of sound," in which Galileo tells the story of a man who hears a certain sound and tries to discover its origins. "Each time he thinks he has found its real cause, he hears that same sound again and realizes that there was yet another way by which nature produced it." Eventually the man finds a cicada and, thinking he can "find the *real* cause," dissects the cicada further and further until there is neither life nor sound:

> Urban liked this parable not only because it was the literary high point of the *Assayer* but because it was the epitome of court culture itself. It showed that pleasure was in the appreciation of the virtuosity of nature, that is, in the multiplicity of the causes by which nature (and, therefore, God) could produce a given sound. By trying to find the real, unique cause of the sound, one not only fails, he also kills the cicada and the pleasure of the inquiry with it. By seeking necessary causes rather than enjoying the novelties encountered along the way, one is a deluded philosopher *and* one who does not know how to play the courtier. . . . Great patrons could not put their status and power on the line by siding with a client whose claims might be judged to be wrong or too controversial. . . . Those who made strong claims were not represented as court virtuosi but as technicians—uncivil people who did not appreciate the elegant play of alternative views. . . . To be mentally enslaved by a philosophical system was similar to belonging to the lower classes. . . . Nature's ability to produce a given effect in a number of different ways . . . allowed Urban to enjoy Galileo's spectacular discoveries and controversial hypotheses without having to take a stand about how they fitted the Scriptures.[5]

As in the case of Copernicus, we can now see that Galileo subscribed to a more complex view of how science related to the "real" world. Science could describe multiple, probable, but not necessarily true causes. Galileo's "neutrality," his ability to engage in the "elegant play of alternative views," was another highly charged expression of cherished values of seventeenth-century Italian courts. As a court virtuoso, Galileo could use these aesthetic standards of court culture to validate his new ideas and methods, and to make himself one of the most well-financed and popular practitioners of the new breed of scientists.

Well, if all that's so, what happened at the famous trial? Didn't Galileo *really* defiantly teach the Copernican system as the truth about the world, and write about that truth in his *Dialogue on the Two Chief World Systems*, audaciously challenging social power with scientific truth? And didn't he, overpowered, *really* recant and end his days under house arrest for such a violation of cultural norms?

We'll try to keep our answers complicated.

When *The Assayer* was in press, Gregory XV died and Cardinal Maffeo Barberini was elected Pope, taking the name of Urban VIII. Barberini was a friend of Galileo's, and had even dedicated a poem to him three years previously; the poem celebrated Galileo's observations of the Medicean stars, Saturn, and the sunspots. The new Pope Urban's secretary and lord chamberlain were colleagues of Galileo's in the new "scientific society" which they had formed, the Accademia dei Lincei (named for the lynx, who can see in the dark). In turn, Urban's nephew was elected to the academy. "In a matter of weeks," writes Biagioli of this changing political scene, "the Lincei found themselves closer to the center of power than any other cultural faction in Rome." Galileo moved to take advantage of the new developments. When his book *The Assayer* went to press, it was dedicated to Cesarini, the friend who had become Urban's lord chamberlain. Galileo ordered late printing changes so that when finally published, *The Assayer* was dedicated to the new Pope Urban and formally presented to him in Rome in the fall of 1623.

Rather than defying power, Galileo courted it. Even the publication of the controversial *Dialogue* reveals his strategic savvy, and the complex charges of the field in which he moved. Galileo wanted that book to be printed in Rome, and received a provisional imprimatur from the Vatican's Master of the Sacred Palace, Father Riccardi, allowing Galileo to negotiate with printers. The Vatican wanted a few changes made, and Pope Urban was especially eager to see something like the fable of sound incorporated near the end, so it would be clear that Galileo was indeed a virtuoso who could court a number of cosmological hypotheses, rather than a boring dogmatist committed to the reality of one. There were a number of delays; an outbreak of the plague, for example, made it difficult to ship the revisions from Florence back to Rome. (Shades of the spacecraft *Galileo* making its way along Interstate 10, lacking the necessary lubricants.) Galileo began to arrange publication in Florence rather than Rome. Using his friendship with Father Riccardi and his connections to the powerful Medici court, Galileo managed to get the final checking and printing of the manuscript transferred to the Florentine Inquisitor. So not only was Galileo actively working with the Vatican and other powers to get his book published, but the final version was reviewed and approved by the Florentine Inquisitor before appearing in print in February 1632.

URBAN PROBLEMS

Events accelerated. That summer, Urban ordered the book taken out of circulation, and a special investigation was started. In the fall, the Pope handed the case over to the Inquisition; Galileo arrived in Rome in February 1633; the trial began in April; and by June he was beginning his sentence of confinement and weekly recitation of penitential psalms.

How did things turn around so quickly and dramatically? To avoid the simple certainty of power smashing knowledge, we have to exercise some patience with uncertainty, complexity, and possibility. The fact of the matter is that among people whose job it has been to pore over the documents, weigh the possibilities, and mull over the various interpretations, the question of what *really* happened to Galileo remains—as it probably always will—a question. You'd have to be pretty seriously deranged to deny that Copernicanism, Scripture, and their differences are important to this question; that the authorities of (a new and nebulous) science and (an old and rigid) religion were at stake; that differing standards and methods of knowledge were being publicly displayed. But you'd have to have a pretty serious agenda to think that explained everything.

On the "legalities" alone, reading the trial of Galileo is a challenge. The Vatican "leveled a number of different accusations against Galileo," Biagioli reminds us, including violating the publication agreements, insulting the Pope by putting his views in the mouth of the aptly named Simplicio, and violating an injunction against teaching the Copernican system supposedly placed on him in 1616. In the end, Biagioli writes, "it is difficult to assess which could have legitimately led to Galileo's sentencing and which were little more than juridical pretexts."[6] The question becomes even messier when you take into consideration the personal relationship between Urban and Galileo, the personal views of the Pope's advisors and colleagues, and the larger political and religious tensions which Urban was trying to mediate. (And were those scientific and religious rivals, the Jesuits, scheming something too?)

To elaborate: The Spanish were accusing Urban of being soft on heretics, favoring the French politically, and not supporting Spain's military efforts against the Protestants in Germany. In the spring of 1632, when Galileo's book came out, the cardinals came to blows over these issues, and Urban "was facing a serious and delicate political crisis." He became a little paranoid, shutting himself up in Castle Gandolfo, having visitors searched, fearing poisoning, patrolling the road to Rome, adding to the military forces at the border. He fired his secretary Ciampoli, a key supporter of Galileo, on the grounds that Ciampoli had tried to improve on a letter Urban had written, but rumors of Ciampoli's sympathies and contacts with the Spanish antipapal forces also circulated. In short, 1632 was not shaping up to be a good year for Urban, and he "needed to show he was a firm, decisive, and great papal prince."[7] Galileo was one of the bigger targets around.

It's also possible that Urban had to draw attention away from his own involvement in the book's publication. Urban declared at the trial that he had never been informed about Father Riccardi's provisional imprimatur, let alone his granting of the final license. Yet letters and documents now coming to light suggest otherwise. One letter from an Italian count to Galileo congratulated him on getting one of the book's referees to negotiate successfully with Urban to have the argument of the ocean's tides included in the book. (The tidal analysis was one of the stronger arguments in favor of the Copernican system.) Another referee wrote directly to Galileo, about how Riccardi liked his book and was going the next day to speak to Urban about the preface. And when the papal ambassador to Florence pointed out to Urban that "Signor Galilei had not published without the approval of his ministers and that for that purpose I myself had obtained and sent the prefaces to your city [Florence]," Urban raged about being deceived by Galileo and his friend Ciampoli, and how even the Master of the Sacred Palace (Father Riccardi) might have been a part of the cabal.

Then there was the problem with one of the key documents. The Holy Office expected that showing an *unsigned* 1616 injunction from Cardinal Bellarmine ordering Galileo to "abandon completely" his Copernican teachings would go a long way toward conviction, with no chance of easy response. The inquisitors appear to have been somewhat taken aback, then, when Galileo produced a *signed* certificate given to him by the same Bellarmine, in the same year, granting Galileo the right to discuss the Copernican doctrine in a hypothetical form. That move made the inquisitors squirm a bit; one of them, Father Maculano, wrote to a cardinal about how "the Most Eminent Lords of the Holy Congregation . . . approved what has been done so far, and then they considered various difficulties in regard to the manner of continuing the case and leading it to a conclusion. . . . [T]he Holy Congregation grant[ed] me the authorization to deal extrajudicially with Galileo, in order to make him understand his error and, once having recognized it, to bring him to confess it. The proposal seemed at first bold, and there did not seem to be much hope of accomplishing this goal as long as one followed the road of trying to convince him with reasons; however, after I mentioned the basis on which I proposed this, they gave me the authority."[8]

Whatever was decided in private, "extrajudicially," behind the more public scenes of the trial appears to have unknotted this potentially knotty problem. Two days later, Father Maculano could write to the cardinal that "the case had been brought to such a point where it may be settled without difficulty" and, adding just the right hint of intrigue and narrowly avoided disaster, "the Tribunal will maintain its reputation. . . ." Galileo's confession followed immediately, and instead of bogging down, the trial quickly concluded. In the judgment of the historian Guido Morpurgo-Tagliabue, "the spontaneous confession of Galileo did not save him but rather the judges from a very delicate situation."[9]

Most current stories about Galileo involve only crude displays of power, and unidirectional ones at that. While reinforcing convenient myths about the sciences, their autonomy, and their dedication to the real truth, these conventional understandings are at the same time quite unscientific. They gloss over details, they omit complications, and they ignore real subtleties in favor of preestablished assumptions about what the sciences and scientists are. With pinpoint historical inquiry, informed by the concepts and methods of cultural and social analysis, a new world opens up. It's a world in which the exercise of power can be both crude and delicate, the cosmos is both "real" and "our construction," the old and the new are continually rearticulated, and scientists work in the impure, powerful, and magnificent in-between.

Every anxious defense of a Galileo reconstituted as the lonely, misunderstood, genius-at-odds-with-the-establishment, is a way of affirming an easy and simple distinction between *power* and *knowledge.* By the end of this chapter, we hope that such a simple distinction will no longer be easy to make. We will have to get used to the uneasiness that comes with that hybrid entity of modernity/postmodernity, best articulated through Michel Foucault's hybrid term *power/knowledge.*

Clothing Naked Truth: California and Tuskegee

Power/knowledge. It is perhaps the philosopher and historian Foucault's best-known neologism. That hybrid term upsets purists, who like to think concepts or things should be neat and well ordered, one thing or another—or two things identified and joined with a transparent "is." Their discomfort is in some ways quite understandable, since such "slash terms" are supposed to upset our habits of thought, which depend so heavily (yet so subtly) on the metaphysics undergirding the sciences: power is X, period; knowledge is Y, period; X is Y or X is not Y, period. For the purist, the proliferation of these slash terms in the writings of the people often lumped together as postmodernists can only be a sign of muddle-headedness, willful or not—they simply can't make up their minds, or think straight. But since realitty is far from straight, and exhibits all manner of indirect and muddy connections and combinations, we have to learn to think more "crookedly," which is what Foucault encourages us to do. There is no doubt that Foucault's extensive body of scholarship has had far-reaching effects on historians, sociologists, and anthropologists of the sciences.

Why use the combined term power/knowledge? Because "there can be no possible exercise of power without a certain economy of discourses of truth which operates through and on the basis of this association. We are subjected to the production of truth through power and we cannot exercise power except through the production of truth."[10] We have to learn how to question these conjunctions, and that takes some patience and openness, just as working creatively in the sciences does.

Patience can be hard to come by for a purist. One such purist, Simon Jackman, quotes Foucault's remarks about how "power and knowledge directly imply one another," and how there is never "any knowledge that does not presuppose and constitute at the same time power relations." He then immediately goes on to argue that such a philosophy reduces to the "argument that knowledge *is* power," which "amounts to nothing less than the collapsing of the fact-value distinction. And the prescription for social-scientific practice that typically results is *not* to do 'better science,' but to give up on the endeavor altogether."[11]

That purist's rearticulation of Foucault is a bad one. As anyone without purist blinders can see, Foucault is in fact quite careful not to collapse anything, or equate power and knowledge, but instead demands that we articulate how they implicate, or fold into, each other, or how they depend on each other for their constitution. The slash between the terms doesn't represent a collapse or an identifying "is," but a question, a challenge to investigate the ways in which power and knowledge, in a particular set of circumstances, get muddled together in their presuppositions and in their effects.

Power/knowledge, then, signals a messy confluence that has to be thought about rather than assumed, an impure compound requiring analysis. But not in the way that one would analyze, say, a chemical compound: 60% power, 40% knowledge. The slash denies any such clean and easy breakdown.

Foucault offers another way of thinking about this: "There is a battle 'for truth,' or at least 'around truth'—it being understood once again that by truth I do not mean 'the ensemble of truths which are to be discovered and accepted,' but rather 'the ensemble of rules according to which the true and the false are separated and specific effects of power attached to the true'."[12] Hence the importance of always asking that other question of Foucault's with which this section opened: What is this reason that we use? And further, what are its rules at a specific moment in history, in a specific domain? How is reason, truth, or science "accounted for," and what are the consequences of such accountings?

Galileo was articulating a new set of such rules, building (along with others in seventeenth-century Italy) a new "power/knowledge regime" that was of course going to clash with the old ones. (Even as it had to partially depend on those old ones to get itself going.) The authority of scripture, whether religious or Aristotelian, was going to play a smaller role in defining the rules for producing truths and power. In its stead (although not totally) would be the mathematical, conceptual, and instrumental rules and tools for reading truths from The Book of Nature—all mediated not through church figures, but through the new class of philosopher-scientists, the cultural identity that Galileo was so expertly creating. It was indeed a new world system, a new way of articulating reality. If it had been neutral, it never would have worked. The new rules for articulating "truths" about celestial and earthly bodies had to be muddled together with new sources of power, and new power effects.

We're going to add more layers to this discussion of Galileo's new power/knowledge system later in this chapter. Some other, briefer stories here may help you get a better handle on this elusive and difficult slash word of power/knowledge.

Consider a recent move by California to castrate—either physically or "chemically"—men convicted of child molesting. In criticizing the proposal, one law professor used phrases such as "the dynamics of ignorance in action," and "this is 'don't bother me with the facts' legislation." Unlike the eugenic measures earlier in this century to sterilize criminals and the insane, where "the horror was what we would do in the name of science," he considered the current proposal "a celebration of not needing any scientific information or controls on punishment policy. And that naked aggression is much scarier."[13]

It could be looked at that way, and in the adversarial systems of law and politics, it is often tactically necessary to look at it that way: opposing myth to science, naked aggression to the civilized clothing of reason. But we would argue that, as Foucault taught us how to see in his landmark book *Discipline and Punish*, the prison system since the Enlightenment has not exercised "naked," pure power, but has always depended on a dense and confounding mixture of power/knowledge. It's simply no longer legitimate for the state of California—other nation-states are another question entirely—to practice "naked aggression." The identity as well as the "treatment" of the child molester today is established by the knowledge practices of a system of medicine and science, which as this example makes clear cannot be separated from questions of power. To say, as California Assemblyman Bill Hoge did, "Why not give these people a shot to calm them down and bring them under control or, alternatively, give them the option of going under the knife?" depends on a confluence of power/knowledge, not on ignorance. It depends on biochemical profiles that define normalcy and deviance, criminological data on recidivism rates, social scientific distinctions among acts of aggression and acts of sex, and medical or scientific studies that articulate the connection, however statistically uncertain, between a decrease in testosterone and a decrease in sexual, violent, or sex/violence acts. All these form the ensemble of rules for producing truths about who and what sex offenders are, and what should be done with them. Establishing these different kinds of knowledge takes hundreds of scientists, doctors, and technicians, laboratories scattered around the country, supported by state and federal grants, police personnel and protocols from the local community to the FBI, and dozens of other factors. It takes power to produce that kind of knowledge which produces and legitimates power which . . . means we're always in the midst of power/knowledge.

Also in the news recently has been that historical episode known as the Tuskegee Syphilis Experiment or, more "neutrally," the Tuskegee Study. In a White House ceremony, the federal government formally apologized to the few remaining survivors from this study that ran for forty years, from 1932 to 1972. Over that period, the Tuskegee Study withheld treatment for syphilis from nearly 400 black men in Macon County, Alabama, to learn more, it was said, about the progression of that disease.

The book *Bad Blood: The Tuskegee Syphilis Experiment*, by James H. Jones, provides a much fuller history than we can here. We're only going to draw out one particular thread which is often overlooked, a thread which emphasizes the necessary (and in this case horrible) collusion involved in power/knowledge. The accounts in magazines, newspapers, and on television which accompanied the White House media event almost made it into an episode that involved only the "power" side of the equation. To be sure, such accounts had plenty of material to draw on: how the study had violated many norms of informed consent; how the poor and illiterate men were treated as subjects or guinea pigs, and not human beings; how even the small payment for burial which was the study's only recompense to these men was gratefully accepted; how there was not just one, but a series of decisions to deny treatment; and of course the fact that the Public Health Service doctors were white and the men being studied were black. It all seemed like an exercise of gross, blatant power. But it is better viewed as a much more complicated exercise of power/knowledge, which was nonetheless gross for having its less blatant aspects.

Going back to look at the original motivations for the study and its design returns us to the difficult muddle of power/knowledge. At the time the study started, it was simply a matter of scientific "truth" that blacks were frail and inferior in general, more susceptible to infections, and more prone to sexual indulgence because their smaller brains had not developed a center that would inhibit such behavior. As a team of physicians for the U.S. Medical Corps theorized, "the negro's well-known sexual impetuosity may account for more abrasions of the integument of the sexual organs, and therefore more frequent infections than are found in the white race." [14] We would now call such a statement "pseudoscience," but at the time it operated—forcefully, unquestionably—as science. It fell within the current rules for producing truth, one of which was something like "Scientists can rely on stable, real definitions of different races and their hierarchical placement on an evolutionary scale." Those articulations were further articulated with other statements about social and sexual behavior which were not merely powerful cultural beliefs, but thick knots of power/knowledge involving culture, social theory, physiological theory, and evolutionary theory all at once. For all these reasons and more, syphilis was thought to develop differently in one race than in another, and the Tuskegee study was meant to highlight those differences. All of which is why the statement made by Dr. John Heller, who ran the study from 1943 to 1948— "There was nothing in the experiment that was . . . unscientific"—is, in some partial, unwelcome, and chilling sense, true.[15]

Clearly, we need better, more reliable and more *just* accounting mechanisms for producing truth. Given these kinds of dense tangles, the question is not how to get rid of "myth" and "pseudoscience" so that we have only secure knowledge which would allow power to be exercised rationally and, by implication, justly. The questions are: How can we grapple with the subtle and dense collusions of different kinds of power/knowledge? How can we judge among them without always having the solid and perfectly stable place that the sciences were supposed to provide?

But once more we're getting ahead of ourselves. Since evolutionary theory has come up again, let's turn first to examine some of the rules which were used to produce it.

Evolving Darwin

In the previous chapter we touched on evolutionary theory in the work of primatologist Sarah Blaffer Hrdy. Evolutionary theory is without a doubt one of the greatest accomplishments of the modern sciences. It continues to be a generative, and largely stable, framework for experimenting and rearticulating within the life sciences, but is also becoming important for generating new questions, ideas, and research directions in computer sciences, in medicine, psychology, cosmology, and in the study of language. So it's a particularly good time to return to an inquiry into Darwin, and those questions of ours for rearticulating such geniuses and their theories. From what infrastructural support did Darwin benefit? What charges did his work carry, and how did the highly charged fields of Victorian Britain make that work jump and swing?

Just as Darwin himself found both a new wealth of empirical data and new theoretical tools to work it into new evolutionary models, we now have similar resources available to rethink Darwin and the evolution of evolution. Because he is so justly revered as a scientist, the Darwin Industry (as it's called in academia) has produced in the last few years voluminous published works of carefully preserved documents, a staggeringly rich deposit of historical fossils for us to puzzle over and try to fit together. There is his complete correspondence, now available on CD-ROM; a *Calendar* of 14,000 letters to and from him; full transcriptions of all his notebooks; and a guide to Darwin's marginalia—the notes scribbled on the edges of the books he was reading in the middle of it all.

Drawing on this recent explosion of primary documents, and coupling them with analyses from historians, philosophers, and literary theorists, two eminent

British Darwin scholars, Adrian Desmond and James Moore, have recently written a new biography that is as simply titled as it is provocatively subtitled: *Darwin: The Life of a Tormented Evolutionist*. While not without its flaws and overreaches, it is an excellent reentry point for reconsidering Darwin and the transformations of both science and the social world which he accomplished—and which could also be said, in a clumsy reversal, to have accomplished him.

Darwin's heroic stature stems not from simple genius, meticulous methodology, or pure scientific clarity—although such attributions are not at all unjustified. We of course prefer to think of his greatness as a product of his complexity, contradictions, and ability to hold multiple forces together. As Desmond and Moore write in their introduction:

> Irony and ambiguity shrouded Darwin as no other eminent Victorian. He hunted with the clergy and ran with the radical hounds; he was a paternalist full of *noblesse oblige*, sensitive, mollycoddled, cut off from wage labour and competition, who unleashed a bloody struggle for existence; a hard-core scientist addicted to quackery, who strapped "electric chains" to his stomach and settled for weeks at fashionable hydropathic spas. . . . He thought blacks inferior but was sickened by slavery. . . .
>
> And how did grave Victorians observe the observer? . . . The butt of jokes, yes. The godsend to cartoonists, of course. Yet his science became a pillar of late-Victorian liberalism. How else to explain the earl and two dukes, representing Gladstone's government, acting as pallbearers in his Westminster Abbey funeral? How, indeed, to explain the body ending up there at all? Or *The Times*'s comment that "the Abbey needed it more than it needed the Abbey"?[16]

Having summoned up that funereal image, let's make something clear: We're not here to bury Darwin; nor, for that matter, are Desmond and Moore, or any other scholar who has tried to think critically about Darwin, his achievements, how those achievements were accomplished, the powers they exerted, and the powers that were exerted on them. As with Galileo, and all the other scientists we mention in this book, we're pursuing a richer, denser, and more complicated account than "he was smart and he was right." We're at least in pursuit of what made him smart and what, under particular circumstances of time and place, counts as right.

If Darwin was able to produce—in his materialist, chance-driven theory of natural selection—what philosopher Daniel Dennett has recently dubbed "the most dangerous idea," it was in no small part because he lived at a time when a lot of dangerous ideas were floating around. Darwin rearticulated a number of these—adding original contributions and taking them in new directions, to be sure, but nevertheless not starting from a blank slate. He had studied the evolutionary thought of French scientists Jean-Baptiste Lamarck and Geoffroy St. Hilaire, drawing some strength of conviction from their emphasis on biological change over long periods of time, evidenced in both the fossil record and anatomical comparisons between living species (although Darwin's own theories ended up differing significantly from theirs). Darwin first encountered these authors in his brief career as a medical school student at Edinburgh, in classes and long discussions while collecting sponges and other specimens from the sea with Robert Grant. Grant was a "freethinker," whose materialist and evolutionary views dovetailed with a political radicalism; he was of the minority opposition, challenging scientific, religious, and state authority all at once. Indeed, it was in social and academic life at Edinburgh that Darwin first experienced the force with which evolutionary views were denounced by clergy as well as scientists.

Among the Darwinian documents now available are minutes from the Plinian Society, whose meetings he and Grant both attended. At the top of one page of minutes from a meeting in March 1827 is a recounting of Darwin's presentation on the ova of the *Flustra*, a sea-mat which he and Grant had collected and Darwin had dissected; at the bottom are a dozen or so lines of notes, stricken over with the heavy black line of a censor. The presentation following Darwin's innocent piece of marine biology had argued for a strictly materialist view of life and mind; the ensuing discussion escalated into a vituperative argument, and the preceding remarks were stricken from the record. (Even the announcement of the presentation in the preceding week's minutes was deleted!) The modern viewer can't help but make the leap to the all too familiar image of the FBI or CIA document released under the Freedom of Information Act, but with most of the substance blacked out in the name of national security. And in 1827 such views *were* a kind of national-security threat; to espouse a self-evolving, materialist nature amounted to both blasphemy and nonsense, and undermined the moral fabric of society and civil order. Understandable, then, that for twenty years Darwin kept his evolutionary ideas to himself, penned in special separate notebooks, imagining the responses from Victorian society, fellow scientists (many of whom were also clergy members), and his wife, Emma (a devout Christian who feared for Darwin's soul). He was also mulling over his own doubts and convictions.

But while Darwin may have been familiar with and sympathetic to the lines of evolutionary thought already threaded through the scientific culture in 1827, it was his famous five-year voyage on the *Beagle*, beginning in 1831, that presented him with the extraordinarily rich diversity of new empirical encounters that started to really crack the old world and its truths wide open.

What was Darwin—a 22-year-old gentleman studying at Cambridge in preparation for a career as an Anglican minister as per his father's wishes—doing on the *Beagle*? And what was the *Beagle* doing in the first place?

The voyage of the *Beagle* was part of the Royal Navy's coastal survey of South America, where the early winners in Britain's newly industrializing economy were investing in markets and raw materials. To compete with Spain or the United States, coastlines and channels had to be mapped, local social and political conditions accounted for, tides and weather understood in more detail. (Francis Beaufort, the Admiralty's man in charge of the survey, had just devised the Beaufort wind scale, and the *Beagle* would be the first ship to employ it as part of its mapping methodologies.) The years at sea on these commercial/military missions had proven to take their toll on navy captains, who had to isolate themselves from the crew to preserve a commanding authority; the *Beagle*'s former captain had lost his own bearings and shot himself off the South American coast he was mapping. Darwin was along on the latest voyage, paying several hundred pounds of his own money, in large part to provide the new captain, Robert FitzRoy (only four years older than Darwin), with dinner conversation and companionship. That Darwin was a Whig and FitzRoy a Tory mattered less to FitzRoy than that Darwin was a "gentleman, mannered and cultivated."

From the edges of empire, this gentleman-scientist shipped back to the museums and scientists of England extensive specimens and reports. Darwin collected spiders, molluscs, birds, beetles (a favorite of his since his "beetling" days at Cambridge, where it was a faddish hobby), plants, coral, flatworms, fossils, and rocks. He catalogued and tagged meticulously, sketched geological formations, and mulled over animal behavior and physiology. He read Milton and the pamphlets of Harriet Martineau criticizing England's Poor Laws and the way the welfare system only exacerbated social problems, encouraging the poor to breed and further strain the limited food supply. A firsthand encounter with the devastation in Concepción, Chile, following a tremendous earthquake solidified his belief in a geological world of continual but gradual transformation, in which strata of sea fossils high on mountain sides were the trace not of a catastrophic Biblical flood, but of slow crustal upheaval. ("Gradual" and "slow" by geological standards, although devastatingly sudden by the human and social standards of Concepción.) And he saw people the likes of whom he had never seen before, and was left with the residing challenge of making sense of them in both moral and scientific terms.

There were no easy answers here, but only nagging contradictions that sparked further questions. Darwin had become quite expert at the sciences of geology, comparative anatomy, and the other areas of natural history that he had been schooled in; he was compelled by and committed to a purely material world, while retaining a sense of religious awe—having sat, as he wrote, in the "temples filled with the varied productions of the God of Nature."[17] But how or why did that Creator make humans so different, including everyone from the most savage Fuegian or Tahitian to the most civilized naturalist or captain? What scale could encompass these differences?

How had *one* God produced this cultural spread? Had he personally locked the Fuegian into his miserable environment? Surely he could not intend that man remain a savage? How much better to see the one God using evolution to spread the human races naturally. And how much more reassuring; evolution for Darwin posed none of Lyell's bestializing threat—the gentlemen were at the top in their rightful place. They were the evolutionary successes.[18]

When Darwin returned to England late in 1836, the words and things he had been sending back had secured him a new and solid scientific reputation and popularity. But the major scientific institutions and society were themselves in upheaval. Radical democrats, among them Darwin's former teacher at Edinburgh, Robert Grant, had won a new museum for the Zoological Society, and were now demanding that the society itself be run by their new class of "paid experts" rather than the old breed of parsons, noblemen, and other dilettantes. They also wanted to get rid of the British Museum's trustees, and turn it into a research institution styled along lines made popular in revolutionary France. While Darwin might have had some sympathies with their scientific views, he valued his "dilettante" status and the research freedom his privileged economic status afforded (Darwin's father now providing him an allowance of about £400 per year), and he was repulsed by their "mean quarrelsome spirit," "snarling at each other, in a manner anything but like that of gentlemen."[19] Other scientists, too, particularly those associated with the more conservative Geological Society, found the democratizing, anti-Church, anti-Cambridge radicals like Grant more trouble than they were worth. Darwin associated more with members of the Geological Society, including its president Charles Lyell, whose new *Principles of Geology* was both enormously popular and scientifically influential, and Richard Owen, a comparative anatomist who shared Darwin's enthusiasms for fossils and molluscs, but was deeply opposed to evolutionary thinking—and to evolutionists and political radicals like Grant, whom Owen helped vote out of a post at the Zoological Society.

In the spring of 1842, Darwin pressed beyond his anxieties and sketched out (in his secret notebook) a draft of his evolutionary theory. He kludged together his fossil evidence, his geological theories, his conversations with farmers and animal breeders, and his ideas about overpopulation and competition into a 35-page argument for descent by "Natural Selection." But the harshly naturalistic argument at this point was muddled with a relatively benevolent theological one—indeed, what today we would call a kind of modified creationist view. The natural world and its display of biological transmutation obeyed grand laws, laws that "should exalt our notion of the power of the omniscient Creator." And the messy, vicious natural world served to exculpate the Creator from directly causing evil: "It is derogatory that the Creator of countless systems of worlds should have created each of the myriads of creeping parasites and [slimy] worms which have swarmed each day of life on land and water. . . . We cease being astonished, however much we may deplore, that a group of animals should have been directly created to lay their eggs in bowels and flesh of other [parasitic wasps]—that some organisms should delight in cruelty. . . . From death, famine, rapine, and the concealed war of nature we can see that the highest good, which we can conceive, the creation of the higher animals has directly come."[20]

Still Darwin knew that such transmutationist views, the geological theories which they depended on, and the theological issues which they inevitably raised, were controversial and highly charged, and so he still refrained from publishing his ideas. Meanwhile, it wasn't only scientists who were articulating different linkages between natural, social, and political ideas, and coming into conflict over those different articulations. Atheist and socialist revolutionaries were using the widely read "penny papers" such as the somewhat paradoxically titled *Oracle of Reason* to circulate a revolutionary Lamarckism, in which biological change, driven from below, accumulated toward a higher and *more cooperative* future, with less hypocrisy, injustice, and economic exploitation—and with no help necessary from a postulated God. Desmond and Moore describe the tensions between these different articulations of evolutionary theory, each carrying a slightly different set of charges:

> *Their* evolution was a world removed from Darwin's. His suited the rising industrial professional classes. Theirs was for socialist workers. His was stabilizing, theirs revolutionary. And yet, they would have relished seeing simians substituted for souls. Nothing could have stopped them from pirating his book and playing up a monkey ancestry. . . . Darwin's lawful chains would be even better to truss up the Anglicans' meddling God.
>
> *Of course* Darwin could not publish.

But he continued to tinker with his ideas, in private, elaborating the mechanisms of selection until, as he wrote to his new friend Joseph Hooker a few years later, he was "almost convinced (quite contrary to the opinion that I started with) that species are not (it is like confessing a murder) immutable." As he was sending his extended manuscript to be copied by a local schoolmaster in July 1844, a pamphlet entitled *Conversation on the Being of God* appeared on the scene. In it, the feminist Emma Martin argued that evolution needed no Creator—a message she toured England with, speaking in the "Halls of Science" then popular with the working classes. In October of that same year, Robert Chambers's book *Vestiges of the Natural History of Creation* created even more of a stir. His impressionistic, popularizing account of nature's self-development was intended to reassure "ordinary readers" that everything in nature was geared to moving up, in keeping with the age of progress then in full swing. Enormously successful, *Vestiges* drew the fire of Anglican ministers who predicted "ruin and confusion in such a creed." Darwin continued to lie low, tinkering, kludging, rearticulating. For fourteen more years.

As many readers will know, when Darwin received a manuscript from the self-educated, socialist-leaning Alfred Russel Wallace in 1858, detailing a theory of evolution strikingly similar to the one Darwin had been assembling for twenty years, Darwin saw the value-producing labor of those decades "smashed." Only then was he prompted to publish. Well-placed friends helped manage the delicate priority and credit issues, seeing that Wallace's paper was published together with selections from Darwin's 1844 sketch and an 1859 letter to the Harvard biologist Asa Gray, in the proceedings of the Linnean Society. (The other choices, the Geological Society and the Zoological Society, were theoretically and politically out of the question.) A year after Wallace made his work known to Darwin, *On the Origin of Species* appeared in print. Darwin could mail a copy of his book to Wallace, who was then on the other side of the globe, with the comment "God knows what the public will think."

Some of what "the public" thought—and even the part of the public that might be said to know what God knows—made it into later editions of the book. Darwin received a letter from an obscure country rector (and a socialist) Charles Kingsley, who wrote that the book "*awes* me." But even "if you be right I must give up much that I have believed," Kingsley wrote. Other articulations were still possible: it could be "just as noble a conception of Deity, to believe that He created primal forms capable of self development . . . as to believe that he required a fresh act of intervention to supply the *lacunas* which he himself has made." Those lines Darwin quickly inserted into the last chapter of the next edition, obscuring the obscurity of their author and strategically attributing them to a "celebrated author and divine."

We end these stories of Darwin and the evolution of his evolutionary theory abruptly here, and somewhat unjustly. To go on to write of the *Origin's* tumultuous insertion into wider scientific and social discussions would require another chapter. We would only note that the remarkable rearticulations of the obscure/celebrated clergyman, realigning religious faith with materialist science, suggest a line of response to the question raised by Desmond and Moore: How did it happen that Westminster Abbey (what rhetoricians would call a metonym for both Church and State) "needed" Darwin's corpse as one of its legitimate heroes and saints, when Darwin had been working and playing with things and ideas so apparently dangerous to the social order? How did a corpus of scientific work that so challenged widespread and deeply embedded cultural beliefs, social practices, and institutions, that it provoked decades of intellectual anguish and possibly even physical illness in its originator—how did that revolutionary body of knowledge come to work as a conservative, legitimating power? How, in other words, did the positive and negative charges of evolutionary theory undergo such a reversal?

To answer that, you have to know that it wasn't a matter of knowledge challenging or legitimating power, but of complex moves of power/knowledge. And that it wasn't a matter of changing charges so much as of "contingent affinities within an assemblage" to which we alluded before, and which we now assemble.

Realitty=Assemblages of Power/Knowledge

If the sciences are a system, is it possible for them to be a nonsystematic system? Or is that merely a contradiction in terms, a product of muddleheadedness?

One could look to contemporary chaos theory, or the sciences of complexity as they are called by some, to begin to see how the sciences themselves are today encountering and/or building new kinds of systems that, if not thoroughly nonsystematic, at least involve, in very precise and odd ways, elements of nonsystematic randomness and sensitivity to "noise." But chaos theory gets enough attention, so we'll be taking a different tack.

Over the latter part of the twentieth century, many people from many disciplines have made efforts to develop new concepts for recognizing and working with events, phenomena, and structures that challenge conventional ideas about systems. Just as mathematicians and computer scientists have had to invent new languages for these new phenomena, so theorists from the human and social sciences have had to coin new words and introduce new concepts or usages. When trying to articulate the articulations of power/knowledge among institutions, scientific practices, and social discourses, for example, Foucault employed the French word *dispositif*. No good English equivalent exists for this word, although "aggregative machine" would be a good enough kludge job. And we are once again talking about kludging things together—things, words, and practices. (Or, about things, words, and practices being kludged together, since it's not always "us" doing the kludging—a crucial point, as we'll see.)

But there is another French word similar to *dispositif* but with a more direct English cognate: *assemblage*. One set of connotations for the word "assemblage" comes from the art world, and describes the sculptural equivalent of the surrealist collages previously mentioned: formerly unrelated and often "found" fragments are brought together, stitched, glued, or welded into a whole that doesn't quite add up to a whole. And in archaeology, assemblage is defined as "the aggregate of artifacts and other remains found on a site, considered as material evidence in support of a theory concerning the culture or cultures inhabiting it."

Walter Benjamin and Theodor Adorno, the German philosophers and social theorists associated (more or less) with the Frankfurt School of critical theory, invented the concept of the "constellation" to address the same sorts of issues. As the philosopher Richard Bernstein elaborates, their intent was to convey "a juxtaposed rather than integrated cluster of changing elements that resist reduction to a common denominator, essential core, or generative first principle."[21] They wanted to avoid recreating the kind of badly "scientized" social theory that neatly systematized "structural" and "functional" elements and relations, where everything fit perfectly with everything else, and social science explained all the determinations clearly and systematically.

In an assemblage, nothing explains it all: not the sciences, not the social sciences, not the human sciences. There isn't anything that is first or fundamental in an assemblage—nature, language, culture, institutions, whatever—it's all at once, and we with our questions come *after* it. Meaning that we are both assembled by it, and in pursuit of it. Even though we're consigned to *come after* the assemblage has been assembled, both with and without our intentionality, that doesn't stop us from *going after* it, too. (It's the particular style of going after it that's at issue, and the style we're advocating throughout this book is the one called muddling through.)

To make the concept of an assemblage more concrete, imagine an Olympic athlete. Every four years the media, the International Olympic Committee, and athletes themselves pay a lot of attention to the question of anabolic steroids. It's easy to gain the impression that steroid use alone can make you run faster, jump higher, or build endurance and strength. Drugs seem to have a magical, primary power. But what is easily forgotten is that steroids, if and when they are used, are still just one part of a larger assemblage, one node in a constellation of productive forces that included rigorous training regimens, scientifically defined and carefully monitored diets, submission to technological monitoring and imaging devices, highly motivational and productive cultural beliefs involving nationalism and individual excellence (and much in between those two semicontradictory poles), and a great deal else. Steroids aren't any more fundamental to athletic performance just because they're biochemical. They will always be one force among many—which certainly shouldn't be taken to legitimate their use. But by making them exceptional and demonizing them, we keep the athlete "natural," and allow the computerized training equipment to be naturalized as well: "Pay no attention to these multimillion dollar machines and resource-intensive regimens; behold the individual athlete!" Similarly, we have also taken too long to acknowledge the insane cultural forces that lead to anorexia and other physiological disruptions in female gymnasts, since only drugs and not culture are supposed to have real, material effects.

The sciences are like steroids: It's a mistake to think they act alone, or are a more fundamental part of an assemblage—the assemblage we're calling realitty. They can have pretty profound effects—and "side effects"—but they don't tell the whole story.

But we need something still more complex than that image of an Olympic athlete. The first approach to understanding an assemblage should be to make a map of it, as precisely as possible. In the previous chapter we sketched out part of a web of connected Peircean triads, to suggest how language worked to create meaning and realitty via an extensive but always incomplete nest of articulations. We're going to shift that diagram slightly, so that it can map not just language, but the more heterogenous elements—newly catalogued objects, newly articulated concepts, rising scientific geniuses, vast social movements, and many others—that are articulated into a new assemblage. To convey something of its irregularity and vitality, we need to shake up the orderly honeycomb of Peircean triads we drew previously, and give assemblages the form of a living, moving, grasping lobster. Let's start with the "Darwin-assemblage":

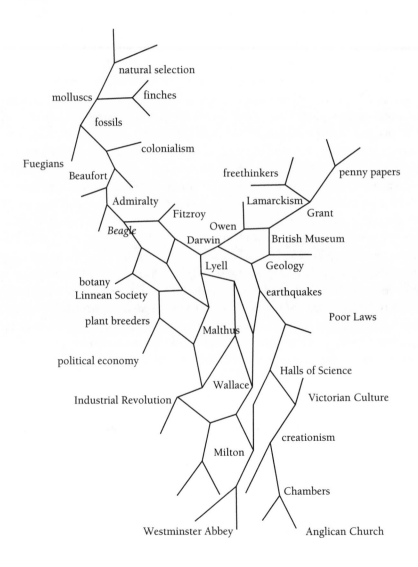

Darwin was a great scientist, and the theory of evolution by natural selection a great scientific achievement, because of these complex intersections. Darwin muddled persistently with specimens that he gathered on his voyages (courtesy of the British Admiralty), with fossil bones (collected, prepared, and catalogued at the British Museum), and with finches and molluscs (in part enabled by the freedom afforded by his socioeconomic position to roam around the English countryside, as one of the last of the gentlemen scientists). He built meaning from the muddle through precise description of minute physiological differences. He articulated all that work with observations of geological formations, and new articulations of geological theory produced by other gentlemen scientists. His own thinking and observations were shaped by the thought-styles of previous theorists of evolutionary change—articulations which Darwin rethought, and restyled.

But he also theorized by cobbling together reflections on the wrenching demographic shifts of the Industrial Revolution; fears about the food supply, population, and the Poor Laws; reading political economists like Adam Smith and Robert Malthus; and Victorian beliefs about "savage" and "civilized" races. He made productive, if not always conscious, use of the metaphors and allegories he found in his reading of Milton and other popular literature.[22]

When scientists remember Darwin today, they often do so in the standard terms of science and the truths of nature overcoming dogmas of culture and religion. At the same time, some social historians and other analysts of Darwin and his work have argued that Darwin was really a kind of Malthusian in scientist's clothing, merely garbing political and social beliefs about inequality, competition, and so on in scientific law and fact. In this view, science is really nothing but ideology; there is no recognizable difference between them. Science is a weapon, or a tool to be employed by the capitalist system to legitimate its needs and desires.

This is why we're better off thinking in terms of assemblages. Darwin went up against religious forces all right, but that's far too simple and historically inaccurate. He was also up against many of his fellow scientists, particularly those in the Geological Society in which he shared membership. It's misleading to ask whether Darwin was going against religious beliefs or scientific beliefs because those two categories (as well as others, like political economy) were already kludged together. Charles Lyell, Adam Sedgwick, other geologists, and other scientists (a brand new word at that time, coined to describe this nascent professional identity) were operating within specific articulations of both "scientific" and "religious" beliefs, articulations in which study of the natural world was inseparable from the study of the Christian God's creation. That linkage was both productive and incomplete, and yielded both insight and blindness. And even if Darwin might have eventually disarticulated his theories from religious beliefs, there was enough play in the assemblage of linkages to allow others to maintain connections between God and Nature (a fact that Darwin could make strategic use of, as we saw above).

That play in the sometimes tight, sometimes loose assemblage of kludged elements also means that we have to think in more complicated terms about the "social implications" of the "scientific theory of natural selection" (separating out those two areas for the time being). Remember that Desmond and Moore pointed to the "irony and ambiguity" characterizing Darwin, and his work; he "hunted with the clergy and ran with the radical hounds," as they so colorfully phrased it. If the theory of natural selection reinforced social hierarchies by "naturalizing" the differences between civilized British captains and primitive Fuegian savages—and at an even finer grain of analysis, the differences between the new captains of industry, the old nobility, and the men and women of the English working class then in the making—it also allowed for another series of articulations to be made. The possibility of revolutionary, "leveling" evolutionary materialism being articulated by the radical hounds did not sit well with Darwin, and he was powerless to stop it. No single logic could characterize the whole nonsystematic system.

Darwin assembled a new assemblage—sort of. (It's the subject-verb-object problem of language again.) Even as his work took advantage of the cultural forces around him, it fit uneasily within them and their implied changes. The inquiry infrastructure which he did so much to extend led to difficult places, and he was profoundly troubled by the uncontrollably multiple religious and social implications of his work. His assemblage was reassembling him, others were actively assembling their own articulations, and it could be said that the assemblage was even reassembling itself. Which is part of the reason why Westminster Abbey could "need" the remains of someone who thought his ideas would be so disruptive of the social order: the social order was in the process of reordering itself. The assemblage called "England" was already mutating from a primarily agrarian, religious, class-bound society to a more industrial, more secular, and more socially fluid and "evolutionary" one. (All relatively speaking, of course.) Natural selection was, in a sense, selected by a changing environment, surviving very well in the new milieu which it was helping to shape at the same time.

Assemblages have, in effect, a life of their own. They are monstrously complicated Leviathans. They are fantastically productive because they are wildly impure. Removing some pure "kernel" of evolutionary theory from its "application" in explaining social and political problems has been a popular move by later commentators on Darwin, particularly scientists. But things are never so neat with assemblages. Desmond and Moore argue, as do many other analysts, that "'Social Darwinism' is often taken to be something extraneous, an ugly concretion added to the pure Darwinian corpus after the event, tarnishing Darwin's image. [H]is

notebooks make plain that competition, free trade, imperialism, racial extermina-
tion, and sexual inequality were written into the equation from the start—
'Darwinism' was always intended to explain human society."[23] True enough, as
long as we remember that the equation is a very complex one indeed, with more
than one solution. And it's continually being rewritten.

Let's return now to Galileo—the seventeenth-century human Galileo—and
quickly sketch out that assemblage:

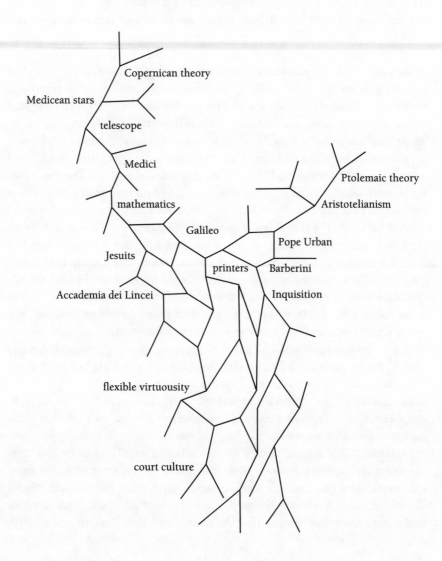

Rather than read this diagram in the same kind of detail as the Darwin-assemblage, we're going to use it to emphasize a few important things about assemblages generally. First, assemblages are a kind of infrastructure—a complex, crazily reticulated transportation system. You can move, but always in certain directions. There's no absolute freedom. As the stories of Galileo and Darwin demonstrated, even the flashiest of vehicles will be caught on roads for which it was not designed. They encounter roadblocks of many kinds; signals that resist them are material, to be sure, but also social and cultural and interpretive and political. And all of those muddled together. Molluscs and moons, foreign peoples and exotic mathematics, subtle shadings of Victorian values and Italian court etiquette—any and all of these will be sign-forces that compel the thoughts and acts of scientists in unexpected directions. It's this complex infrastructure that carries our inquiries and experiments, allows us to articulate all "really?" questions, and their tentative answers. It's the impure, heterogenous multiplicity of assemblages that makes them a challenge to work with and within, but which also makes them productive.

Second, assemblages nevertheless allow for a kind of agency, or rather, various kinds of agency and sources of power. Galileo was a virtuoso who could rearticulate existing nodes and elements, and invent new ones. Great scientists are never fully determined by the assemblage, nor is their work. They can build off-road vehicles, or initiate a kind of chain reaction of self-organization of new roadwork. By working on and working out new observations, new experiments, new kludge jobs of material-social artifacts, they create not "neutral" knowledge, but new forms of power/knowledge. Which is why they can be such a threat to the old forms of power/knowledge, as Galileo discovered. He might be said to have focused his efforts in that part of the diagram labeled "mathematics," "telescope," "Medicean stars," "Copernican theory," and so on—a sector which, when "purified" from the rest of the assemblage, we conventionally call science or nature. But we hope to have shown that such a purification is artificial, that Galileo couldn't have so focused his efforts without the other articulated elements, that the distant and indirectly linked "court culture" was as necessary to the realitty of Copernicanism as "mathematics" was.

What powered Galileo and his new sciences? Many things, which is the best kind of power source to have. Certainly the accurate observations, the new technology of the telescope, the new emphasis on mathematics were each powerful and compelling, and together even more so. Those can in part be attributed to Galileo's genius, but also to a growing social resource of mathematicians, and other scholars trying to join math to physics. Galileo's salary from the Archduke, and his affiliations with the Medici and papal courts, were another source of strength. There are his powerful texts, written boldly in the vernacular, employing rhetorical devices brilliantly and deftly. There's no question that Galileo was a great astronomer, or that his physical-mathematical tools for analyzing the motions of falling bodies were far more generative of new knowledge, new tools, and new questions than the Aristotelian concepts that they challenged—and that those tools still work, reliably, today. But why contend that those tools were neutral, or that they were all Galileo had to rely on?

Such creative agency is not found only in individual geniuses. Just as the federal government encouraged the construction of the interstate highway system, new traffic patterns can emerge at the behest of collective work and institutional prodding. The Medici and the Accademia dei Lincei functioned in part as support for creative, collective endeavors. The Jesuits, the *Society* of Jesus, were becoming an effective and enthusiastic institution for the development of the sciences. The construction crews of the sciences may be many, large, and only partly organized, but it is always possible to reconstruct the broader system of transportation—although such reconstruction is a slow, piecemeal, and often unpredictable process.

That suggests a third point about assemblages. The suggestion can take the form of a kind of reprogramming of Galileo's apocryphal exclamation on leaving the Vatican trial: *Eppur si muove!* (*And yet it moves!*) The lobster form is not entirely whimsical, but a deliberate reminder that the sciences are in motion and, indeed, composed of linked motions. The sciences aren't a frozen mirror image of a preexisting world; the sciences are things and thoughts and people and institutions in movements that are both concerted and noisy, coordinated and clumsily contingent.

Local pockets in an assemblage can indeed be stabilized; that's why the sciences can be "science," and allow us to do so many things. But such stability comes not from reality, but from realitty: the collective effort, thought-styles, metaphors, and all the other articulated elements of an assemblage. The pocket called electron spin, for example, is now quite stable and real. But that's because of all the mathematical equations, interpretive symbols, productive instruments, institutional support, and other parts of the assemblage that are all kludged, half-assedly but indissolubly, onto the Secondness of "real" electron spin. In the Galileo-assemblage for creating a new science and new scientific identities, the earth really does go around the sun. In the NASA-assemblage for getting Americans to the moon and *Galileo* (spacecraft) to Jupiter and its satellites (a.k.a. the Medicean stars)—the earth really is at the center of the universe.

There is movement in another sense as well. These diagrams can be read as diagrams of contingency—not in the sense that they're the product of chance (although they are in part), or that the sciences and their connections to the "social" are arbitrary. That would return us to a kind of neutrality, in which anything could go anywhere along with anything else. The sciences are contingent in the sense that they always touch and *depend on* many other points in the assemblage. This is where "charge" comes back into our picture. The dependency among the elements, the forces exerted by its charged components and relationships is always felt somewhere, somehow. But the dependencies can be quite distant and indirect, the attractions and repulsions subtle and far from unbreakable. Yet they still have to be accounted for, rather than neutralized or purified. The charged affinities between evolutionary, political economic, and social theory were contingent ones, quite forceful yet still movable in different directions by radicals and Tories alike. One-way relationships or direct causalities are rarely at work in the assemblages of the sciences. The trap of an easy kind of functionalism, in which charges line up and lock into place, should generally be avoided.

Finally, remember that these diagrams are themselves a kludged tool for our inquiries into the sciences. They suggest questions, tell us where to look. They're a map of possibilities that have to be analyzed in detail, different assemblages demanding different kinds of questions and approaches, and allowing for different kinds of inventions. Whether in the race to reach the moon or Jupiter, or to map the human genome (where new assemblages of evolutionary theory can be found), the sciences move in directions we can learn to chart—with precision, care, openness, and attention to complexity and our own explanatory limitations. The challenge is not to know where the sciences are going, once and for all. For now, our goal is more limited and experimental: to learn to map the complex highway systems which provide the sciences with ways to move, and to learn how to question the contingent affinities permeating our material and cultural assemblages. Instead of an itinerary telling us where we will arrive when, we want better navigational tools. Instead of a clear and unilaterally decided destination, we want more tinkering devices, and more people involved in their making. And instead of either the pure neutrality of discovering reality, or the pure sociality of overarching values or power, we want better power/knowledge systems for working out the assemblages of a better realitty.

That's why these stories about Galileo and Darwin still matter today. Their analysis can provide new ways of questioning and understanding what has charged the sciences in the past, so we can better imagine the worlds to which our future sciences could take us. When scientists from Merrell Dow publish peer-reviewed studies proving that the drug Bendectin has no adverse effects, how should we evaluate those studies? *Who* should evaluate those studies? Is it merely a matter of "what power wants, power gets," or is there something more complicated going on? Are people just supposed to trust, as transparent truth, statements made through the sciences that alcoholism "is" genetic, or at least has a strong genetic component? Or do we want a citizenry capable of asking about what experiments are articulated deep within such articulations, what metaphors are used and for what effect, and what contingent affinities might operate between the worlds of the laboratory and the worlds of culture? When a group of citizens living around a military installation want to know how clean and safe their drinking water is, what kind of assemblage—of instruments, concepts, coalitions, and whatnot—has to be built? And who's going to pay for it? How do we know which realitty of Multiple Chemical Sensitivities, and other emergent disease categories like Gulf War Syndrome, is the right, true realitty? How do you create sciences in such highly charged fields? Why should people learn to think critically and creatively about quantum physics, rather than simply enjoying the wonderful, dazzling, even mysterious show?

These examples, which we'll be taking up in more detail in the rest of the book, may seem a far cry from the cases of Galileo and Darwin, but they can be thought of in similar and productive ways. Current scientific traditions—their standards of evidence, their cultures and subcultures, their thought-styles, their language and experimental practices—have to be examined just as these classic examples of heroic science were. What we find is that while science and scientists may never be neutral, such a view doesn't rob them of all authority, power, or of our fascination and respect. By learning to map the assemblages of the sciences more completely, we can ask a different set of questions about how the multiple sources of their authority and power—in nature, in a social and intellectual infrastructure, and in current values and interests—and the multiple charges they carry. We can ask where and how democratic interests might require that we question such authority and power in the name of better sciences, even as we remain fascinated and impressed.[24]

The biggest challenge, of course, is in delineating exactly what will count as "better" in these processes. Despite our many attempts to make our articulations careful as well as playful and provocative, qualifying our arguments with painful contortions, there will no doubt be readers who can only see in our metaphors of changing realitties, thought-styles, and sign-forces, the haunting, horrifying specter of scientific—and therefore historical, social, and political—relativism. And that leads, they might argue, to moral relativism, where judgments cannot be made of "better" or "worse," either in terms of scientific theories, or who we are and what we should do. If the sciences aren't going to be grounded in transparent truth, or don't have a ground in reality but only in the articulated realitty that they set up and work within, then how are we to judge what's better?

Patience . . .

Chapter 4

Judging

The Fruitfulness of Uncertainty

A Paris-assemblage, 1901: The Eiffel Tower. A new aesthetic of speedy, sleek aerodynamics. War and colonialism. An urban sprawl of grimy slums and green boulevards, populated by *chiffoniers* and *flâneurs*, rag-picking beggars and strolling artists and intellectuals. An imminent future of ethereal radio broadcasts of entertainment, advertising, and propaganda. Transparent glass arcades displaying shiny new consumer goods, and in a dark laboratory somewhere else in the city, another new glow, almost unnatural: radium.

In the same year and city, a possibly apocryphal story (since it appeared under the title "Truth or Fiction?") placed the French socialist Jean Jaurès standing and watching an "airship" (a blimp) flying overhead. Two other gentlemen standing with him were not impressed; one denounced it as a glamorization of the scandalous Eiffel Tower, "built with stolen Panama money," while the other thought it a cheap bourgeois trick, the circus part of the old bread-and-circuses routine. The latter proposed establishing "a group of Revolutionary Aeronauts . . . who will wait until the revolution is accomplished before they invent balloons." Against these two summary judgments, which subsumed the new within knee-jerk, established social and political categories, Jaurès found the scene wonderfully interesting. Even if imperialism was involved, even if a new class of "technocratically correct" engineers was now necessary, there remained the simple fact and remarkable engineering/social feat of *flying*.[1]

Massive change and realignment, with no smooth fit between spheres and ideals, and time itself may be out of joint. All disjunctive elements of a new, shifting assemblage that can't be evaluated by reducing it to one of its components, or to an old social ideal that dealt with them individually. How can the whole mess be analyzed, considered, *judged* as an ensemble?

Jaurès, who coined the uneasily juxtaposed phrase "revolutionary evolutionism," has been placed within the context of the *via media*, an intellectual and political tradition articulated by historian James Kloppenberg. Around the turn of the century, American pragmatists, British Fabian socialists, German social theorists, revisionist Marxists, and French socialists such as Jaurès were imagining and enacting the *via media*, or middle way. All were trying to invent new ways of being *in between*: of developing an epistemology between idealism and empiricism, of imagining an ethics between intuitionism and utilitarianism, and of making a politics between revolutionary socialism and laissez-faire liberalism.

> Although things have not turned out as these generations hoped, they accepted the unpredictability of the democratic project as an integral part of its value in a world where truth and justice are to be carved from culture rather than found already etched in reason. They understood that such indeterminacy leaves open the possibilities of liberation or control, community participation or domination by elites, and they did not take for granted the outcome of such struggles. If the failures of the welfare state have mocked their confidence and provided melancholy confirmation of their doubts, the last sixty years have also failed to provide better answers to many of their questions. In philosophy and in politics these thinkers proclaimed the fruitfulness of uncertainty; in both spheres their victory has itself proved uncertain. While I would not suggest that we ignore the gulf of time separating our world from theirs, it may be time to build bridges back to their ideas.[2]

It should be obvious by now that we too respect the "fruitfulness of uncertainty," indeterminacy, and unpredictability, especially as it applies to questions of the sciences in the democratic project, and the need for a critical approach to science literacy.

We've seen how dense articulations compose the sciences, and their objects. But in our lobsterlike assemblages for Galileo and Darwin, anyone can see that the structure of kludged elements necessarily traces out its own gaps. Every bridge spans an abyss, and something always remains unspoken, unarticulated, left out of the system (especially the nonsystemic kinds of systems). What causes these holes—forgetting? error? the future? the inexhaustible plenitude of the world? "Finitude" would be a decent catchall term—the inability of language, thought, and practice to exactly match the remarkable plenitude of the world.

While all the sciences in some sense struggle to overcome this finitude, perhaps nowhere has the promise of infinitude been clung to more intensely than in the zone where particle physics meets cosmology. Writing in 1993 in the pages of *Physics Today* for his own "community," the physicist and historian of science Silvan S. Schweber (whom we encountered tangentially in Chapter 1, interviewing

Richard Feynman) provided an eloquent and insightful analysis of the intellectual and social crises in which particle physics and cosmology now find themselves.

Schweber articulates a widely felt sense of interminable fragmentation that has yielded no intellectual unification, a shaky social infrastructure of employment and equipment with the end of the Cold War, and a feeling that the exciting questions, perspectives, and opportunities were now to be found elsewhere, especially in the biological sciences. Then, at the end of this lengthy, carefully detailed account of the current intellectual and social state of particle physics, he puts in a conclusion that appears somewhat out of joint. Just as it was not readily apparent how airships and the Eiffel Tower were to be thought of together, Schweber's conclusion isn't easy to integrate with the rest of his article:

> . . . [T]he goals of most of the scientific enterprise are no longer solely determined internally; other interests come into play. The scientific enterprise is now largely involved in the creation of novelty—in the design of objects that never existed before and in the creation of conceptual frameworks to understand the complexity and novelty that can emerge from the *known* foundations and ontologies. And precisely because we create those objects and representations we must assume moral responsibility for them.
>
> I emphasize the act of creation to make it clear that science as a social practice has much in common with other human practices. . . . I believe that in the reconstruction we are engaged in we must accept that the separation between the moral sphere and the scientific sphere cannot be maintained.[3]

Nor can it simply be collapsed. But it's nevertheless becoming increasingly clear, to more and more scientists like Schweber, that spheres which were once separate and "pure" are now muddled together. We can no longer escape the questions that arise from the moral and the scientific getting articulated, kludged together in a dizzying array of specific circumstances. How can we begin to reconstruct the inquiry infrastructure, and the charged cultural and political territory through which it moves, so that it better serves a system of democratic values? And how do we engage in that reconstruction of a charged assemblage, when it engages and catches us in its own unexpected reconstructions? How do we pursue a shifting realitty that also pursues us? How should those reconstructive pursuits be judged, and by whom? And finally: in the play of other interests which determine the sciences in addition to their internal logics, in the play of loosely fitted, kludged elements of nonsystematic assemblages that now have to be judged very seriously, do the sciences still get to be fun?

Fun, Responsibility, and TNT

"Fun" used to be a basic principle in the defense of pure science in the modern era, particularly among physicists. And particularly among physicists in the post–World War II era, as historian Paul Forman has noted, when their means of social and fiscal support depended on that key articulation, national security. In such a situation, "it became imperative for scientists to fashion a self-image that allowed them to close their eyes to the real basis for the generous social support of their knowledge-producing activities and to maintain the illusion of personal and disciplinary autonomy. That new self-image, originating with the American physicists and then spreading to other disciplines, projected 'fun' as the predominant feature and leading attraction of 'doing science'. . . . [It] was meant to trigger the fantasy of perfect autonomy: any eudemonic activity is an end in itself, and as such wholly autonomous—but also wholly irresponsible. Moreover, putting 'fun' forward trumps any question of the end for which the scientist's knowledge-making is sustained, and thus supports from this side as well a stance of radical irresponsibility."[4]

That way of talking about and legitimating the sciences seems to have passed, Forman observes, and now the talk everywhere is of "responsibility." Even though the term is used loosely and left relatively empty—the National Academy of Sciences 1992 report called *Responsible Science*, for example, leaves the word undefined—that does not mean, contends Forman, that "its use is insignificant." Something about the sciences and the ways in which we need to think about them *has* changed, however hard that change is to grasp, and the words "responsible" and "responsibility" mark this new space, and keep it open for inquiry. It marks a middle space, where bridges might be built over chasms (without, of course, making the chasms disappear). As the Chairman of the Board of the Nobel Foundation said in his opening address at the 1989 awards: "[t]he steadily increasing demand for responsibility on the part alike of the researchers and the humanists is now bridging the gap between the two cultures."[5]

That's a nice thought, but still pretty vague. But the vagueness and emptiness often associated with "responsibility" isn't entirely a failure of thought or will. It also recognizes the monstrousness of the problem, the almost overwhelming demand entailed in acting responsibly in and toward the sciences. The word responsibility should evoke, as Kierkegaard knew, more than a little fear and trembling.

In dealing with situations of responsibility and judging in the sciences, we need to recognize the enormity and explosiveness of those situations. The previous chapters have built up a characterization of the sciences as complex assemblages, charged by all manner of sign-forces, constituting nonsystemic systems of what Foucault called power/knowledge. If we take that last hybrid term, and rewrite it somewhat glibly as "Truth 'N' Transactions (of power)," we could kludge together a new metaphor for the assemblages of the sciences: TNT. The sciences and their assemblages are dynamite, a volatile mixture with powerful uses. They require special handling—subtle, ginger encounters. Almost any proposal for "doing something with the sciences"—purifying them of values and power, adding values to them, democratizing them, making sure they're responsible, even just judging them—should evoke a sense of unease and impending disaster.

Science-assemblages have never been purified of power or values, even when those admittedly imprecise terms are scattered toward the frayed ends of our lobster diagrams. And since science-assemblages are already bound up with a whole set of cultural norms and values, pleas to add values to the sciences make no more sense than the pleas to purify them. It is because they are densely articulated and tightly packaged assemblages of power/knowledge that the sciences are so extraordinarily potent, useful and awesome. That's also what makes clumsy attempts to control them so dangerous.

We've highlighted the need to recognize complexity, to appreciate ambiguity, to avoid easy oppositions and abstract idealizations, and to stay close to questions. In this chapter, we show how these demands and others combine under the umbrella of *judging*, the processes required in all muddling attempts to reconstruct the infrastructures of our volatile power/knowledge assemblages. We start by considering two particularly explosive historical episodes involving science-assemblages—the cases of eugenics (particularly in Germany under National Socialism), and the "Lysenko affair" in Soviet genetics—which are frequently used to excoriate the intrusion of values and power in science, and to defend the necessity of the freedom and purity of scientific inquiry. We then look at some court judgments that have recently been handed down to the worlds of the sciences, and the people affected by them. The issues become even more complicated when we turn to a more recent and highly publicized case of alleged scientific fraud, and how different kinds of "judging" are tied up with the sciences at all levels, from the laboratory bench to institutional oversight. At each step along the way, we describe a number of benchmarks for judging, questions and suggestions to help us fill in some of the gaps in those hole-ridden ideals of democracy and responsibility—or to at least keep us from blowing ourselves up.

Some Myths About Eugenics

The history which hangs heavily over questions of biology and social policy today is, of course, that of eugenics, the word British scientist Francis Galton coined in 1883 to denote the "science" of improving humankind by biological means. Without providing a full discussion of this extensive, troubling, and complicated history, we at least want to revisit a few myths (as we did for Copernicus, Galileo, and

Darwin) that have proven to be both remarkably persistent and unhelpful for thinking about the sciences and democracy today.

The historian of science Mark Adams has identified four such myths about eugenics.[6] The first is that the popularity and social power of eugenics can be restricted to any one country or culture, whether (most notoriously) Nazi Germany (as detailed in such books as historian Robert Proctor's *Racial Hygiene: Medicine Under the Nazis*, and the geneticist Benno Müller-Hill's *Murderous Science*) or the Anglo-American context (as portrayed in Daniel Kevles's *In the Name of Eugenics*). While the history and memory of the eugenical and racial policies of Nazi Germany continue to provide a powerful and important example of the dangers of biological fundamentalism, that example has often been used to obscure a much more extensive historical record. German racial hygienists, including natural scientists, social scientists, and doctors (vocal and influential, by the way, before Hitler and the National Socialist party came to power), were quick to justify and legitimate their own efforts with the scientific theories and laws generated in the U.S. and Britain. (It was the state of Indiana, invoking the authority of objective science, that passed the first law allowing for the sterilization of the mentally ill and criminally insane in 1907. Twenty-eight other states had followed Indiana's example by the late 1920s; 30,000 U.S. citizens were sterilized on eugenic grounds—groundless grounds by today's standards—by 1939, nearly 13,000 of them in California alone.) And beyond these two well-documented national examples, historical research is now revealing the extent of eugenic ideas, social movements, and government policies in the Soviet Union, France, Brazil, Norway, Denmark, Yugoslavia, Turkey, Finland, Sweden, and other countries. Eugenics can't be easily "othered," as something that just the Nazis believed in and practiced. No nation's hands, nor the hands of some of their best scientists, are clean here.

The second myth is that eugenics was essentially tied to the rapid scientific development and spread in the first part of the century of Mendelian genetics, which provided a basis for biological determinism in the gene—a basis which "softer" versions of heredity (such as Lamarckian inheritance of acquired characteristics) lacked. But as studies of eugenics movements in France, Brazil, and the Soviet Union have shown, a scientific commitment to hereditary theories that incorporated environmental influences like culture and education (whether those influences resulted in the inheritance of acquired characteristics or not) did not necessarily lead to kinder, gentler social policies. Lamarckians could advocate sterilization of the "unfit" just as enthusiastically as Mendelians; whether alcoholism or a propensity for crime came from the supposedly manipulable "nurture" or environmental side of the equation, or from the supposedly unchangeable "nature" side of the genes, people had to be stopped from "passing it along." There was no hard and fast connection between a scientific world view and state policies; there were "contingent affinities within charged assemblages," as we put it in the previous chapter. The affinities could be quite powerful indeed, but with room for a variety of combinations and movements according to changing historical and social situations.

These issues lead to a third myth about eugenics, that it has been essentially right-wing or "reactionary." "According to this view," suggests Adams, "depending on who you are talking to, eugenics variously grew out of and supported racism, sexism, anti-Semitism, or capitalist exploitation of an oppressed working class, and led naturally to fascism, Nazism, and ultimately (either directly or by natural extension) to the 'Aryanism,' barbarous human experimentation, genocide, and the death camps of the Third Reich." But the history is more complicated. Some scientists were subtle, thoughtful, and modest, while others were crude, unthinking, and brash—and while most of the ardent eugenicists fell into the latter category, subtlety and intelligence were no guarantee that one stood on the correct side of these issues. Some scientists wedded biological theories to reactionary, fascist politics; some to principles of "scientific management" and the hope for a technocratically achieved and maintained future; still others to progressive and socialist agendas. Diane Paul has shown how the goals and ideals of eugenics were part of a tradition of leftist scientists, particularly Marxist ones such as American Nobel Prize winner H.J. Muller and British geneticist J.B.S. Haldane.[7] Birth control and women's rights advocate Margaret Sanger was a strong supporter of eugenics, and women in the American eugenics movement combined its language and ideas with demands for improved educational opportunities for women. While miscegenation was taboo in the United States, racial mixing was valorized in Mexico, on eugenic grounds, for its production of "hybrid vigor" that would improve the national stock. So again, we find an intricate set of political articulations being produced "in the name of eugenics," under specific historical and social circumstances.

The fourth myth exposed by Adams, and from our perspective the one most in need of debunking, is that eugenics was essentially a pseudoscience, or the misapplication of a crudely formulated science contaminated by personal bias and social prejudice. The corollary historical myth here is that once genetics became less crude, more "up to date," and more objective, scientists saw the light and abandoned their beliefs in and commitments to eugenics. "This myth may once have had a certain utility as a way of acquitting the science of genetics and freeing it from any unsavory eugenic associations, but it can find relatively little support in the historical record," writes Adams. Ideas about and policies regarding the "genetically inferior"—criminals, drunkards, morons, racial degenerates—operated at the time as expert, reliable, *objective* science. They followed the rules for making articulations within the power/knowledge systems at the time. And many re-spected, rigorous, and productive scientists, particularly in the U.S. and Britain, remained committed to various eugenics measures until well after World War II. (And in some cases, until *now*.)

To these four myths, we add a fifth: that eugenics involved science and scientists being (mis)used by the state. Such a model of the relationship between knowledge and power, as we argued in the previous chapter, just doesn't match the historical reality. In Nazi Germany, as in the U.S. and Britain earlier, and as in many other nations, scientists, psychiatrists, doctors, and anthropologists could be found ahead of the curve of state policies: proposing eugenic measures which might then be adopted by the state, testifying before government officials for immigration restrictions, rationalizing forced sterilizations, devising new scientific and medical procedures to better classify or "treat" the "genetically inferior." We do not have a history of innocent scientists being corrupted by the powerful and nefarious state. We have a history of the interaction and mutual accommodation between forces of sciences and scientists and the forces of power and statesmen. Putting it baldly: Scientists *were* the state, and their knowledge *was* a form of power.

What can we learn about the demands of judging from this history? First, of course, is that there is no purity. If the history of National Socialism in Germany reminds us of anything, it should be that any articulation that depends on "purity"—racial purity, ideological purity, and even, as we hear today, "pure foods" like tomatoes uncontaminated by the genes of other species—deserves the harshest skepticism. So when we're judging sciences, we have to give up the equations that all good science will and must be pure, and all bad science impure. Since there is no such thing as pure science, judging first involves a mapping out of the complex, articulated linkages that always exist among the sciences, their culture, and social institutions.

If you do that mapping with sufficient complexity and subtlety, you will then see that there are also no easy equations between good and bad science and good and bad politics. There are no necessary, essential, or easily predictable connections between the sciences of genetics and the practices of power, but there are always contingent affinities. Such contingency doesn't make those links or articulations any less real, any less strong, or any less demanding of reflection and judgment—in fact, just the opposite. History shows us that the charged affinities that exist between repressive forms of state power and various kinds of biological determinism are indeed powerful and persistent, in part because states (especially democracies) want to appear rational and objective, and articulations about essential biological qualities fit that bill nicely. Indeed, the charges here are strong precisely because of the reciprocal productions of meaning and significance:

Articulations about essential biological qualities such as idiocy, schizophrenia, or "Jewishness" appeared rational and objective at the time; they operated so forcefully because such categorizations drew on simultaneous cultural and political articulations.

History confirms this reciprocity among scientific, political, and cultural productions again and again. But the nexus itself is always specific to time and place. Any secure, preformed analysis of the way power and the biological sciences go together will be only an imperfect guide, and you will always have to think and rethink in terms of specific historical and social conditions. Judging requires a pursuit of specificity—scientific, historical, political—and a pursuit of complexity, to map the affinities among these specific domains.

A social scientist recently remarked to one of us, "The Human Genome Project is just a new version of nineteenth-century craniometry," referring to an earlier "science" of biological differences in brain size that also legitimated violence and discrimination against women and certain racial and social groups. We regard that as a failed judgment, and failed inquiry. It fails to pursue both the specifics and complexities of craniometry and genetics, their different methods, materials, and results, and the different ways in which each is linked to different institutional, cultural, and political elements of an assemblage. Can scientists today who make strong claims about the biological basis of certain human qualities be fitted into the same power/knowledge assemblages in which physiologists of an earlier era worked? Or have the articulated elements so changed that we face a radically new challenge? The judging of that and other questions is reserved for Chapter 7.

Judging one part of the assemblage or another isn't good enough either. We don't deem eugenics "bad" because it depended on insufficient sciences, or because scientists and politicians actively sought linkages with each other. Insufficient knowledges and their collusions with social interest will *always* be with us. We can, however, judge the various "eugenic assemblages" as horrible practices which disguised contingent articulations as necessary truths of biology and state, disguised political interests as rational policies, shunned both empathy and uncertainty, and had awful effects.

We can also judge eugenics as a failure to recognize differences among people as a dynamic social and scientific resource, rather than something to be feared and eliminated. Judging the sciences will always depend on the assemblage of different perspectives which can interrupt and supplement each other. As we've seen with primatology in the previous chapter and with other cases, one criterion of excellence within the sciences is a plurality of articulations. That kind of pluralism is realized, in part, through constant attention to the ways every categorization, every theory, and every fact is charged by multiple sign-forces. At their best, the sciences themselves are pluralistic, and are best judged from a plurality of vantage points

Lysenkoism: Beyond the Nucleus of Power

The other historical episode that makes a spasmodic appearance whenever some-body questions the impermeability of the boundary between power and knowl-edge is "the Lysenko affair." Until the ardently hoped-for day when science purists cease advancing tendentious arguments based on distorted historical reconstitu-tions of these events, any book that tries to open up an inquiry about the conflu-ences of politics and science will have to deal with the question of Lysenkoism. It's not a particularly appealing job; most analyses of Lysenkoism are so rigidly shaped by simplistic oppositions that new questions and attempts to recollect complexity are all but barred. But it is precisely because available articulations of Lysenkoism are so barren that we have to confront the subject here. The challenge is not to fig-ure out how the sciences then and now are the *same*. The challenge is to turn his-tory into a set of subtle heuristics which suggests routes of inquiry that will lead us to judgments benchmarked for the specific ensemble of our own time and space.

Often it is nonbiologists who are guilty of taking the most simplistic lessons from this historical episode, as Peter Huber does in *Junk Science*: "Whatever the so-cial purpose of the Marxist state, Mendel was right and Lysenko wrong; when so-cial purposes decree otherwise, the upshot is Stalin's Academy of Junk Science Genetics."[8] When power dictates truth, in other words, you get both bad science and bad politics. The episode known as Lysenkoism or the Lysenko affair in the history of the former Soviet Union is undeniably tragic, with geneticists exiled or killed and an atmosphere of scientific inquiry poisoned. But Lysenkoism was also, as biologists Richard Levins and Richard Lewontin maintain, "a phenomenon of vastly greater complexity than has ordinarily been perceived."[9] It's a complexity that any judgments should take into account.

To be sure, there are some agricultural scientists today who can remain neces-sarily critical of the Lysenkoist efforts, yet still respect the complexity of this his-torical episode sufficiently to keep from falling into unthinking, knee-jerk dismissals. Wes Jackson, whose reflections on Lysenkoism can be found in his book *Becoming Native to This Place*, draws on the writing of those other historically and philosophically attuned biologists Levins and Lewontin to remind us that "though some Lysenkoist ideas were absurd, that does not mean that philosophy as such should or can be kept out of science."[10] Not only philosophy, we would add, but the exercise of power—political, social, and cultural—cannot be kept safely outside of the sciences, either. One way or another (actually, one way *and* another), these things will get muddled together, so scientists and citizens alike had better be prepared to think about which particular mixtures might be more productive and less deadly than others.

To provide the most skeletal of historical and scientific frameworks: T.D. Lysenko was an agronomist from "peasant stock" (as it is often said, although we would prefer "peasant culture") who managed to dominate Soviet agricultural science and practice, as well as genetics, from the late 1930s to the early 1960s. The standard explanations of how he accomplished this feat give central placement to the authoritarian power structure of the Soviet Union (especially under Stalin). There is no doubt that this power structure was crucial to Lysenko's championing of Lamarckian theories of inheritance (in which environmental influences can be passed on to the next generation) over Mendelian genetics (in which only the unchangeable chromosomes were involved in heredity). Lysenko and his followers came to dominate Soviet research establishments, purging them of geneticists such as Nikolai Vavilov who held to the tenets of Mendelism. Mendelian genetics held on by the barest and most clandestine of threads, while the Lysenkoists enforced their theories of acquired inheritance in the laboratories and attempted to substantiate them and put them into practice in the vast, collectively farmed wheat fields. The Lysenkoists' wheat never sprouted earlier and never yielded the predicted increases, the story continues (how could it, since it was based in knowledge that simply wasn't "true" and was only brutal ideology?). Although state power kept Lysenko in charge for decades, eventually even that could not withstand the force of scientific truth and failed experiment. Lysenko fell, and "real genetics" returned to the Soviet Union.[11]

While "dogmatism, authoritarianism, and abuse of state power helped propagate and sustain an erroneous doctrine and even established its primacy for a time," as Levins and Lewontin readily concede, "a theory of 'bossism' is not sufficient to explain the rise of a movement with wide support nor to explain its form and content."[12] Levins and Lewontin, and Wes Jackson after them, provide excellent summaries and reflective analyses of the "form and content" of the Lysenko period in Soviet agricultural genetics. We encourage readers to read their work for a bracing immersion in the social, political, scientific, and philosophical complexities of this episode in the history of science: the climatological, social, and experimental challenges specific to the context of Soviet agriculture; the bearing social class had on scientific research and popular understanding and uptake of science; the role of peasants in the collective farms; and other factors which complicate the clean and comforting story about the eventual triumph of good truth over bad power. Rather than resummarize and reanalyze their articulations of this history, we'd like to focus on a few more tangential events at the margins of the Lysenko story as a different way of sketching out many of the same complex connections among powers and knowledges.

The American geneticist H.J. Muller (whose work using x-rays to induce mutations in fruit flies would win him the Nobel Prize) released his popular book, *Out of the Night: A Biologist's View of the Future*, to critical acclaim in 1935. As we saw in the previous section, historians like Diane Paul have detailed how eugenics was not simply a conservative movement in the U.S. and Britain, but was actively supported by a number of leftist biologists such as Muller. In his book, Muller advocated a plan for eugenics based on artificial insemination that had been laid out some years previously by the Soviet geneticist Serebrovsky. In May 1936 Muller sent a copy of his book to Stalin, accompanied by a long letter advocating his eugenics plan. "It is quite possible," he wrote to Stalin,

> by means of the technique of artificial insemination which has been developed in this country, to use for such purposes the reproductive material of the most transcendently superior individuals, of the one in 50,000, or the one in 100,000, since this technique makes possible a multiplication of more than 50,000 times. . . . A very considerable step can be made even within a single generation. . . . After 20 years, there should already be very noteworthy results accruing to the benefit of the nation. And if at that time capitalism still exists beyond our borders, this vital wealth in our youthful cadres, already strong through social and environmental means, but then supplemented even by the means of genetics, could not fail to be of very considerable advantage for our side. . . . We hope that you will wish to take this view under favorable consideration and will eventually find it feasible to have it put, in some measure at least, to a preliminary test of practice.[13]

Our point is not to disparage Muller as a scientist, but only to show that relationships to power were cultivated by all parts of the geneticist spectrum, from staunch Lamarckians like Lysenko to ardent Mendelians like Muller. And truly scientific Mendelians could be just as prone to exaggeration, extrapolation, wishful thinking, and other nonrational behaviors as the ideological and pseudoscientific Lamarckians.

Moreover, as Mark Adams argues, Muller's enthusiasm may well have helped Lysenko's rise to power. In December 1936, the same year he had sent his book to Stalin, at a meeting of the Lenin All-Union Academy of Agricultural Sciences, Muller flouted the political advice of his scientist colleagues and attacked the Lysenkoists, who of course immediately returned the attack. The Lysenkoists "drove home the argument that genetics, eugenics, and fascism were all of a piece . . . one remarked that Soviet women would never forgive [Serebrovsky] for his human breeding scheme." Muller had to be hurried out of Moscow after the meetings; murders and arrests began almost immediately thereafter. "However inadvertently," Adams concludes, "Muller had played directly into the hands of the Lysenkoists. By resurrecting Filipchenko's and Serebrovsky's forgotten arguments—

so fitting in the 1920s, so inappropriate in 1936—he helped reestablish the ideological links between eugenics and genetics. In the process he compromised the Soviet Union's leading geneticists . . . [and] inadvertently undermined the field he had helped to create—medical genetics."[14]

The articulations possible between the sciences and various forms of power are not only diverse, they can also be quite subtle. Not all power comes from the barrel of a gun, or from the connection of that gun barrel to the pen of a Lysenkoist autocrat. Nor is power only exercised in the personal, social, and political spheres. As we argued in the previous chapter, our inquiries into power need to include the ways in which the sciences are always a form of power/knowledge that works even at the levels of metaphors and thought-styles. In far less brutal but ultimately no less significant ways, the Mendelian "truth" of an unchanging chromosome as the sole agent of heredity was also mixed up with operations of power.

In their account of Lysenkoism, Levins and Lewontin take care to document an "impressive body of experimental data" circulating in the 1930s which, at the very least, *held open the possibility* that things like disease resistance, color patterns, and other traits could indeed be environmentally modulated and then inherited. "Thus when Lysenko and his followers began to put forward claims of directing hereditary change in the 1930s, Lamarckism was not a dead relic dredged up from the past," they argue, but in fact persisted, quite reasonably, in fields outside mainstream genetics, with "an extensive literature of experimental results that had never been refuted."[15] To believe in the possibility of acquired characteristics in this period, then, while certainly a marginal and risky position, was not simply an "ideological" one.

The idea that biologists the world over had all come to the only rational conclusion possible—that genes controlled for Mendelian traits were the only reproductive or developmental mechanism worthy of examination, and that anyone who questioned that was an ideological robot or crackpot—is not supported by historical evidence. The history of inquiry into "cytoplasmic inheritance" and embryology in this period is too large a topic to open up here. Jan Sapp's *Beyond the Gene: Cytoplasmic Inheritance and the Struggle for Authority in Genetics* provides a good account of the various resisting "sign-forces" which biologists encountered in the cellular, intellectual, and political domains as they tried to experiment and articulate outside the "Mendelian paradigm."[16] Gregg Mitman, a historian, and Anne Fausto-Sterling, a developmental biologist, have examined the history of the flatworm *Planaria* as a research tool in the life sciences, and the work of Charles Child at the University of Chicago.[17] Mitman and Fausto-Sterling's collaborative work details how research organisms like flatworms or fruit flies produce specific, limited, and different kinds of knowledge, and how an eminent scientist like Child could pursue "non-Mendelian" investigations through the entire first half of this century. Their account shows not only the personal and professional politics of the relationship between Child and T.H. Morgan, the towering figure of fruit fly genetics, but how very different assemblages of personal, social, cultural, and institutional interests could accrete around different research organisms and their equally sound sciences. And Evelyn Fox Keller's biography of the corn geneticist Barbara McClintock, *A Feeling for the Organism*, documents the often subtle, often overt ways in which McClintock's painstaking work on transposition (when genes change their location on a chromosome) was ignored or dismissed as heresy because of certain "powerful" intellectual and social forces in the American climate of biological research. "Even though McClintock is not a Lamarckian," Keller comments, "she sees in transposition a mechanism enabling genetic structures to respond to the needs of the organism. . . . [T]ransposition . . . indirectly allows for the possibility of environmentally induced and genetically transmitted change. To her, such a possibility is not heresy—it is not even surprising. On the contrary, it is in direct accord with her belief in the resourcefulness of natural order."[18]

"Neo-Darwinists should not be allowed to forget these cases," stated John Maynard Smith in 1983, referring to the inquiries of Child and others into cytoplasmic inheritance, "because they constitute the only significant experimental threat to our views."[19] If you're not a research scientist, all we're asking you to remember is that Mendelian genetics was not the only rational pursuit in the time of Lysenko, and that power works in other ways besides gulags and overt repression. The combination power/knowledge can be much more subtle, and harder to track, query, and judge. Non-Mendelian, non-fruit fly geneticists were not exiled or shot in the U.S.; it does not follow that American genetics was free and pure.

Double-Crossing Corn

In a later chapter we'll return to T.H. Morgan and the "discourse of gene action" which he was so successful in formulating and propagating. But first, a quick look at corn:

"The development of hybrid corn was no simple matter of the transfer of theoretical science from an elite academic to an applied commercial context," write Diane Paul and Barbara Kimmelman. The Mendelian theories of genetics on which hybrid corn was supposedly based were already an applied science, which "ensured that fundamental problems in genetics would be addressed within institutions oriented to practical ends—and that the subsequent development of genetic research would often reflect dominant social and economic interests in American agriculture." Paul and Kimmelman support this argument by analyzing the role of such organizations and institutions as the American Breeders Association, the U.S. Department of Agriculture, and the state agricultural colleges and experiment stations in the development and spread of Mendelism in America in the first few decades of this century. We're going to lift just one of their stories.

The three main actors here are George Shull, working at the Carnegie Station at Cold Spring Harbor (still one of the premier institutions of genetics and, for a time, a central institutional resource for the American eugenics movement[20]), and Edward East and his student Donald Jones, at the Connecticut Agricultural Experiment Station. Working independently, East and Shull had each developed certain inbred lines of corn which were at first superior to the open-pollinated corn plants for which they were derived, but in successive generations declined in vigor. Despite that degeneration in the hybrids over time, they developed and stayed close to a theory of hybrid vigor. "The object of the corn-breeder," Shull would write, "should not be to find the best pure-line, but to find and maintain the best hybrid combination." Shull thought that "hybridity itself" had a "stimulating effect upon the physiological activities of the organism." Research was focused on the development of new hybrids, rather than on developing different pure lines.

The problem, as Paul and Kimmelman point out, is that "the concept of physiological stimulation arising in some unknown way from heterozygosity was both distressingly vague and unsupported by any evidence. Even East was later to admit that it was 'an assumption for which there was no proof, and which was not illuminating as a dynamic interpretation.'" When East and Jones published their influential book *Inbreeding and Outbreeding* in 1919, they would conclude that pure lines were indeed more desirable than hybrids. By that point, however, hybrid corn was all over the place, and since it was also effectively sterile, farmers were obliged to return to the seed purveyors year after year to get the same hybrids to achieve the same yields. In their 1919 book, East and Jones characterized the brief period in which the unproven and unilluminating theory of hybrid vigor ruled as truth, as a well-timed "happy result," since it "locked the door on any hope of originating pure strains having as much vigor as first generation hybrids." Their book was quite open about the ways in which politics, economics, personal values, and science were combined here. The "double-cross" method of hybridization that Jones had developed was a complex one, they noted, not suited for everyone:

> It is not a method that will interest most farmers, but it is something that may easily be taken up by seedsmen. . . . The man who originates devices to open our boxes of shoe polish or to autograph our camera negatives, is able to patent his product and gain the full reward for his inventiveness. The man who originates a new plant which may be of incalculable benefit to the whole country gets nothing—not even fame—for his pains, as the plants can be propagated by anyone. . . . The utilization of first gen-

eration hybrids enables the originator to keep the parental types and give out only the crossed seeds, which are less valuable for continued propagation.

The Soviet Union wasn't the only place where power collided and colluded with knowledge. In a different way, it happened here too. "East and Jones knew that, in theory, hybrids were not the only, or even the best route, to improvement of corn," conclude Paul and Kimmelman. "They themselves identified the alternative. But that method [of developing pure strains], as a breeder associated with Jones wrote, would probably 'spoil the prospects of any one thinking of producing the seed commercially'.. . . East and Jones held a scientific theory according to which pure lines should produce maximum increases in yield. But they also held a social theory (reflecting the facts of their actual world), according to which improvement of corn required an incentive for commercial producers that only hybrids could offer."[21]

The benchmarks for judging in these cases are similar to those in the previous section: Pursue specificity and complexity before offering a judgment. Don't judge on the basis of a purity that will never be there, but on the basis of impure assemblages with impure effects. That doesn't mean that all assemblages are equal— "everything's relative," or some such imprecise formulation. When you map the complex specificities and specific complexities, you can begin to see how different assemblages kludge together different realities. Or, different realities emerge with the help of different assemblages. Assemblages and their emergent effects can and must be judged.

The assemblage of (in shorthand)

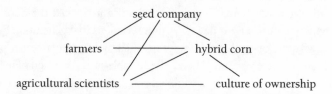

had a set of effects with which we are all too familiar: dependency of farmers, concentration of both farms and capital, higher socioeconomic status for plant breeders, certain lines of scientific inquiry pursued while others languished. Scientists *knew*, at some level, that creating hybrid corn and adding it to the muddle would help that particular assemblage emerge and stay strong. They knew that another assemblage of

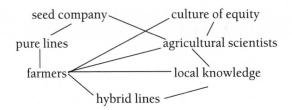

was, at the very least, possible to work on, and could further a different set of effects: what would later be called agricultural sustainability, farmers less dependent on seed companies, more farmers getting a better shot at modest prosperity.

Another, related benchmark for judging: power/knowledge assemblages involve both gross and subtle linkages. Sometimes economic and political interests will be apparent and forceful in their collusions with the sciences, and sometimes they won't. People need to be able to ask how gross interests might come into play, but also need heuristics for provoking subtle, seemingly insignificant linkages into visibility. Pursuing and judging the sciences democratically will depend on developing the literacies necessary to inquire about the most subtle articulations within the sciences: Why are some people working on the margins of a "dominant paradigm" like Mendelian genetics characterized as misguided, bad scientists, or even "heretics"? What all is involved in *that* kind of judgment? "Just the facts"? Or a more complicated and perhaps more ambiguous combination of resonances among words, beliefs, and interests?

Let's emphasize again that what has to be judged is the entire, messy assemblage and not simply one part of it. Which also means that no single perspective can dominate the entire assemblage. Here, strangely enough, we find ourselves in *partial* agreement with those science purists Paul Gross and Norman Levitt, referring to the question of Lysenkoism in their book *Higher Superstition*:

> Dialectical materialism [is supposed to have] the power to oversee and indeed to rectify all the other sciences. Supposedly, it is uniquely situated to escape the illusions in which ordinary science, because of its historical and social situations, may become entrapped. Marx himself took these questions quite seriously and was eager to sit in judgment on scientific questions despite his modest and sometimes absent competence to do so. This sorry tradition was carried on by Lenin in a number of notorious pamphlets; under Stalin's regime in the Soviet Union it was at least a factor in the wretched Lysenko affair, which crippled Soviet biology for decades.[22]

As we stated at the beginning of this chapter, we think that monologic solutions to the problem of judging the sciences are recipes for trouble. No perspective or no discipline—not Marxism, not feminism, not social constructionism, but also not the sciences themselves in all their supposed purity—can oversee or "sit in judgment" over the entire assemblage. The worlds of the sciences are just too complicated and potentially explosive for that. Gross and Levitt are right to criticize such tendencies. But they're wrong to think that they invalidate dialectical materialism or any of these other approaches to the sciences; they're wrong to conclude that "totalizing" is the only way to pursue such inquiries. (Indeed, it's their own totalizing, purist instincts that lead them to such a totalizing conclusion.) Each of these forms of inquiry offers different partial perspectives, sets of questions and insights that together can inform, without overseeing, our much-needed process of judging.

Toxic Assemblages

Now we consider a particular set of judgments, made by flesh and blood jurists, about science, the scientific method, and the role of experts in a democratic legal system, which, in our opinion, has to be founded on juries.

A COURT SUPREME

The case of *Daubert v. Merrell Dow Pharmaceuticals, Inc.,* decided by the U.S. Supreme Court in 1993, has been discussed in detail by Shana Solomon (a law student at the time) and Ed Hackett (a sociologist of science); our account here has been greatly informed by their work.[23] The *Daubert* case illustrates very well the kind of situation we're in now regarding the sciences, a situation requiring multiple judgments, at every step of the way. It will help show that rather than subverting the possibility of making judgments, our perspective of a world in which things and ideas, language and facts, nature and culture are all assembled together in fact demands judgments more than ever.

Far in the background of this case is the drug Bendectin, one of those "better living through chemistry" drugs which promised to relieve the effects of severe morning sickness in pregnant women. Bendectin was used widely in the U.S. throughout the 1960s and 1970s; the plaintiffs in *Daubert* had birth defects which were alleged to be caused by their mothers' taking the drug during pregnancy. A few reports in the medical literature suggesting links between Bendectin and birth defects began to appear as early as 1969, and in the fifteen years beginning in 1977 close to 2,000 legal claims were filed against Merrell Dow, the drug's manufacturer. Only about thirty of those cases went to trial, and the verdict in those was almost always overturned, either by the trial judge or on appeal. At the center of such tort cases is the issue of causation (an issue we encounter again in Chapter 6)—and in the kinds of complex scientific assemblages we've been talking about in this book, causation is a very difficult thing to prove. The first step in proving causation to a court of law is getting the judge to accept evidence as admissible.

Knowing that causality in torts involving toxicity is very difficult to track and prove, the plaintiffs in *Daubert* attempted to have expert testimony in four different areas admitted: in vivo animal studies, in vitro laboratory studies, chemical analysis comparing Bendectin to other chemicals known to cause birth defects, and a reanalysis of the published epidemiological studies of Bendectin. Court after court ruled this testimony inadmissible or insufficient, on the basis of "the overwhelming body of contradictory epidemiological evidence" marshalled by Merrell Dow—over thirty studies, none of which showed a statistically significant increase of birth defects among the children of women who had taken Bendectin.

The plaintiffs didn't dispute the lack of a statistical link. They were arguing that if you considered *multiple* lines of reasoning—the in vivo and in vitro studies and the other analyses—then it appeared that Bendectin was more likely than not responsible for the birth defects. At the very least, it could appear to a jury that it was more likely than not to have caused birth defects, and the plaintiffs thought a jury should have the chance to decide.

When the District Court dismissed the case, by insisting that epidemiological data were the only admissible evidence of causation, the plaintiffs appealed. The Ninth Circuit Court of Appeals upheld the District Court's judgment, and added a stricter kicker: "reanalysis of epidemiological studies is generally accepted in the scientific community," the Court noted, "only when it is subjected to scrutiny by others in the field. Plaintiff's reanalyses do not comply with this standard; they were unpublished, not subject to normal peer review process, and generated solely for litigation. . . . [Three federal appellate decisions] recognize that 'the best test of certainty we have is good science—the science of publication, replication, verification, the science of consensus and peer review.'" When the case subsequently went to the Supreme Court, it was these broader issues of peer review and the apparently almost oxymoronic "science of consensus" which came to the fore, rather than what Bendectin could or couldn't cause.

One of the reasons the Supreme Court gave for its decision to review the case was the existence of "sharp divisions among the courts regarding the proper standard for the admission of expert testimony." Federal courts had been inconsistently invoking one of two such standards, the 1923 *Frye* rule or the 1975 Federal Rules of Evidence (FRE). The *Frye* rule is the more restrictive of the two, demanding that the data and methodology underlying scientific evidence and arguments must be "sufficiently established to have gained general acceptance in the particular field in which it belongs." (We'll come back to some of the problems with this standard in just a bit.) The Federal Rules of Evidence say nothing about "general acceptance," only offering the following description of who can act as an expert witness: "If scientific, technical, or other specialized knowledge will assist the trier of fact to understand the evidence or to determine a fact in issue, a witness qualified as an expert by knowledge, skill, experience, training, or education, may testify thereto in the form of an opinion or otherwise." Some courts were employing *Frye's* "general acceptance standard" across the board, others employed it only for forensic evidence in criminal cases, and other courts used the FRE's "helpfulness" or "relevancy" rule. The Supreme Court was supposed to clear this mess up, or at least make another judgment.

And indeed, what's interesting about the *Daubert* decision is that it's hard to tell what exactly, if anything, was decided. The Supreme Court's judgment in *Daubert* was not only short and sweet; it also didn't clear anything up, "raising far more questions than it answers," according to Solomon and Hackett. Although its decision on the surface marks the end of the more restrictive *Frye* standard of "general acceptance," and encourages the FRE standard, the Supreme Court emphasized that the "subject of an expert's testimony must be 'scientific . . . knowledge.'" What the Court left out with that ellipsis are the other two terms from the FRE, "technical" and "specialized."

And what does the Supreme Court think can count as "scientific . . . knowledge?" For that accounting, "an inference or assertion must be derived by the scientific method . . . the requirement that an expert's testimony pertains to 'scientific knowledge' [here the ellipsis has disappeared entirely] establishes a standard of evidentiary reliability." And just how does the Supreme Court think a trial judge will evaluate the "reasoning or methodology underlying the testimony" to determine if it was derived by the "scientific method," or was a product of so-called junk science? Although it noted that "many factors will bear on this inquiry" (indeed!), the Court listed four in particular that trial judges should bear in mind.

The first "key question to be answered in determining whether a theory or technique is scientific knowledge" that will assist the trier of fact will be whether it can be (and has been) tested. Here the opinion quotes philosophers of science Carl Hempel—"statements constituting a scientific explanation must be capable of empirical test. . . "—and Karl Popper—"the criterion of the scientific status of a theory is its falsifiability, or refutability, or testability." (Not the most helpful or definitive statements in the world, but let's move on.)

The second factor concerned peer review and publication, and here the Court displayed some subtlety. Instead of the hyperbolic statements about the "science of publication" and the "science of consensus and peer review" that came from the Ninth Circuit Court of Appeals, the Supreme Court classified peer review and publication as a "relevant, though not dispositive, consideration in assessing . . .

scientific validity." The Court added that publication "does not necessarily correlate with reliability." (Again, questions arise here, but let's continue.)

Third, judges should consider the "known or potential" error rate in the "scientific . . . knowledge" in question. The Court doesn't offer much help in the way of specifics here, either.

Fourth, finally, and most paradoxically: After rejecting the *Frye* rule, the Supreme Court held that "'general acceptance' can yet have a bearing on the inquiry" which judges undertake into the admissibility of evidence. Solomon and Hackett note that "allowing that a determination of general acceptance is permitted, though not required, the Court resurrects the spirit of *Frye*, if not the corpse. But the Court offers little guidance about what exactly 'general acceptance' means and how to find it. The opinion invites further confusion among lower courts by noting that without 'widespread acceptance,' evidence may 'properly be viewed with skepticism.' The Court does not resolve what role general acceptance should play, how it can be determined, or even why it matters."

"The inquiry envisioned by Rule 702" of the FRE, the Supreme Court concluded, "is . . . a flexible one." That much is certainly clear. But where most commentators might lament that, wanting better "guidance" from the Supreme Court on these matters, we have more patience with their lack of clarity. Not that lack of clarity is simply good, or doesn't create some problems, but by and large we think it is indeed better to be somewhat muddled on these issues—to allow room for judgment. Like the punchline about ancient cosmology—"It's turtles, all the way down!"—what we're saying about the sciences, and where the sciences intersect with laws, is: "It's judgments, all the way down!"

The Court's decision explicitly acknowledged the need for judging in the matter of peer review, which went from a rigid standard to one criterion among many to be considered and weighed. That's a good thing; the simple fact of publication in scientific journals, as we'll see later in this chapter and in Chapter 6, is hardly free of either personal or political influence, or of scientific uncertainty.

At other points in the decision, the need for judging remains more implicit. Judges, the Court says, will have to make a ruling on just how general the general acceptance of a particular method or theory is. But we would like to focus on two other implicit and related judgments here, concerning the criteria of falsifiability and relevant specialized knowledge.

Justice Rehnquist stated in his dissent: "I am at a loss to know what is meant when it is said that the scientific status of a theory depends on its 'falsifiability,' and I suspect that some [trial judges] will be too." As anyone who has ever read Sir Karl Popper's essays knows, what exactly constitutes "falsifiability" and "falsification" is a matter of judgment, even for philosophers (especially for philosophers!), let alone trial judges. Even the relatively idealistic Popper found himself trying to make ever-finer distinctions among "naive falsificationists," "sophisticated methodological falsificationists," and other categories which better captured what scientists actually believe and do in different situations and at different times than the unmodified "falsifying" does.

More important still is the hidden judgment being passed off here: How were Popper and Hempel judged to be the appropriate authorities here? Who judged that their perspectives provided the unshakable ground on which to judge whether inquiries were scientific or not? Why not any one of Popper's critics, whether the more conservative like Imre Lakatos, or the more radical like Paul Feyerabend? These rhetorical questions are meant to suggest only that certain judgments are lacking here, or at least hidden. Citing the authority of Popper and Hempel was less of a judgment and more of an assumption, less of a question and more of a culturally authorized answer. It's interesting that the Court doesn't even refer to scientists for their characterization of "the scientific," but to philosophers. We wonder what they might have made of the physicist David Bohm's comments on Sir Karl Popper, uttered during an interview with the more straightlaced physicist Paul Davies:

> *Davies:* This raises Popper's idea about what we can regard as scientific. He argues that you have to be able to show the theory is potentially falsifiable, and that depends on being able to make observations which could contradict the theory.
>
> *Bohm:* That's Popper's idea. I'm trying to say, why should we take him as the authority? There are all sorts of ideas which people have had, and Popper has proposed an idea which has some merit, but it needn't be the absolute truth. If one says that Popper has given the absolute last word as to what science is, then why should I accept that?[24]

Judgments about "falsifiability" presume the authority of certain philosophers—that they have specialized knowledge about the sciences, we might say. Recall that "specialized knowledge" was that ambiguous term used in the less restrictive Federal Rules of Evidence as to who would be allowed to offer expert testimony. Our question here is: What would happen if the Supreme Court judged the judgments of scientists like Bohm, philosophers like Feyerabend, or any of the historians, sociologists, anthropologists, and other scholars of the sciences to be "specialized knowledges" which, in our opinion, they indeed are? What would it take to make their presumption into a judgment, a question of judging whose view of the sciences should be deployed in the courts?

The question is neither ridiculous, nor so far in the future that our society will be able to neglect it. Consider who filed *amicus curiae* briefs in the *Daubert* case: not only organizations such as the American Association for the Advancement of Science, the American Medical Association, and the National Academy of Sciences (all in support of the defendant, Merrell Dow), but also the American Society of Law, Medicine and Ethics; a group of four scholars with expertise in epidemiology, medicine, and history of science and medicine; and another group of twelve comprised of physicians, scientists, and historians and sociologists of science (all in support of the petitioner Daubert). The question of who has specialized knowledge of the nature and practice of the sciences is one that has clearly begun to open up, and needs to be opened up still further.

TRIAL BY JURY

How might different judges judge differently on some of these matters of admissible evidence, causation, and the nature of the scientific enterprise? And how might other judges decide who, ultimately, should have the authority to judge these matters—judges or juries? We're fortunate that the brother of one of us, Mark Bernstein, was sitting on a judicial bench in Pennsylvania when a related case was brought before him, *Blum v. Merrell Dow Pharmaceuticals, Inc.* A 1986 jury had awarded $2 million in compensatory and punitive damages to the Blums, whose son Jeffrey had been born with club feet; his mother had taken Bendectin during her pregnancy. The Pennsylvania Supreme Court reversed that verdict, arguing that Merrell Dow had been deprived of its constitutional right to trial by jury when the judge had allowed the trial to proceed with only eleven jurors after the twelfth had become ill. In the retrial, a new jury heard seven weeks of evidence and awarded the Blums $19 million in compensatory and punitive damages. Merrell Dow was back in court, seeking a new judgment or a new trial.[25]

This case is distinct from *Daubert* in two important ways. First, where *Daubert* restricted the relevant scientific evidence to statistical epidemiology, *Blum* admitted the importance and validity of chemical structural analysis and in vitro and in vivo studies in addition to epidemiological evidence. And second, the Pennsylvania court placed the process of judging firmly within the purview of the jury, rather than of the judge.

The scientific and medical experts called by the Blums established that many of Merrell Dow's experimental designs had too few animals, and too low experimental dosages of Bendectin. Scientific studies performed outside of Merrell Dow showed that Bendectin was indeed a quite powerful teratogen (a substance which causes birth defects). One of Merrell Dow's expert witnesses conceded under cross-examination that his own epidemiological study was "less than good" (he interviewed 1,427 mothers, of which only 122 had taken Bendectin). He also testified that such "less than good" scientific studies nevertheless passed the highly touted scientific standard of peer review and appeared in print. Another of Dow's own expert witnesses had to concede that his study on Bendectin included women at later stages of pregnancy whose fetuses could not have been affected by Bendectin, and that "this illogical grouping resulted in an underestimate of the risk of club feet in offspring."

It also came out during the trial that one scientific study "proving" Bendectin's safety was rejected by *The New England Journal of Medicine, The Lancet,* and *The British Medical Journal,* only to appear in the journal *Teratology* (the editor of which had been retained by Merrell Dow as a scientific expert for eighteen years). In that instance, the study's author had written what the Court called a "grovelling letter" to Merrell Dow, soliciting funds for his research since, as he put it, "much clearly depends upon the value of this publication to Merrell Dow National Labs . . . I should appreciate any gesture Merrell felt inclined to make, but I imagine that if we are able to give [Bendectin] a clean bill of health with regard to teratogenesis, this would be of substantial help in the courtrooms of California." Other experts in "teratology" (a specialty that is not certified by any professional review board) routinely submitted drafts of their scientific articles to Merrell Dow's attorneys for editing. Perhaps most egregiously, Merrell Dow had failed to report abnormalities in test animals to the Food and Drug Administration. In other words: basic violations of both scientific methodology and the sociopolitical process of trust and justice had occurred. They manipulated experimental design and reporting, and they lied. Sometimes the operations of power are quite direct.

This is not even half the sorry story. But we now cut to some of the court's conclusions. As Mark Bernstein wrote:

> By this appeal, defendant asks that the law of Pennsylvania be transmogrified so that each trial court can preside over a "scientific court" whose primary function is to embody, as precepts of law, the current "generally accepted" opinion of any self-identified scientific establishment. Counsel claims that only generally accepted scientific principles and only subjects having "general agreement" should ever be permitted in court. By this view, the trial judge becomes the courtroom door guardian for scientific conformity, and each trial judge creates, as precepts of law, his or her own individual determination of proper scientific orthodoxy.

Bernstein rejected such an appeal on at least two counts. "What does history reveal about orthodoxy and the establishment consensus of scientific principles? . . . Spontaneous generation was the scientific consensus for the existence of vermin. A clear, scientific consensus confirmed that people of color belonged to an inferior class of mankind. . . . Is our collective memory so short that the Judiciary cannot remember the horror of the consensus in the 'relevant scientific community' that women were unfit to be lawyers or doctors . . . or that genocide of the Jewish race improved racial purity?"

"By defendant's legal theory," he added, "judges, as doorkeepers, must seal the courtroom until 'science,' itself, reaches a new consensus. Defendant's principle would have precluded testimony by every seminal thinker in the history of the world, including Newton, Pasteur, Freud, Darwin and Einstein. . . . The courts, and thereby all society, would be locked into outmoded thought, erroneous principles, and false 'truths.'" The appeal by Merrell Dow's lawyers, as well as other judicial decisions made along similar lines, "never permits minority opinion to be expressed in court. No new opinion; no, as yet, unaccepted opinion; no changing opinion; no minority view could ever be exposed for factual determination in a court of law. Only the scientific establishment can open the courtroom door."

Inside that door were not only judges, but more importantly, juries. Their job would certainly not be made any easier by the admission of multiple "specialized knowledges" for them to weigh and judge. But that is both the system's pitfall and its most redeeming quality. "[C]astrating the fact-finding role of the jury, the judiciary becomes an absolute bar to legal inquiry, until a new 'scientific consensus' claims the mantle of divine revelation required to open the courtroom doors, but only to let in the new established orthodoxy. The testimony in this case demonstrates how 'scientific consensus' can be created through purchased research and the manipulation of a 'scientific' literature, funded as part of litigation defense, and choreographed by counsel."[26]

There are many benchmarks for judging to be found here. We've encountered a number of them already: Use history as a guide. Pursue complexity—ask about the linkages between a scientist and a corporation, and between an article in a peer-reviewed journal and "the truth." Demand specificity—ask what criteria were used to construct a control group in an epidemiological study.

Judging also requires accountability, where rules of evidence and discourse are laid out and enforced. The courts are, for better and for worse, one of the primary sites where such accountability can be maintained in our society. They are a form of that public space for witnessing that Boyle and the other founders of experimental science at least said they valued for its ability to guide the production of truths and facts. These public spaces of accountability force people to articulate what might otherwise be forgotten or hidden, to account for their methods, to say where and how they supplemented their methods and calculations with their own processes of judgment, to state as clearly as they can—*under questioning*—where they bridged certain gaps, and where holes remained.

Judging requires expert witnesses, people who have immersed themselves in one pursuit or another: controlled experimenting with substances and animals, careful statistical and epidemiological analyses, reading and assessing a burgeoning scientific literature containing not only numbers, charts, figures, and method, but articulations of meaning as well. Many questions of justice today, toxic torts being just one prominent example, are addressed through the methods and articulations of the sciences. We can't escape that fate, nor should we necessarily want to; the sciences can indeed be one important form of knowledge with which to make decisions. But as any good trial lawyer will tell you, the plural term "expert witnesses" is absolutely necessary. As we saw in the Blum case, multiple sciences have to be considered. To elevate any one, such as epidemiology, is to deny that realitty is too complex to be captured by one kind of pursuit—especially in the complex chains of causality that constitute toxins and their effects.

The gentlemen scientists of the Royal Society knew that when such expert knowledges are presented in public, they are best presented as *opinion*: a professionally rendered judgment call on probabilities, which judges and juries will in turn weigh before rendering their own judgments. That entails another kind of literacy campaign, geared to both judges and citizens everywhere who will be called upon to serve on juries. That is a daunting and mammoth task, but we do believe that it is possible to cultivate a citizenry for juries which are quite capable of judging not only complex scientific issues, but who are also able to inquire into the sciences themselves as the kind of complex entities we've been describing here. Without being antiscientific, they should be able to ask critical questions about evidence, relevance, and the possible ways in which social, economic, or political interests are articulated with the most seemingly objective matters of fact.

A more open, more multiple admissions policy to our courtrooms is certainly not without problems. But it is better to err on the side of openness and multiplicity than on the side of monologic consensus. We therefore judge the criteria in the Federal Rules of Evidence to be a better error to live with than the criteria in *Frye*. The FRE criteria are better for addressing and engaging with the complexities, ambiguities, and changing "consensuses" of the sciences.

Another Affair

In Chapter 1 we used the story of Robert Millikan's determination of electron charge to show how scientists often work in ambiguous circumstances, making a series of interrelated judgments: whether they should hold onto their theories or their experimental results, which observations were "good ones" and which might be discarded, what was signal and what was noise, when a piece of equipment was working "properly," and so on. The Millikan story is a popular one with scientists, because it shows a creative reading working out for the best, i.e. the truth. The Millikan story has a happy ending.

Another story with a less happy ending (not to mention middle) for most of the people involved in it, is what has often been referred to as "The David Baltimore

Affair"—although a better headliner for it would be "The Margot O'Toole–Thereza Imanishi-Kari–David Baltimore Affair." (Better still would be something like "The O'Toole–Imanishi-Kari–Baltimore–altered mice–MIT–Rockefeller University–Rep. John Dingell–Secret Service–Office of Research Integrity–etc. etc. Assemblage Affair." But in our stargazing culture, we tend to name our scandals according to the one who has the most fame at stake, which in this case was the Nobel laureate and, at the time, Rockefeller University president Baltimore.) It is an affair which began in 1986 and only reached official closure in 1996, that has left damaged careers and reputations in its ten-year wake, and that will no doubt continue to serve as a key reference point for all manner of contentious discussion about everything from ambiguous data and questions of fraud, to the perils of whistleblowing in the lab and what public oversight mechanisms—if any—are appropriate to the sciences.

We start with the briefest of synopses of the relevant events.[27] In the summer of 1985, Dr. Thereza Imanishi-Kari hired Dr. Margot O'Toole to work as a postdoctoral researcher in her lab at MIT; the two had met a few months previous at a party at the house of Henry Wortis, O'Toole's thesis advisor. Working in collaboration with Baltimore, Imanishi-Kari was using a breed of transgenic mice (mice which had a gene spliced into them that produced a certain antibody) in immunological experiments; what Imanishi-Kari and Baltimore seemed to have found was that "normal genes" in the transgenic mice had somehow become able to produce the same antibody that came from the gene that had been deliberately spliced into these special mice. O'Toole was to work on a series of experiments that would substantiate and extend these findings—experiments which involved not only these highly specialized technomice, but also highly specialized chemical reagents (particularly one called BET–1) in very demanding and delicate experimental protocols.

After some initial success, O'Toole could not get the experiments to work, despite months of struggle. She went back over the notes of other researchers in the lab, and found what appeared to be discrepancies between those notes and data that had appeared in a paper authored by Imanishi-Kari, Baltimore, and several other researchers in the journal *Cell* in the spring of 1986. O'Toole felt that there were serious errors in the *Cell* paper, but not necessarily amounting to fraud. She brought her concerns to a senior colleague outside the lab, who consulted with other scientists, including O'Toole's former thesis advisor Wortis. This informal, "third-party" group of scientists thought there were some "mistakes" and "overstatements" in the *Cell* paper, but that these did not significantly affect the final argument or result. O'Toole thought these should be corrected publicly, in print, but the others could see no benefit or need for that.

O'Toole was not happy with that judgment, but things were about to go from bad to worse. Her postdoc finished and no other positions having arisen, O'Toole went to work for her brother's moving company. She had discussed these matters with another scientist who, unbeknownst to O'Toole, informed two rather self-aggrandizing, "fraud-busting" scientists at the National Institutes of Health (NIH), Walter Stewart and Ned Feder, about the whole business. Stewart and Feder in turn contacted O'Toole, and after their repeated requests for the pages from the laboratory notebooks, O'Toole eventually provided them. Stewart and Feder concluded there was a serious problem, and wrote to Baltimore asking for more laboratory records.

Now it was Baltimore who was not pleased with this judgment from out of the blue, and things went from worse to worst. NIH convened an investigative panel; Stewart and Feder met with aides of Congressman John Dingell, who chaired the Energy and Commerce Committee (which had authority over the NIH budget), and they scheduled a hearing. Baltimore sent a letter to hundreds of colleagues, warning that "a small group of outsiders" was threatening to "cripple American science."

The NIH investigative panel found no fraud or misrepresentation (although they did find, as the informal third-party scientists had, "significant errors of misstatement and omission, as well as lapses in scientific judgment"). Meanwhile, clarifications and additional data had already been published in *Cell*. But Stewart and Feder still felt there was a conspiracy afoot. The Secret Service was called in to analyze the laboratory notebooks. NIH, through its newly established Office of Research Integrity (ORI), reopened the investigation. In 1991, after only a year and a half as president of Rockefeller University, Baltimore decided to resign. The ORI issued its final report late in 1994, upholding a preliminary finding of guilt for Imanishi-Kari, and making her ineligible for government grants for ten years. Imanishi-Kari appealed; the Research Integrity Adjudications Panel issued its findings in June 1996, clearing her of the fraud charges and making her eligible once again for federal research funds.

Our goal here is not to argue who was right or who was wrong, who were the heroes and who were the goats, or what the immune cells of transgenic mice "really" produce. As we will see shortly, those kinds of judgments are fraught with complexity, mounds of empirical data from both the laboratory and the hearing room, and multiple demands for interpretation. As it stands now, the official judgment is that the preponderance of evidence indicates that no fraud was perpetrated, and both Imanishi-Kari and Baltimore have been cleared of any charges of falsification or fraud. (Baltimore is now the president of The California Institute of Technology.)

But that does not mean that we think there are no interesting questions or problems remaining, nor that we agree with many of the lessons or conclusions being drawn from this case. What needs to be judged now is not the outcome of the affair, but what the very complexity and messiness of the affair itself means for the sciences today.

Consider some statements from the final report of the Research Integrity Adjudications Panel (when we downloaded it from the NIH Internet server, this document ran to over two hundred pages), the report that "vindicated" Imanishi-Kari and Baltimore:

> The *Cell* paper as a whole is rife with errors of all sorts . . . some of which, despite all these years and layers of review, have never previously been pointed out or corrected. . . .
>
> Responsibility for the pattern of carelessness in writing and editing of this paper must be shared by all the participants, including the main drafter (Dr. Weaver), the leading collaborators who shaped the communication and drafting process (Dr. Baltimore and Dr. Imanishi-Kari), the contributing authors who failed to catch errors in their areas of expertise, those who read the paper in draft form (including Dr. O'Toole and Dr. Wortis), and the reviewers and editors who failed to pick up errors in the original submitted text of the paper. While a high rate of careless errors is no defense to intentional falsification and fabrication, the presence of so many pointless mistakes at least raises a question whether the mistakes singled out as intentional (because they arguably favor the authors) really represent conscious efforts to deceive.[28]

The original *Cell* paper, in other words, went through multiple authors and readers, passed through a peer-review process, and still emerged riddled with errors. Some of these were corrected, and some were only found after years of intense review. Indeed, the "high rate of careless errors" is half-seriously offered as a sign that there were no "conscious efforts to deceive." It can certainly be argued that such errors were "insignificant," as they well appear to have been in this case. But such is the daily reality in the fast-paced, high-stakes fields of contemporary biomedical research that scientific papers—even those with prestigious authors, recounting experiments done at prestigious institutions—can be "rife" with "obvious" errors.

The panel also pointed to questions about more subtle matters:

> All of the scientists who looked at the questions raised about the *Cell* paper over the preceding decade (at Tufts, MIT, and on the NIH Scientific Panel) found no evidence that scientific misconduct had occurred. While they found errors in the paper, and the authors published corrections, the dispute appeared to center on differences about how judgment was exercised, how experiments might be interpreted, and whether phrasing in the paper was correct.
>
> Although ORI raised some new or differently presented scientific issues before us, these also proved to be largely matters of interpretation, judgment, and confusing laboratory jargon.

Judging, interpreting, phrasing: these less-than-scientific processes are always being "exercised" by working scientists—including those trying to judge and interpret questions of scientific fraud. At the core of the most exact experimental sciences is a certain inescapable fuzziness; the most apparently factual statement about antibodies in transgenic mice is intimately entangled, in some way, with interpretations and assessments. Tracing those entanglements can be a demanding, delicate, and prodigious task:

> The sheer volume of material presented by ORI was enormous. The number of exhibits and reports generated over the many years of investigations and the lengthy hearing process required a massive amount of time and resources to evaluate fairly. Weighing evidence is not a mechanical process comparing the number of pages or hours of testimony for each side; rather, the fact-finder must assess the quality and probative value of all of the evidence. This approach is in accord with that of courts confronting large records.

Here was a case of absolutely "normal" science, facts that appeared in print quickly along with thousands of other normal facts in hundreds of other normal papers in dozens of other normal journals that same month. After the fact, however, and only because of questions initiated by a well-versed postdoc exceedingly familiar with the research, it took "a massive amount of time and resources to evaluate fairly," to judge if the rapidly produced fact was indeed a fact.

Before we offer our own summary opinions on these events, we want to turn to an interesting exchange of opinions in the pages of *Science* magazine. Coming as it did in the late spring of 1996, the exchange was perhaps prompted by news of the impending final report on the Baltimore affair. The subject of the exchange—ambiguity in the sciences—was certainly relevant to these events.

The Haunts of Ambiguity

Frederick Grinnell's editorial in *Science* titled "Ambiguity in the Practice of Science" highlighted these same questions of judging and assessing responsibility that we've been staying close to here. Grinnell, of the Department of Cell Biology and Neuroscience at the University of Texas Southwestern Medical School, referred to two recent reports from science institutions concerned with problems of fraud and misconduct in scientific research. One of these was a 1995 report issued by the Department of Health and Human Services' Commission on Research Integrity (CRI), which was anxious to establish "the fundamental principle that scientists be truthful and fair in the conduct of research and the dissemination of research results." One of the problems with this platitude was that, like all platitudes or "fundamental principles," it glossed over the ambiguities, uncertainties, and complexities encountered in actually *doing* science. In support of the latter, Grinnell also referred to a report previously published by the National Research Council (NRC) in 1992 titled *Responsible Science: Ensuring the Integrity of the Research Process*. Even this premier establishmentarian scientific institution recognized instances in which "the selective use of research data is another area where the boundary between fabrication and creative insight may not be obvious."

The very existence of such reports points to both a problem in the social worlds of the sciences, where questions of fraud and misconduct are becoming increasingly visible and problematic, and to the urgency of having some kind of institutional mechanisms and established procedures or principles that address these kinds of situations. Everybody wants the sciences and scientists to be "responsible." But where the CRI or NRC thought it necessary to lay down rules, clear standards of behavior, and fundamental principles of what science is and how scientists should be expected to act, Grinnell thought efforts like these were "inappropriate," and that they would actually "add to rather than resolve what has been an ongoing controversy."[29]

"Fundamental principles" about truth and fairness are "inappropriate," Grinnell argued, because they don't recognize the actual complexity and ambiguity involved in doing science. Grinnell looked to the real rather than the idealized working strategies of scientists; he's studied what eminent scientists like François Jacob (whom we've quoted earlier) and Rita Levi-Montalcini have written about their own thinking and doing. In her book *In Praise of Imperfection* (a muddling-affirmative title if there ever was one), Levi-Montalcini writes of the work that led to the isolation of nerve growth factor (and a 1986 Nobel Prize, shared with Stanley Cohen); Grinnell quotes approvingly her postulation of "the law of disregard of negative information . . . facts that fit into a preconceived hypothesis attract attention, are singled out, and remembered. Facts that are contrary to it are disregarded, treated as exception, and forgotten."

Grinnell uses such stories to remind us that the sciences are an unwieldy fusion of "experience, intuition, and creative insight," a contradictory game combining chance and passionate commitment with method and detached analysis. "By blurring the boundary between creative insight and scientific misconduct," he writes, "ambiguity will frustrate any attempt to deal with misconduct through the application of fundamental principles. . . . Unless we understand that ambiguity is an inherent feature of research, we may find the practice of science restricted in ways that make creative insight far more difficult."

That last sentence should remind us that Grinnell is playing a necessary but risky line, and we are in treacherous and complicated territory here. There is a conservative response to these arguments with which it is easy to imagine many readers of *Science* being sympathetic: Yes, scientific practice is riddled with ambiguity, and only the best, most intuitively creative scientists at the top of their fields are able to appreciate the subtleties involved, and therefore only scientists should be authorized to judge matters of responsibility, fraud, or misconduct. The conservative response to ambiguity is to argue for an even more insular, self-governing, elite scientific community.

Indeed, the Deputy Assistant Inspector for Oversight at the National Science Foundation, Donald E. Buzzelli, articulated just such an argument in a letter to *Science* responding to Grinnell's editorial:

> The important thing is that scientists should not be held to a standard of perfection; rather, a misconduct case should be initiated only when a scientist's departure from truthfulness or fairness has been egregious.
>
> Ambiguity haunts misconduct cases at least as much as it affects the normal practice of science. Even the relatively clear-cut examples Grinnell cites (data fabrication, plagiarism) raise questions such as intent and whether the action in the actual circumstances was serious enough to be misconduct in science. Simple rules do not resolve such questions. The ambiguity of actual cases calls for the application of judgment based on the standards of ethical practice of the relevant community of scientists. Any definition that does not recognize the need for that kind of judgment in individual cases will in the end subject the accused scientist to legal standards taken from some other source. One of the major issues . . . is whether scientists can continue to be judged according to the scientific community's own standards of practice.[30]

Grinnell in turn responded that "we have yet to adequately describe those standards" of what constitutes "ethical practice." "Instead of appealing to theoretical, fundamental principles," he went on, "we need to articulate what doing science actually entails and to explain how the ethical practices of science fit into the context of everyday research."

The sciences, which were supposed to get rid of all ghosts, now recognize themselves as haunted by ambiguity. That's a valuable recognition. It would be a shame, though, if we recognized ambiguity only to have it policed by scientists themselves—or *only* by scientists themselves. We would have failed to recognize the ambiguity, then, in deciding who is in the best position to be involved in Buzzelli's abstract, faceless "application of judgment" (which sounds suspiciously unambiguous, like "application of paint").

We could make ambiguity, and the questions of judgment and interpretation it entails, into a new rationale for defending the autonomous self-policing of the sciences. But what would it mean to say that because the actual practice of the sciences is complicated, creative, selective, and ambiguous; because the sciences rely on "blurring the boundary" that is conventionally thought to separate power/knowledge, reason/imagination, "the law of disregard of negative information"/fraud, or any other opposition you can think of; because reality is absolutely "other" *and* a matter of our own judgments—that for precisely these reasons, we need better, more complicated, and more pluralistic social mechanisms of accountability?

With that very big question in mind, let's return now to the Baltimore affair and try to extract some more benchmarks for judging. The first thing that should be said is that it was indeed a muddled, messy, complicated set of events at every level: at the immunological and laboratory level; in the social dimension of scientific research, careers, and the writing of papers for publication; in the institutional politics where the sciences meet the Office of Research Integrity and powerful congressional committees. And in the face of such complications, we should be very loath to make quick and hard judgments, even though judgment is precisely what is demanded.

Judging assemblages requires that we respect ambiguity and uncertainty. The problem here is not fraud, or at least not simply fraud, but something much more subtle and ambiguous. A thousand things can go wrong with an experiment, a hundred holes can plague explanation, a dozen interpretive possibilities can suggest themselves. When we look at the specificities and complexities of actual cases, as both Grinnell and Buzzelli rightly demand that we do, we will better respect what a difficult, arduous, and creative craft the sciences are. Margot O'Toole phrased the risk in terms of the ever-present possibility of "self-deceit": being so wrapped up in your theories and results and hopes that you lose critical distance. And yet being so wrapped up is in fact essential to doing the sciences well.

But there were many more things in addition to personal theories or hopes in which the various actors in the Baltimore affair were enwrapped. These other parts of the assemblage also demand judging, again while respecting ambiguity, because they reflect important elements of the social worlds in which all the sciences are done today. There are personal allegiances and informal obligations (remember that O'Toole got her job as a postdoc in Imanishi-Kari's lab at a cocktail party) that can make it hard not only to become a whistleblower, but even just to carry on open discussions about ambiguity, interpretations, and other haunting gaps encountered every day. Real differences in power in research institutions everywhere, along gender lines and along seniority lines, only make it harder to discuss, think about, and judge complex muddles. The intense specialization that has characterized the sciences for quite some time now, and is only intensifying, compounds the difficulty: Baltimore didn't know how to do many of the experiments that had his name on them. And all researchers, senior and junior, are under extraordinary pressure to publish often and quickly. This is not simply a question of commercial pressures, although as more and more university biologists affiliate with biotechnology and pharmaceutical companies, those pressures increase. But even within the supposedly sacrosanct academy, scientists are under pressure to publish new and useful results to keep their grants coming in, to keep donors happy, and to keep their careers going.

All in all, this is not a social system which promises to handle ambiguity patiently, kindly, or subtly. Within such a highly pressurized container, it's no wonder that the Baltimore affair escalated so insanely, with congressmen, self-appointed fraud busters, and the Secret Service getting called in. We should not conclude, however, that an organization like the Office of Research Integrity has to be "Orwellian," a kind of "Star Chamber" of faceless judges that is both unnecessary and harmful to science. (These two metaphors were employed by Daniel Kevles in his *New Yorker* article on the Baltimore affair.) That's not to say that the operations of the ORI or its precursor did not need reform; Kevles notes the way it denied certain basic rights that apply in other legal proceedings, to know what the accusations are and to cross-examine one's accusers, for example. But those are questions of procedure, and can be subject to change and debate without calling the whole enterprise into question.

This situation escalated to almost unmanageable levels, but that doesn't mean that you get rid of the structures you have, in the name of scientific purity, autonomy, and "fun." It means you figure out how to do it better, how to have a social structure of collective accountability that works not just to address the relatively easy cases of fraud, but those much grayer and more difficult cases shot through with ambiguity. We need to imagine how these important institutional mechanisms of oversight and judging could be made better, more effective, and less invasive.

In our usage, "judging" oscillates between verb and adjective. We have to learn how to judge assemblages: examine them in their full complexity, ask about the motley connections that give them their strength and allure, attempt to make evident what might be hard to see or conceptualize. But read another way, we're pointing to the need for new kinds of "judging assemblages": institutional arrangements, social conventions, ethical expectations, and conceptual tools that together make a space in which all the judgments which are so necessary to doing the sciences well can proceed.

That kind of space need not only be institutional. It can also be a more diffuse cultural space, which is in some ways more suited to these kinds of ambiguities and complexities. To return to the words of Silvan Schweber: If the separation between the "moral sphere" and the "scientific sphere" in particle physics "cannot be maintained," then what do we do? *Don't* install phys-ethicists or social constructivists or cultural anthropologists *over* the physics community, as some sort of moral and social police force. These people are no less (and no more!) muddling through than the particle physicists. *Do* find better ways of fitting those people into the assemblages of the sciences, where they can ask "third party" questions, and bring different perspectives and insights. *Do* invent new curricula, textbooks, and other teaching strategies that provide the questions and analytic tools for pursuing the social, historical, and moral dimensions of the physics-assemblage so that, little by little, a culture that promotes and respects multiple forms of inquiry, acknowledged uncertainty, and social and political engagement can emerge.

The task, however, is not simply to "make a space." As these few examples show, building a judging assemblage is perhaps even more urgently a question of structuring time. The ambiguities that are the unavoidable accompaniment to the sciences today—the gaps, voids, and lacunae that are always a part of our densely reticulated experiments with realitty—require often extraordinary amounts of time for inquiry and judging. Examinations of the damaging "side effects" of pharmaceuticals, or of environmental toxins, require months of testimony detailing the experiments which produced those chemicals or established the parameters of their effects, the articulations involved, the possible ways in which powerful economic interests or social beliefs have become entangled within the "simplest" matter of fact, and a patient, attentive jury to listen and think critically. When a question arises about the kind of scientific claim that our laboratories routinely produce every day in vast quantities, as with the Baltimore affair, it can take years of investigative labor, producing thousands of pages of evidence—and still never manage to eliminate the various ambiguities involved, but perhaps only shift them to a less shrill register. Extremely muddled sciences require extensive time to mull. And time, as no one needs to be told, is something that is in very short supply these days.

Game Time

We've been handling explosives here, assembling our own volatile mixture of examples drawn from the negative side of the sciences. There is something unfortunate about that selection, since we could have pursued these questions of judging through the creative, joyous, wonderfully imaginative and productive, positively charged side of the sciences as well. Indeed, in a chapter opening the image of Jaurès marveling at the achievement of flight, it would seem that we've stood closer to his companions who groused about the sorry state of politics and society. And it looks like the answer to one of our early questions, "do the sciences still get to be fun?" is an emphatic "no." The sciences now have to be responsible.

Our own judgment call is that the demand for judging is heard more urgently when things go awry, and if we can respond to these kinds of events, people could

figure out the less ugly ones on their own. If we had made that judgment differently, the questions raised and responses offered in this chapter would have been
different, in perhaps subtle but no less significant ways. The book would have
been different, and would have had different effects on readers bringing who
knows what variety of their own interpretive assemblages to bear on it. Many scientists would no doubt be happier, right now, if we had used more upbeat stories
here and elsewhere. (And not merely happier, but perhaps more willing to think
through the rest of the book.)

Which is precisely the point: The smallest judgments affect what will be
known, written, and created. Decisions which are anything but strictly rational,
decisions taken far in the past or hidden almost inaccessibly in the densest of assemblages, shape an emergent realitty that is always full of ambiguities. In their
complex, faltering, and extremely effective manners, all of our inquiries will set up
the terms, concepts, and practices that will fundamentally define what we judge as
natural, reasonable, and real. Scientific inquiry is not a means to an end, the dis-
passionate extraction and application of certain answers from and to a given reality. Scientific inquiry is the simultaneous production of means and ends, the
invention of capacities to frame "really?" questions along with the creation of a realitty in which those questions will find answers. The sciences are a kind of performance, and what they perform is realitty.

"Performance" is a better metaphor for thinking about the sciences for a number of reasons. First, "performativity" is not just a trendy term from cultural studies—it's a real characteristic of the sciences now. To reiterate what Schweber
pointed out at the beginning of this chapter: "The scientific enterprise is now
largely involved in the creation of novelty—in the design of objects that never existed before and in the creation of conceptual frameworks to understand the complexity and novelty that can emerge from the *known* foundations and ontologies."
Representation as a goal of the sciences is no longer the dominant concern; performance or production of the new is: new drugs, new ideas, new businesses, new
mice with new genes.

If we continue to think that the goal of the sciences is to represent reality, then
we will run into the same problem which the science writer John Horgan encounters at the end of his book, *The End of Science*: Confronted with changing, multiple
scientific representations, we know now that the sciences don't really deliver the
truth, so we can take their theories and pronouncements as kind of ironic fictions,
with a wry laugh, a nudge and a wink.

It's an interesting suggestion, and since it might seem close to the kind of analysis we are suggesting here, we'll point out some important differences. First, Horgan interviews the big names, the superstars of science who can sell books. He
pays little attention to what happens outside that tight focus—to where the sci-

ences really move and shake the world. Second, as long as he was going to do that, and end his book suggesting irony as the appropriate attitude to take toward the sciences now, he should at least have interviewed a couple of superstar literary theorists, who would have informed him that irony can be just as complicated as a transgenic mouse.

Charles Baudelaire was one of the great practitioners and theoreticians of irony in the nineteenth century. (The solution of ending something—realist literature, realist painting, realist truth—by resorting to irony is hardly a new proposal.) Irony depends on a knowledge of one's fallenness from a state of grace, a reminder that one has only imperfect, instrumental knowledge about transcendent nature. But instead of ending anything, either by self-deprecating humor or some other mechanism, irony sets off what Baudelaire called *vertige de l'hyperbole,* a cascading dizziness. "It may start as a casual bit of play with a stray loose end of the fabric," as Paul de Man comments on Baudelaire, "but before long the entire texture of the self is unraveled and comes apart. The whole process happens at an unsettling speed. Irony possesses an inherent tendency to gain momentum and not to stop until it has run its full course; from the small and apparently innocuous exposure of a small self-deception it soon reaches the dimension of an absolute."[31]

Irony, then, is not a good solution to the question of the sciences. It's much too absolute in its own insistence on nonabsoluteness. Moreover, irony depends on an outdated model of the sciences, as an all-too-human representation of a preexisting, transcendent reality. Finally, irony leaves no space or opportunity for judging, only dizziness.

No, irony won't see us through. As they perhaps always have been, the sciences are now pursued for new performings of realitty, and performances can and must be judged for their different effects.

The sciences, then, are something of a game, albeit a very serious one. But if we are in a time in which responsibility has become a key word for the sciences, that doesn't mean that having fun at this game will or should go away. We need a new aesthetic for *performing* sciences that includes both the pursuit of responsibility and the preservation of the joy, exuberance, and creative *affirmation* that the sciences have always provided for their practitioners—and sometimes for the rest of us. Such a juxtaposition will not always be easy, or in perfect synch; airships will undoubtedly continue to precede the creation of the class of revolutionary aeronauts who will know how to use them responsibly. We need to pursue sciences that are both serious *and* playful. Since we don't know how to do that very well, maybe we can start by thinking of them as a game, or rather, as a game of games.

"Game of games" is what one scientist, the biophysicist Henri Atlan, suggests as a metaphor for thinking about the contemporary sciences. (Playing the professional and scientific credential game for a moment: Atlan is professor of biophysics at the University of Paris, and Isaiah Horowitz Hadassah Scholar in Residence for studies in Philosophy and Ethics of Biology at the Hadassah Hebrew University Medical Center in Jerusalem, and a member of the French National Presidential Commission for Ethics of Life and Health Sciences.) The sciences are a very serious game of games—indeed, "nothing is more serious than the games of knowledge, for their stakes are not beliefs—beliefs are not serious and do not require research programs—but our very existence as living beings, members of the human species."[32] Like a game, the sciences work by delimiting a space where specific rules will be observed, specific things go together with other things, and specific moves are defined as reasonable or unreasonable. Within those specified conditions, specific performances can take place. Like a game of games, the sciences change and grow by simultaneously getting outside that delimited space, changing the rules, props, and dialogue which the performance depends on—and playing with ideas and things once again.

Thinking of the sciences as a game of games doesn't mean that you can play any way you like:

> The game of games is not devoid of all rules. In particular, it includes negative rules such as: do not mix up the rules of the different games; be careful with analogies and use them only to make differences stand out more clearly; steer clear of the temptations of all-embracing fusions; do not give in to the fascination of grand cosmological syntheses or make them operate in the mode of legend and (science) fiction; finally, and above all, do not pretend to ground ethics (and even less, politics) on some objective knowledge, established scientifically (or otherwise), that is supposed to disclose the Truth of Nature. For what distinguishes game from ideology is that the latter is *believed*, while the former is played.[33]

Does acknowledging that the sciences are a game of games, in which realitty is performed, lead one to a relativism of knowledge that's nothing more than sterile skepticism or irony?, asks Atlan. Far from it. It "actually permits a critique that is even more fruitful because it is not based on a metatheory that could produce a theory of critical thought. Recognizing the unbridgeable distance and the incommensurability of two modes of knowledge allows us, by jumping from one to the other, to have a radical exteriority toward each of them. The result is a critique that is not simultaneous, but rather alternating and reciprocal, and thereby always able to refresh its praxis, because it avoids setting up an autonomous and overarching domain that would be the locus of a necessarily noncritical theory of critical thought."[34]

In describing the sciences as a game of games, Atlan is also describing a game of judging, which is different from a game of policing. Jean Jaurès's companions, gazing at the balloon-Eiffel Tower-Paris assemblage, wanted a kind of police force, a new overarching authority to control the strange new, commingled traffic. It's an understandable demand that can still be heard from many quarters today. But heavy-handed proposals for sociocultural value-policing of the sciences are as ill suited to their complexities and muddlements as demands for laissez-faire purity or for total, autonomous fun. Any such simple solution or program is far too jarring and inept, and threatens to blow the whole volatile enterprise sky high. The charged, contingent juxtapositions of the sciences, and their multiple incommensurabilities, require something more akin to the reciprocal and even contradictory alternations among games suggested by Atlan—something we're calling muddling through.

The New Rules: Soft and Slow

The cases we've juxtaposed here—eugenics, Lysenkoism, toxic torts, questions of fraud and ambiguity—are themselves muddled, contingent, and complex juxtapositions of natural and social elements. Like Tansey's fractal shoreline, we've staged some of the layered iterations of the sciences: the same complex structure, from the macrolevel to the microlevel. (Perhaps you should break the oft-repeated rule of not judging a book by its cover.) Valuable work is nevertheless possible and even necessary, in this case the work of judging.

Since the sciences are such a dense, intricate, and volatile assemblage of practices, metaphors, articulations, and other kludged-together elements of nature, culture, and power, they have to be muddled through. Judging is key to this process of muddling through, in part because the sciences already depend on judgments at every level. In the Baltimore and Bendectin cases, scientists (as well as judges and juries) had to make judgments about the significance of ambiguous data, and about the trustworthiness of experiments and experimenters. These cases also involved judgments about the durability of a certain theory's truth, and how theories might be linked to economic, personal, and professional interests. In the cases of eugenics and Lysenkoism, the judgments involved all those issues, and their entanglement with larger historical and cultural factors. In all the cases, our judging had to respect the complexity and contingency of the multiple entanglements, while trying to show how judgments could still be made—indeed, how judging might be enabled by recognizing a more muddled state of affairs.

Judging is not a science. The sciences can, however, serve as an important tool for judging a variety of empirical, social claims—as long as we remember to simultaneously judge the sciences, too. It's a tricky, double operation: using a tool for inquiry and judging, while being ready to question and judge that tool at the same time. This is the refreshing alternation between contradictory "radical exteriorities" for which Atlan argues. Rules may not be hard and fast, but soft and slow rules can nevertheless be quite effective, humane, and just.

"Judging assemblages" is another way to phrase the challenge we call muddling through. The stakes, as we've seen in this chapter in particular, couldn't be higher: bodies damaged, careers shattered, lives destroyed, hierarchies reinforced, imaginations quashed, wonders deadened, societies realigned . . . realitties made. But at the same time, remember also those qualities and effects of the sciences that we judged to be covered well elsewhere and so have tended to downplay here: bodies healed, careers blossomed, lives saved, hierarchies chipped, imaginations cultivated, wonders produced, societies realigned . . . realitties made.

Judging assemblages means pursuing sciences means muddling through realitty: all processes that are cautious, nimble, and respectful, since they deal with explosive matters. Yet even small movements can carry a bang. Like Max Planck, who introduced the idea of the quantum into physics just to balance a simple equation about heat, we know that the tiniest and most tentative of new ideas or actions can lead to sweeping transformations. The immense scientific and social effects of such small innovations are at once the sciences' greatest beauty and their most disquieting power. Pursuing sciences, muddling through realitty, means creating and judging processes that are more attentive to both specificity and ambiguity; better attuned to their own kludged articulations, and to the gaps between and within them; more cognizant of their charged yet contingent relationships within power/knowledge assemblages; and motivated by commitments to *get through* to a more democratic society and more responsible pursuits of inquiry, while acknowledging the desire for and even the necessity of something outside those noble ideals—something *almost* irresponsible: fun.

This is not "the end of science," but it is the end of Section I. The examples in this chapter and in previous ones were designed to evoke the simultaneous complexity, subtlety, and urgency involved in questions of how the sciences are always kludged together or muddled with cultural values and norms, and the operations of political and other forms of power. Our metaphors—of sign-forces and thought-styles, articulating and kludging, power/knowledge assemblages and TNT, contingent affinities and haunting ambiguities—are intended to set up pursuits of the messier, more contingent, and less self-assured kinds of inquiries that are more suited to the messier, more contingent, and less self-assured kinds of realitties that we now inhabit. For better and for worse, we've avoided as much as possible setting up what Atlan calls an overarching, noncritical domain of theory. Instead of establishing a new safe and autonomous domain, we've juxtaposed stories, new metaphors, and partial insights from scientists and others, to obtain a collage that better resembles the pursuits of the sciences themselves: patterned, but not perfectly ordered; assertive, but open to redefinition; densely linked, but with slips and grinds across the ever-present gaps. We've charged you, the reader, with some of the responsibility for making our assemblage work.

To the extent that these chapters can be said to be foundational, they found not a new system or philosophy of the sciences, but a series of questions with which to experiment, articulations to newly connect, and inquiry infrastructures on which new projects can be driven and judged. The chapters of Section II focus on what happens when these kinds of inquiries are taken into specific pursuits of environmental sciences, biomedicine, and quantum physics: the awful and wonderful muddles encountered, and the experimental strategies for muddling through.

PART II

"...we go by sideroads..."

The truth is that research is muddy, and people need to start
acknowledging that. You can't get good, clean answers; the world does
not work that way. Patients tend to not work that way unless you totally
manipulate them. And this [people with AIDS] is not a population that is
going to be easily manipulated. So you either have muddy research that
you know is muddy, and you can at least say, "this is muddy," or you have
muddy research and you don't even know how muddy it is.

—an AIDS activist

A picture is conjured up which seems to fix the sense unambiguously.
The actual use, compared with that suggested by the picture, seems like
something muddied. . . . In the actual use of expressions we make detours,
we go by sideroads. We see the straight highway before us, but of course
we cannot use it, because it is permanently closed.

—Ludwig Wittgenstein, *Philosophical Investigations*

I failed my way to success.

—Thomas Edison

Chapter 5

Cleaning Up
On (and Under) the Ground with Military Toxics

Areas of Concern

A deer and a couple of wild turkeys crossed our path as our half-dozen cars drove across the northern end of Westover Air Reserve Base in Chicopee, Massachusetts, on our way to "touring" the most recent Area of Concern (AOC) for the base's environmental engineers. Someone later joked that the Air Force had arranged this symbolic display of wildlife to demonstrate a serene, unpolluted environment. Arriving at AOC1, a former jet test stand area, did little to threaten the perception of harmony.

If you've seen Richard Misrach's photographs of Bravo 1, the bombing practice range in Nevada, where the occasional carcasses of animals and trucks break up an austere landscape colored crazily by toxic chemicals, you have one set of rather majestic images of the toxic legacy of the U.S. military. AOC1 at Westover is nowhere near as remarkable. No one would ever dream to suggest it become a national park, a visual reminder of power's "side effects," as Misrach has for Bravo 1. AOC1 is just an old, cracked concrete apron, dominated by a large, hollowed-out rectangle of concrete in which jet engines were mounted for testing following maintenance and repairs; it hasn't been used for over twenty years. A few red pipes lead to a nest of pumps and gauges sidled against the small noise-reduction building, where the test operator would have taken shelter. The gray paint on the massive walls is chipped and peeling, the thick glass broken in several places. Strewn around the area is a small wheelbarrow-sized tire, a burned-out smoke canister, some wooden crates shipped from Italy marked "Glass" and "Unisys Defence Systems" and filled with Styrofoam packing, a pile of ruptured bags spilling sand and gravel, a cardboard cutout of tank treads from a practice target. At the edge of the asphalt, the slender alders and beeches typical of any wooded area in New England. There's nothing at all striking about AOC1: no stench to outrage the nose, no muck to make eyes pop, no panorama of twisted metal and spent shells. Just the kind of boring, mundane crap you see driving down any road, any day.

Why are we here? Why are we supposed to be interested and concerned about this incredibly banal Area of Concern? Questions like that must have been running through the minds of the twenty or so people that had come to AOC1 as part of a meeting of the Restoration Advisory Board (RAB), the assembly of base personnel and local citizens established by federal law, and charged with oversight of the environmental cleanup at this base and the hundreds of others across the country. Cyndi, the official from the Massachusetts Department of Environmental Protection, tried to figure out where the former fuel, hydraulic oil, and water tanks had been placed, using the map from the contractor's report that marked out their "suspected former locations." A few of us tag along with Ron, an environmental engineer from the University of Massachusetts who collaborates with our organization, the Institute for Science and Interdisciplinary Studies (ISIS), on this project and other efforts, as he walks through the "rip rap" area of scattered stones directly behind the test stand to a small marsh that eventually drains into Stony Brook. It's here that runoff from the tests would have gone, and where the contractors had detected levels of tetrachloroethane, toluene, 1,2,4-trimethylbenzene, and TPH (total petroleum hydrocarbons—oil, in so many words) at concentrations two to six hundred times the levels that regulations require to be reported. Paul, the local civil engineer who is a civilian member of the RAB, checks out a few of the monitoring wells and chats with the base personnel. Sabina, who always introduces herself in a quavering but clear voice as "a 74-year old resident of Fairview who wants to know what's going on with my drinking water," stands quietly by; she lives near the base, has lost a number of neighbors to cancer, and comes to most every meeting. David Keith, whose activist work with Valley Citizens for a Safe Environment (VCSE) had once resulted in an arrest at Westover, is now the co-chair of the RAB elected by the community. He gives a wide berth to Bob, another RAB member with whom David often squares off (Bob would like to see the Endangered Species Act repealed), and poses a few questions to Jack Moriarty, the chief environmental engineer at the base, appointed by the base commander as the other co-chair. After about fifteen minutes, we've seen all there is to see, and the mosquitoes are getting annoying.

On the drive back to the conference center where that evening's RAB meeting will be held, we stop to look at a hole in the ground. Here, at a random place on the side of the road, had been buried a small concrete cube about five feet on each edge, filled with something called "electron tubes" (probably x-ray generating equipment) and designated as low-level radioactive waste. After locating the cube, which had been forgotten and lost for years, experts from Kelly Air Force Base in Texas had come (with their own funding), placed it on a truck, and hauled it away to Hanford, Washington. It's Hanford's problem now. We continue driving along the edge of Landfill B, where most of our sociotechnical work for the last two years has been focused. It too is an innocuous-looking expanse of grass, shrubs, and small trees. But no one is really sure what's happening under the surface. The uncertainties attached to the direction and magnitude of that groundwater flow are interpreted by the Base as "nothing there to worry about, or do anything extraordinary about," and by others (including ISIS) as "a lot to find out about, and a lot of work to do." (We'll be considering some of the problems of Landfill B in the rest of this chapter.)

Following the tour is the actual meeting of the RAB. Over an hour goes by as the final version of the bylaws is discussed, debated, amended, and eventually passed. The two major sticking points: how frequently will the RAB meet, and how many members will get copies of documents related to the cleanup effort distributed to them. The base is balking at mailing out somewhere between 10 and 18 (no one can agree on the exact number called for) copies of documents such as contractors' reports. They compare it to "junk mail," and make a few gibing allusions to "saving trees." Those of us from ISIS and other community representatives remind the base personnel that this "junk mail" has been requested, and that a previous discussion of this exact issue had already led to an agreement to revise the contractors' reporting requirements to include sufficient copies for all RAB members. The base seems to see it primarily as a burden, rather than a necessary mechanism for public involvement. The same holds for the question of meetings: The Base people object to inserting the sentence, "Normally, RAB meetings will be held every six weeks," into the bylaws.

Another difference, on where the meetings should be held, comes to the boil. Once again, we thought this had been settled, thought it was clear that there were many community members who were not comfortable about coming on to the base, who felt it was not neutral territory, and who were happier with downtown venues like the Chicopee Senior Center. It was more comfortable here in their new convention center, the base argued, and easier to make copies of documents that RAB members needed in a hurry—which is true. True, but not the point. (RAB meetings continue to be held at the Chicopee Senior Center.)

There are a number of pointed references to "you people at ISIS" at various points in the meeting, intimating that we have overstepped our bounds, and have tried to control too much, and too much in favor of the community. While we continue to have an excellent working relationship with both the military and civilian personnel at the base, it might be a sign that our perception of ourselves as mediators between the military and the community may be in need of revision.

Finally, with everyone almost totally dulled down by these important but overblown details and differences, Ron, the environmental engineering expert we invited to the meeting, makes a series of comments on "intrinsic bioremediation" that leads to a crucial question. Studies have shown that the process of bioremediation—in which bacteria already present in the soil are allowed to decontaminate the dirt—can be made much more efficient and effective with the inclusion of a little fertilizer with the air that is bubbled through the soil. The contractor doing this work at Westover is not employing that little (but documented and proven) trick, and Ron quickly shows how the data in the contractor's own report substantiate his claim; his finger traces the clear correlations between measured nitrogen levels and the decrease in toxics. How can the RAB, he asks, incorporate this kind of critique and better knowledge into the cleanup program as it proceeds? No one has an answer for him.

As far as twentieth-century sciences go, the sciences involved in helping to characterize and clean up toxic wastes at military installations are not very glamorous. It's not cosmology, where a privileged few get to speak of how they gently lifted the veil to see the "face of God," in Stephen Hawking's quaint and immodest phrase— although we may learn a few interesting things about the other end of God's alimentary canal. It's not high-energy physics, where spending a few billion public dollars affords a glimpse of the fundamental structures of matter, space, or time— although, God knows, the problem of military toxics could use more than a few billion dollars from the state. It doesn't have anything to do with artificial intelligence, artificial life, or chaos theory—although challenges to the intellect, questions of the preservation of nature, and the quality of life are thrown together in a turgid complex of impenetrable equations.

The fact that it isn't any of these noble pursuits is at least some small part of why the problem of the toxic and nuclear waste left behind by the military and its production and support complex accrues little cultural (or other forms of) capital. So we can't stud this chapter with amazing stories of famous scientists doing astounding work around military issues. That's not to say that there haven't been such scientists: Linus Pauling, who won a Nobel Peace Prize for work related to the atmospheric test ban of the early 1960s comes to mind, as does scientist-activist Matthew Meselson (one of the great but lesser-sung scientists from the early years of molecular biology), who was involved in chemical and biological weapons treaties. At Westover, as at any of the other military installations in this country and around the globe, the West is anything but over. Its legacies of toxic chemicals—almost always hidden, often insidious—continue to plague the armed services as well as the communities and environments in which they were, or are, located. The environmental and health problems posed by these contaminated sites constitute an enormous challenge to the sciences. But while they are as unexplored as the edges of the cosmos, as demanding in their intricacies as complexity theory, and as potentially far-reaching in import as the most exalted immunological discovery, it's a safe bet that no one will ever win a Nobel Prize for work done on the ground in the world of military cleanup. Nor is it likely to be the subject of any popular book about the cosmological or cultural uplift provided by the sciences.

The magnitude of that work, its scientific and technical complexity, and its demand for new social forms in which citizens, scientists, and military personnel can work together, present staggering challenges. Our goal here is to describe the kinds of muddling operations we've had to invent and put into action on the ground at Westover. We've had to demonstrate the constructed nature of such facts as which way groundwater and potential contaminant plumes are moving—while trying to construct a different set of facts and theories that can be relied on in a cleanup program. We've proceeded knowing that technical expertise is not enough, and yet it has to be included and improved. We're committed to citizen involvement in and understanding of science, and we know firsthand just how challenging, frustrating, and occasionally annoying that can be in practice. In short, we are trying to do what many other individuals and organizations across the country are attempting: kludging together a complicated assemblage of technical expertise, citizens and community organizations, military personnel and agencies, federal and state regulators, and many other elements, and bringing it to bear on a pressing national problem of human and environmental well-being.

The Whole Shebang

Westover is just one, relatively clean and relatively cooperative military base in the U.S. There are almost 12,000 contaminated sites at 770 active or recently closed bases, and a few thousand more areas of contamination at over 2,600 former facilities. Another thousand former defense sites have yet to be evaluated. Even conservative estimates place the Pentagon as the single largest polluter in the country. The Department of Defense (DOD) spent $9.4 billion on the problem through fiscal year 1996; it is projecting to spend another $30 billion. But with history as a guide, it's safe to conclude that the military rarely overestimates costs, and the actual figure is likely to be much, much higher.

The problem is even bigger when you throw in the over 10,000 sites under the jurisdiction of the Department of Engery (DOE) that are also a part of what it calls the "Cold War Mortgage." (More sobering still, Russia and the former Soviet Union probably have a much worse mess as *their* Cold War legacy.) Payments are past due. Depending on how clean you want how many sites to be, as well as a host of assumptions about technology development, lack of wasteful spending, early closures, budget stability, and so on, DOE has estimated the costs of that cleanup to range from $200 billion to $375 billion over the next seventy-five years. And that 1995 estimate excluded small things like the costs of disposing of surplus weapons-grade plutonium and of cleaning up the large contaminated river systems like the Savannah, Clinch, and Columbia, arguing that "because no effective remedial technology could be identified, no basis for estimating cost was available." Why even try, the Department of Energy seems to be saying, to figure those costs into an already astronomical budget which itself will be an ongoing struggle to secure?[1]

We will not be treating that part of the mess here. The chemical contamination at active, closing, and former military installations is complicated enough. Gasoline, diesel, and aviation fuel are the most widespread pollutants, and present one set of problems. Organic solvents like trichloroethylene (TCE) and carbon tetrachloride are more toxic and harder to remove from groundwater or soil. Lead, copper, nickel and other heavy metals produced over decades of painting, electronics work, and other manufacturing processes pose different challenges. Pesticides, polychlorinated biphenyls (PCBs), and explosives like TNT create still other problems. At firing and bombing ranges, up to ten percent of the bombs, artillery shells, grenades, rockets, and other tools of the trade do not detonate, and become UXO—"unexploded ordnance." UXO can corrode and release toxins whether it lies on the ground or underground; it blows up when souvenir hunters take it home; and the presence of live ammunition can make studying the other contaminants at a site or cleaning them up much more complicated, to say the least.

This is only a small indication of the toxic legacy of military operations nationwide (not to mention overseas). It is in part a product of the military's history of secrecy, hierarchy, and being above the law, both figuratively and literally. The armed services have done pretty much whatever they wanted to in terms of the environment. There was no need for them to think about the long-term effects of dumping fuel or burying barrels of chemicals, no public oversight of such actions, and no regulatory or legal mechanisms by which they could be addressed. The military could and did act with impunity, and very often stupidity. As a result, they've created not only a legacy of chemical toxins, but of social toxins as well. Citizens feel that the military has kept secrets, controlled information, lied, patronized, and ignored—which it has. The result is a persistent background of distrust, suspicion, and outright hostility, which doesn't make experiments in collaboration easy. All that is just now starting to change, for reasons which we'll detail below.

While global and monolithic in many ways, the problem of military toxics is exquisitely sensitive to local history, local geological conditions, and local social organization. Fortunately, the contamination at Westover is apparently nowhere near as bad, in magnitude and complexity, as it is at many other bases. We haven't had to contend with rocket fuel, chemical weapons, extensive radioactive waste, or many of the other horrendously dangerous and complicated problems that communities in other locales face. The problem of military toxics is always "site specific," a phrase that recurs frequently in the vast literature on military waste.

(The Restoration Advisory Board, for example, is just one of many kinds of Site Specific Advisory Boards that have been established for federal facilities.) It is perhaps the most important phrase for us to keep in mind as we go about our work, and for you the reader to keep in mind as you go through this chapter. Almost any generalization—about technical issues, about the military, about citizens, about how they can or can't work together—has to be qualified according to local conditions. "Site specifics" demand the kinds of pursuits that we described in Section I: going after the specific details of actual practices, rather than relying on idealized abstractions; asking about contingent affinities instead of assuming easy "fits" between entities; respecting uncertainty and ambiguity; and making a series of judgments about all of these things at once. To the extent that we can say anything more than we already have about the national problem of military toxics, we can only do so through the lenses of our local involvement. Working "on the ground" is the only way to follow, analyze, and, one hopes, deal with the complexity of realitty. Most of the remainder of this chapter, then, focuses on those "site specifics" that we've encountered in our own work. We do so not because that work is particularly laudable, nor because those local specifics illustrate perfectly the larger picture. Staying close to the ground means staying close to questions. An account of the daily work at the local level can show both the complexity and contingency of the bigger problems, and how it is nevertheless possible to pursue both the sciences and democracy in the face of that complexity and contingency.

Site Specifics I: Some Geology and Some Demographics

Perhaps the most important local condition to take into account when dealing with military toxics on the ground is, not surprisingly, what kind of ground you're on. Where you are in the geological world can often make all the difference in the world between a mess and a disaster. A few hours east of us by car, for example, at the Massachusetts Military Reservation and Otis Air Force Base on Cape Cod, at least ten chemical plumes consisting mostly of jet fuel but including other highly toxic chemicals have spread rapidly throughout the sandy, water-soaked soil, already contaminating 53 billion gallons of the Cape's drinking water and adding somewhere between 3 and 8 million gallons a day to the toxic tally. The aquifer lies just a few feet below the surface, so contaminants can enter it easily and spread quickly. Now a Superfund site, $165 million has already been spent there with little result. The problems remain daunting, the prospects grim.[2]

The geology specific to Westover makes for a different set of questions and potential problems. Seeking input from other sociotechnical experts, Jeffrey Green, who coordinated our work in this area, described it to an electronic bulletin board devoted to the subject of military toxics:

Westover Air Reserve Base is located in the floodplains of the Connecticut River Valley, and is flat as a pancake. The general surficial geology consists of deltaic deposits, over varved [layered] glacial lacustrine deposits, and over glacial till, all over fractured sedimentary and then metamorphic bedrock. There are two aquifers: an unconfined upper aquifer and a deep bedrock aquifer, typically 80–100' below grade, which appears to be confined. At one site (the current fire training area), groundwater is 5–10' below grade. Here there is about 20' of fine-medium sand, over 30' fine silty sand, 25' fine sand, then the varved lacustrine deposits. . . .

Generally, chemicals will have a harder time getting into the aquifer here than they have on Cape Cod, and move more slowly once they do. But there are known exceptions, like the current fire-training area, where the groundwater is much closer to the surface; there, fuel drums are tipped over and set on fire, and base personnel practice putting the fires out. Other sites within Westover, as we'll see, may reveal still more complicated geological particulars. The demand for specificity is high.

Within installation fence-lines, several sites at Westover have extensive and/or high-concentration contamination. The compounds detected in soil and groundwater samples range from less hazardous nonchlorinated hydrocarbons, to perchloroethylene (PCE), trichloroethylene (TCE), pentachlorophenol (PCP), vinyl chloride, and polychlorinated biphenyls (PCB)s. The environmental mobility of various detected contaminants is equally varied, ranging from relatively immobile heavy metals, asbestos, and PCBs, to rapidly migrating TCE, PCE, and vinyl chloride. Other sites contain compounds including BTEX (benzene, toluene, ethylbenzene, and xylene—all petroleum breakdown products), phenanthrene, diethylphthalate, and chlorinated and other solvents. Petroleum hydrocarbons from the fire training area are already present in the soil in Chicopee Memorial State Park, across the fence from that part of Westover, although limited evidence suggests that surface water in the park has not been contaminated. Without effective remediation strategies, contaminants from sites such as Landfill B, the fire training area, and the hangar apron area will—eventually, probably—cross base property lines. Over a quarter of million people live within five miles of the base, about 21,000 within one mile of the base. By the Air Force's own estimates, about 5,000 Granby residents use water from private wells. No one knows how many of these wells tap groundwater that could potentially become contaminated due to Westover waste sites.

Site Specifics II: Some History

If Jeff Green, the young scientist who wrangled many contradictions into a coherent project, emerges in this account as a central figure, it is not because we want to lionize him; Jeff himself would be the first person to resist any such characterization of extraordinariness. Rather, it's because Jeff exemplifies the kind of hybrid figure that scientists today, particularly in endeavors like this, have become almost by default: part hydrogeologist, part computer whiz, part grant-writer, part community organizer, part policy analyst. None of these roles is "central"; each requires the others.

We asked Jeff if he could give a one-line definition of what he thought "science" was. "It's kind of like putting your finger in an electric socket, only noisier," was his immediate answer. So not only was he a solid natural scientist, a questioning social scientist, and committed environmentalist, but he knew enough to answer an impossible question with an oblique metaphor. We hired him for all those reasons, and at least one more: he came cheap.

How did we get involved in this mess? When ISIS first incorporated, building a project that would involve citizens and scientists in addressing the problem of military toxics was high on our list of priorities. It was an urgent social concern, raised a number of immensely difficult scientific and technical questions, and promised to sharpen our thinking on combining scientific and social work. We also thought, rather naively, that funding would be readily available. We were right on three out of four.

We convened a meeting of regional activists, local citizens, and scientists from the neighboring colleges in 1992 to learn about the possibilities for local work. We had few set ideas of our own, and wanted to hear from as many people as possible about what an institute like ours—which so far existed only on paper—could best do. We were, we realize now, muddling through.

It was at that meeting that we learned about Valley Citizens for a Safe Environment (VCSE), an organization of about two hundred local residents. In 1986 Westover changed its basic mission and operations, greatly increasing the number of flights of its C-5A aircraft—mammoth cargo and personnel carriers, the biggest things in the sky. Their engines emit a high, whistling whine. When property values near the base started dropping and local eardrums could no longer tolerate the more and more frequent disturbance, many residents of Chicopee, Granby, Fairview, and other small towns and neighborhoods around the base came together to form VCSE. When we first met the key organizers of the group—Ruth Griffith, who had retired from Monsanto years before, after working in a quality control lab there for decades; David Keith, a local photographer cum historian cum teacher; and Marian Wadsworth, a retired social worker and educator—VCSE had just sued Westover over the noise issue, seeking damages and changes in flight operations. But even as they were preparing to go to court over noise, VCSE members were expanding their attention to issues of groundwater contamination, and to the base cleanup program that was getting underway.

The cornerstone of citizen involvement in the cleanup program at Westover (as well as thousands of other federal facilities) is the Restoration Advisory Board. In the early 1990s, the Environmental Protection Agency (EPA) was being pressured by many governors and their states' attorneys general to develop national environmental and funding priorities, and to suggest mechanisms for achieving them (the contamination at nuclear weapons-related facilities like the Hanford complex in Washington, Rocky Flats in Colorado, and other sites under the purview of DOE having spurred this broader discussion). The EPA asked The Keystone Center in Colorado, a nonprofit mediating and "conflict management" organization focused on environmental issues, to convene a set of dialogues, discussion, and interviews with representatives of federal agencies, the military, environmental organizations, state government, Native American organizations, and labor organizations. They issued a report in early 1993 concluding that "[t]echnical breakthroughs and scientific models alone, will not overcome the current problems with Federal Facility Environmental Restoration decision-making. Indeed, the future viability of federal facility cleanups depends on the ability of the federal government to incorporate the divergent views of all concerned stakeholders into the decision-making process such that all stakeholders can become true partners in ensuring that cleanups are conducted in the safest, most efficient, and most cost-effective manner possible."[3] Thus the Restoration Advisory Boards were born.

One of us (Bernstein) and Jeff had already gotten involved at Westover by joining the Technical Review Committee (TRC). (It helped that Herb could play the physics card to get a position at the table.) So when the draft "Keystone Agreement" was released and the TRC was supposed to evolve into a RAB, we found ourselves right where we wanted to be, in the middle of things. One of the first things we did was organize a tour of the contaminated sites at the base. We assembled a group of geology professors and environmental engineers, officials from state and local government agencies, and some of the community activists with VCSE; the military provided the bus. The civilian base personnel overseeing the cleanup accompanied us and gave a running narrative about the contaminated sites and the current cleanup activities. Here is an excerpt of the conversation at Landfill B among the college professors (Lavigne, Bernstein) and the base personnel (Moriarty, Kelly):

> R. Lavigne: Jack, you say "clean groundwater"—are you saying that the monitoring wells around this landfill are not showing signs of leachate contamination?
>
> J. Moriarty: That's correct.
>
> R. Lavigne: And it's an unlined landfill?

J. Moriarty: That's correct.

H. Bernstein: Are you sure you have the direction of groundwater flow well characterized?

J. Kelly: Yes. No question.

R. Lavigne: Something's not making sense. You've got 1.3 million gallons of water falling on this area per acre, each year. It's passing into the refuse. It's producing leachate. You've got essentially 1.3 million gallons, or three thousand gallons per day, per acre, of leachate leaving the area. You told me how many acres? Ten?

J. Moriarty: Fourteen.

R. Lavigne: Fourteen? So, if you take ten times three, you've got thirty or forty thousand gallons a day of leachate being produced somewhere on this site, and it's got to be going somewhere.

J. Moriarty: Well, we have to look at this—there are other things happening, including the potential for not a great deal of petroleum products in the fill material—that's what we're primarily monitoring for. Although, we're doing a broad brush survey.

R. Lavigne: I thought this was a conventional landfill.

J. Moriarty: Well, it was an aircraft maintenance landfill.

J. Kelly: Primarily domestic.

R. Lavigne: Domestic. So you should have volatile organics, you should have a reducing environment that would be attacking heavy metals and putting them in a mobile, cation form, and you should have forty to fifty thousand gallons a day of leachate going somewhere. If it's not going horizontally, then it should be going vertically; down. . . . No one has asked the question? You've got a water balance problem.

J. Moriarty: We're looking at the physical evidence and seeing what it is, and proceeding accordingly. We'll show you our results. . . .

R. Lavigne: Are you looking at some of the more conventional leachate parameters, or just organics? Let's say, are you looking at iron? Specific conductance? pH?

J. Moriarty: I believe iron was one of the parameters, but primarily it was volatiles, semivolatiles, and base-neutral extractables, and metals—considered a total toxic organic scan, taken as a complete battery of tests. . . . We'll talk about the more technical stuff as we go on. But let's get out of the sun now, and I can tell I'm not going to convince Ron on the site.

We include this lengthy but already heavily edited excerpt from the videotape we made of the entire tour to illustrate the kinds of exchanges that can happen when you open up a situation like this one to more democratic involvement. As you can see, the base personnel were quite open about their study results, and later did provide all the data and the specifics of the methodologies used to create them. But having additional, outside experts present is important for a variety of reasons. First, they *are* experts, and know which questions need to be asked. "Ordinary citizens" can do this too, or can learn how to do it, but the need for specific technical expertise will remain high. Second, good scientists can immediately start making back-of-the-envelope type calculations based on what they're seeing out in the field, and start kludging together other possible perspectives and other questions. In this case, it looked as though there should be a significant plume of *something, somewhere,* and the fact that the military's studies showed nothing was bothering Ron Lavigne. Maybe it would turn out that he was bothered for no reason. But this became the beginning of an extended set of inquiries into what was going on under the ground at Landfill B. If the military had "no question" about what was in the groundwater and in which direction that groundwater was moving, we could think of many questions. Now we had to build an assemblage in which to ask those questions.

Site Specifics III: Trash in Landfill B

If you go to enough RAB meetings, the one phrase you might hear more than any other coming from the mouths of a base commander or engineer is: "We don't know what's really out there." Often this is accompanied by a wry grin, a self-deprecating shrug, and a semijoking allusion to "the next shovelful of dirt"—the chemical nightmare it might contain, and the legal nightmare it might initiate. Until a few years ago, federal facilities such as the national laboratories and nuclear weapons-production complexes managed by DOE, and all the military bases operated by the Department of Defense (DOD), enjoyed "sovereign immunity" from the pollution laws and regulations to which private corporations were subject. The Federal Facilities Compliance Act of 1992 changed all that, and made all federal agencies and departments subject to all the provisions and penalties of the 1976 Resource Conservation and Recovery Act (RCRA), which regulates hazardous waste management throughout the U.S. The "next shovelful of dirt" turned up at a military installation now has the potential to open up a lawsuit. That serves as something of a disincentive, perhaps, in the minds of base commanders who might generally be committed to cleaning up their installations, to doing their jobs as thoroughly as possible.

You might think that our job would be to find what *really is* out there, and that we would abandon all those pretty theories about realitty and get down to work. If so, you should go to more RAB meetings. Because what you'll see and hear are the most detailed, lengthy, and contentious discussions about this method versus that method, what this contractor measured versus what that contractor measured, what this report contained versus what the other seventeen reports contained. The discussion is always about measuring devices and procedures, sampling methods and errors, imaging technologies, statistical analyses, theoretical constraints, and a host of similar realitty-makers. The only way to talk about reality is to talk about realitty: to engage in discussion about what kinds of indirect knowledge our mediating technologies—with all their errors, limitations, and embedded interpretations—can provide. (And as we pointed out when we first introduced that term, when you're sitting around a table with an engineer, a lieutenant commander, and a Chicopee senior citizen, no one can tell if you're talking about reality or realitty.)

Our object, then, is to kludge together more experimenting technologies, more articulations, and the different perspectives, experiences, and expertises of many different people into an assemblage that will make realitty as dense and complicated as possible—and then judge how well that assemblage works. In practice, of course, we run up against all kinds of signals that resist: availability and expense of technology, spotty data, bureaucratic tangles, differing judgments, and severe time constraints made worse by the constant tasks of grant writing and fundraising necessary to handle that other severe constraint, money.

One of our more successful assemblages was made up of local college students, their professors, and some of the best technology that could be begged and borrowed. Jeff collared some of his former geology and chemistry teachers at Smith, Amherst, and Hampshire Colleges, and got them interested in Landfill B. For an entire semester these classes, some thirty students and four professors in all, drove down to Westover on Saturdays and the occasional weekday, and tramped around Landfill B trying to figure out what was happening underground. It's about as far from "transparency" as you can get.

Landfill B has been around for about forty years, which means that no one knows exactly what's down there. "Trash" is the imprecise term that we have to work with, based on not particularly trustworthy records and the memories of base personnel and local citizens. It's approximately fourteen extremely flat acres in surface area and thirty feet deep, which means that it probably extends just into the groundwater.

An early report, based on little more than a consultant's walk-through of the area, suggested that given the flatness of the terrain, any leachate from the landfill would probably be spreading in all directions. Over the years, a realitty congealed in which groundwater flow headed east, meaning that it stayed within base property lines. A series of contractors over those years installed various monitoring wells, clustering to the east and south, which seemed to support the eastward-flowing theory. There are now about twenty "couplets," two wells together with one going down about thirty feet, the other sixty feet.

So the first step in bringing another realitty to the groundwater flow, and to any plume of leaching chemicals that would be carried along with it, has to go through the wells. Even that simplest and most transparent piece of technology, the ruler, immediately becomes a site of complex judgments. The wells consist of a PVC tube extending either thirty or sixty feet down into the earth, and protruding several feet above it; a steel casing sleeve fits over the PVC tube. The students dropped a line with a float attached to it, and measured the length of line from the top of the well to the point where the float hit the groundwater. But what's the "top of the well"? The top of the steel casing, which has moved with every frost heave over the years? Or the top of the PVC tube, whose exposed length above ground might have changed with erosion or human activity? (Measuring according to the PVC tube height turns out to be the better option, since the tubing is pretty stable.) There are disciplined methods to deal with those variations, and of course margins of error are estimated as accurately as possible, but you have to rely on good judgement as well.

And here's where you confront the first layer of uncertainty in answering these questions. When the students tried to combine into a single "grid" their data sets with those of the different contractors that had previously measured groundwater levels, they ran into problems. Remember that this area is very flat; the groundwater level varies by only a few feet. The data constructed by the various contractors were often inconsistent, and when aggregated and coupled with analysis of possible errors, they turned out to be almost useless for showing conclusively which way the groundwater was flowing. Different measurements taken at different times could be connected in multiple ways to construct different models of groundwater flow. The differences were subtle—north? north-northeast? east-northeast?—but serious enough so that we could only conclude that we couldn't conclude. To make things worse, no one had ever performed careful studies on how the water table reacts to a serious run-off event like a thunderstorm or rapid heavy melting of snow, or how the groundwater level varies seasonally (which could explain all those different well levels measured by contractors). All that could be said was that the existing groundwater flow maps were likely to be very flawed.

The students also took water samples from the wells, testing their pH levels and analyzing them for certain ions, the charged forms of atoms like calcium and magnesium. These are a useful tracer for most landfills, and can indicate the presence of dissolved chloride compounds and possibly other substances. The geology classes used chromatographic analysis, and their samples and data were double-checked by Dula Amarisiriwardena's chemistry class at Hampshire College with an atomic absorption spectrometer. (A good realitty-making assemblage always incorporates the disciplined habits of quality control.) Not much turned up. Neither did a terrain conductivity study yield much of interest, other than learning how much work it took to write a grant to get to use some rather expensive and esoteric equipment, and giving an undergraduate geology class the opportunity to use it.[4]

The seismic refraction/reflection studies proved more provocative. The procedure involves "bonking," as Jeff puts it (apologies to our British colleagues), a steel plate at various points on the surface, and measuring how the bonk bounces back up into "geophones" spaced about fifty feet apart. The bottom of the site is supposed to be a smooth shallow bowl of thick varved, or layered, clay, the remnants of the glacial lake bed that was once there thousands of years ago. That's what most of the studies of Landfill B have presumed, for not much more reason other than it's the simplest hypothesis. The Air Force and its consulting firms certainly find the simplest hypothesis good enough for their purposes, including minimizing costs. Geology professors are in a different position and have different purposes, including maximizing curiosity and skepticism. Another is to teach their students how to be curious, and the pleasures of hunting for signals that resist.

And that's what they found. In the range of what was expected to be consistently thick clay, they didn't get thick, consistent readings. There were variations in signals, and while they weren't good enough to be shaped into a definitive and coherent picture, it definitely looked like something interesting, significant, and potentially problematic. There might be a "sand lens" in the midst of the varved clay, that a leachate plume could be escaping through downward, avoiding the monitoring wells to either side. Or there might be channels within the clay, through which contaminants were zigzagging their way around the monitoring wells.

Or there might be nothing at all: no leachate, no plume, no sand lens, no channels, no problem. But that's getting ahead of the story.

The classes were scheduled to present their results at a meeting of the Restoration Advisory Board in December. Their geology professors, Robert Newton from Smith College and Steve Mabee from Amherst College, worked furiously to finish a final few studies, and to help the students compile and analyze their data, and shape it into a polished presentation that would be heard by Air Force officers, the base civilian engineers, and of course the community. That motley crowd filled the Chicopee Senior Center on a cold December evening.

Was it a success? Yes and no. It was clear that the students took pride in having done good scientific work, and relished the opportunity to be speaking both to the military and to the people in the community who were most anxious to have the studies done. The presentations by the students were professional, perhaps too professional. They were already learning how to speak to peers at a professional scientific gathering—although learning may be too strong a word. The subtle and inexorable (and by no means entirely negative) processes of scientific enculturation were already at work. Nervous, struggling to do justice to the specifics and complexities of the geology of Landfill B, they had their hands and heads full, and the "objective professional scientific presentation" is a genre that covers over an excess of anxieties and questions. They were in a difficult situation: Trying to address the people in the community who had no immediate understanding of what "varved clay" was, the students sacrificed accessibility to professional rigor. The power that they were, in theory, to lend to the powerless got lost in a morass of technical data and methodological details. At the same time, the snide remarks of a visitor from the Air Force Center for Environmental Excellence sitting behind one of us gave a clear impression that those at the other end of the power spectrum found the whole endeavor at best amusing, at worst an unwelcome interference from snotty college kids and their elite professors.

Still, some of the students had learned that geologists don't have to work for oil companies, but can use their abilities and skills in the sciences to address a community's more immediate needs and problems. They learned that the hardest work and the best equipment are no guarantee of certainty. They learned that people nonetheless still want certainty.

What we got was another set of articulations, which may not seem like much. It may seem like all we can say is that this set of methodologies, questions, technologies, chemical analyses, water level measurements, and so on, indicates that there *might* be a plume, that *might* be escaping downward or sideways. Or there might not be. Both articulations are indeed possible, and in some sense equivalent. The latter, that there is no plume, is the articulation the Air Force prefers. They can argue that the lack of any data that would indicate a plume of contaminants just shows how *clean* the whole landfill is, after forty years of rainwater dilution took whatever was there in the first place, if anything, out to sea via the Connecticut River. And no one can say for certain that such a view is wrong.

That's one of the consequences about thinking in terms of articulations rather than true or false "theories": there's no final rest in arriving at some objective, settled truth. Everything remains potentially unsettled at the level of articulations: There could be no plume, there could be a plume that we're missing for some geological or technical reasons. But, again, this doesn't make everything relative. It only means that such questions aren't settled by an appeal to an objective reality (which we can't access directly), but through a process of judging indirect realities: accounting for and assessing the robustness of different scientific methods and results, different technologies, different social needs, and economic and political factors. We could be wrong, and the Air Force could be right about what's happening under the ground at Westover, or vice versa. The important thing is having a working assemblage that allows everyone to articulate his or her realitty-producing methods and results, and the residual uncertainties and gaps that accompany those, and to make collective judgments about those results and gaps in combination with many other factors. That's what a Restoration Advisory Board should be for.

Site Specifics IV: The Westover RAB Again

Once again, we're in the Chicopee Senior Center for another RAB meeting. There are no students at this meeting, and very few citizens. There are about fifteen people around the table, only one of whom is a woman (Cyndi, from the Massachusetts Department of Environmental Protection), only one of whom is under the age of thirty (Jeff), none of whom is nonwhite. Those of us from ISIS as well as the four consultants from the engineering firms with which the military has contracts are all technically (i.e., according to the agreed-upon social norms) neutral. Everyone in the room knows that. Everyone in the room also knows—from the way people dress, from what they say, and most of all from history—that that characterization has some "data gaps," to put it in terms that will come up below. Everyone in the room is capable of mapping out the contingent affinities that exist between those two charged poles, the military and the community.

It will be a meeting of some subtle and not so subtle finger pointing. The RAB had learned at the previous meeting that the final report on Landfill B submitted by the contracting firm to the Massachusetts Department of Environmental Protection, as part of the requirements for closing out one stage of the cleanup and moving on to the next, had been deemed inadequate because of serious "data gaps." The base tries to argue that it had contracted with the engineering firm to comply with the Massachusetts Contingency Plan for dealing with hazardous waste; the contractor's representative claimed that's not what they had agreed to, and that they had done the job they were contracted to do as defined in the initial agreement. The contractor admits now that such work was insufficient for meeting Massachusetts regulatory requirements, and proposes to carry out the additional work necessary as part of the next phase of work. Cyndi says that isn't possible; the characterization of Landfill B has to be completed before the next phase could proceed. That was the law. Someone makes a stinging reference to how the law in this case takes up about ten inches of shelf space and requires "about forty lawyers to understand."

The man from the tri-county planning commission finally asks the contractor: Do we have a satisfactory Phase II study? No. Don't you have an obligation to fulfill that requirement? The contractor answers directly: We were contracted to do certain things, we did them, then we were asked to submit a Phase II study only based on what we had; we didn't "blow off" or try to "sneak something through," as some people have suggested; we aren't white knights who go out and sample for free, and run up thousands of dollars in lab costs. The take-away message: "Firm fixed price" contracts don't exactly encourage a contractor to go out and characterize a site any more thoroughly than is absolutely necessary. Of course, the alternative has its problems as well.

David Keith, the RAB's community co-chair, finally suggests that if the cleanup contractor had come to the RAB and said they didn't have enough money to do as thorough a study as necessary, the RAB might have been able to get more money. "I didn't know this board carried that kind of weight," says the contractor. "We don't know either," replies David. "That's what we're trying to figure out."

After close to two hours of discussion, we begin to draw some lessons for the future. As David Keith's remarks implied, the RAB is an ongoing experiment. With gentle and persistent prodding from him and Jeff, the military representatives and the rest of the RAB agree to use the cleanup of the Jet Test Stand Area (which we toured at the opening of this chapter) as a kind of RAB Test Stand. All the relevant documents, draft contracts, scope of work agreements, and so on will be made available to the RAB before any further work proceeds. Contractors will present their plans at a RAB meeting, where everyone will be free to raise questions. Working closely with the DEP officials, the RAB members could ensure that regulatory requirements were being met, that citizens' concerns were being addressed as fully as possible, that the scientific work being proposed was as thorough as it needed to be. All those criteria, of course, could only be answered according to the *collective judgment* of the RAB, the only thing we had to go on in the end.

If we had managed to get slightly, temporarily "ahead of the fact" for the Jet Test Stand Area, we remained after the fact on Landfill B. Another engineering contracting firm, from New Jersey, was at the meeting on a reconnaissance mission: seeing how this particular RAB worked, as they got ready to plan for the capping of Landfill B. A cap is basically a thick plastic sheet (fourteen acres big, in this case) placed over the existing landfill area, mounded so the water runs away from the center of the landfill, and covered with a thick layer of clay and soil. It keeps that 1.3 million gallons of rainwater per year from percolating through whatever is down there and creating a plume of contaminants—if there is one. Capping Landfill B appears to be the best long-term solution—unless a problem with contaminated groundwater is found later, in which case it is very difficult and very expensive to get back in there and try to do something. There's another representative from the Massachusetts DEP at the meeting tonight from its department of solid waste management, which has authority over things like landfills and their caps. He recognizes that the "data to date" suggest that the groundwater is flowing as the military and its contractors argue it is, but he also recognizes there are those "data gaps" left open by the botched Phase II study which complicate the picture. "In hydrogeology," he reminds everyone, "when someone says 'this happens a hundred percent of the time,' you should be suspicious." Jeff in turn reminds him and everyone else on the RAB that our studies showed the possibility of a sand lens or breaks in the varved clay, and that there are at least some grounds for suspicion.

That's all we can do for now. If the question remains somewhat open, the planning for the landfill cap nevertheless has a very real momentum. It will inevitably come down to judgment calls, made on a messy combination of certain and uncertain scientific articulations, social dynamics, political pull, and constraints of time and money. The important questions are again those of who gets to make those judgment calls, on what basis, and through what kind of open, democratic process.

Going Holistic: A Democratic Workshop on Facilities Cleanup

As an experiment in socioscientific democracy, the Restoration Advisory Boards have worked remarkably well. There are, of course, those "site specific" factors encountered at particular military installations that make that general statement questionable: recalcitrant base commanders, blindered contractors, lack of trust, overwhelmingly horrendous contamination, inept state and federal regulators, budget shortfalls, citizen-activists who like nothing better than to hear themselves speak, or who believe their role should only be oppositional and not collaborative. But by and large, the RABs are a good experimental assemblage for creating better accountability, better judgments, and better prospects for cleaner and safer environments. And since they are assemblages, their kludged articulations do not always make for smooth, predictable operations; they always need tinkering and rearticulation.

Those lessons were reconfirmed when ISIS collaborated with the CAREER/PRO (a very kludgey acronym for the California Economic Recovery and Environmental Restoration Project) to present a three-day workshop on RABs and the cleanup process at federal facilities. With a grant from the Environmental Protection Agency, CAREER/PRO subcontracted with ISIS to run the Northeast Federal Facilities Cleanup Workshop in July 1997.

It was a lesson in just how messy, complicated, and oddly effective democracy can be. You just had to look around the room at the fifty-odd people ISIS had organized into this temporary assemblage: active and retired school teachers; physicists, toxicologists, and electronic engineers; real estate developers and a librarian; environmental engineers employed by military bases and retired military officers; a doctor from Physicians for Social Responsibility and an epidemiologist from the Air Force; representatives of local, state, and federal agencies; a state assistant attorney general arguing for the "devolution" of final environmental authority to the states, and a public health specialist with the EPA arguing for full federal oversight; practiced Internet surfers and people who have never gotten or sent a message by electronic mail in their lives; sweet Sierra Clubbish conservationists and confrontational Greenpeacenikky toxic avengers.

The questions, problems, and tasks that came out during the discussions were just as specific and diverse. RABs at active bases are in a somewhat different situation from those at bases which are closing, and both differ from "mixed-use" bases. Army and Air Force installations can pose different challenges and opportunities from those of the Navy. The latter, shaped by a long tradition of the "marry 'em and bury 'em" near-absolute authority of the sea captain, can be either particularly effective, cooperative, and innovative, or especially obstructionist and hidebound, depending on who's in command. (Culture *does* matter.) The specifics of geology and chemistry, as outlined above, make more demands on how the sciences are to be pursued in particular contexts.

These are the kinds of both obvious and subtle differences that people are negotiating somewhere every week at a RAB meeting. It doesn't always work and it isn't always pretty or efficient. In fact, when things start to appear too smooth and easy, we get a little nervous. At our workshop, for example, everyone from the school teacher to the Air Force colonel could call herself or himself a citizen, and it became something of a running joke to remind everyone that "I'm a citizen, too!" The "stakeholder" model of public involvement that the RABs employ, as does the broader environmental movement, evokes a similar worry in the back of our minds: Certainly we're *all* citizens and stakeholders, but at what point do these relatively homogenized terms start to make it difficult to address important differences in perspective, access to resources, economic stakes, and political power?

In a similar vein, the optimism and enthusiasm of the speakers from the Pentagon at our workshop was for the most part encouraging, but also prompted some concerns. By all accounts, the officers and civilians at the Pentagon as well as people lower down the chain of command at the thousands of military installations have seen that "citizen" involvement in the cleanup process has been an asset and not an impediment. In the space of a few years, stakeholder participation has gone from something to be approached warily and somewhat reluctantly, to something that is welcomed, valued, and encouraged. That is at least the official, public discourse, and quite a bit of evidence suggests that such a change in attitude is indeed more than mere words. Perhaps because it is such a relatively rapid and significant shift that leaves one wondering, just a little, if it's too good to be true. At the very least, when the military starts talking about the importance of "strong stakeholder participation" and the need for a "more holistic approach" to environmental cleanup, you know that, as the famous Chinese curse goes, you're living in very interesting times.

Whatever unease a kinder, gentler military might have caused was more than offset by the evident ability of the civilian citizens to continue asking skeptical, insightful, and diverse questions. There are indeed individuals in communities across the country who know how to stay close to questions, who know how to find the holes in holistic articulations, and who are dedicated to building effective assemblages. They can be respectful and critical, trusting and skeptical, and they can show pretty good judgment about when they need to be which. They're willing to sit through long and often dull meetings—not just once, but regularly, for years.

Through persistence and creativity, they have scraped together some of the technical knowledges that they need. They've sought out local experts at community colleges and universities; they've consulted by phone and e-mail; they've taught themselves. There's little evidence of "antiscience" attitudes here. Many of them *are* scientists, but even the ones who aren't have learned how to read toxicological studies, hydrogeology charts, and inches-thick regulatory guidelines and other poorly written government publications—and come up with novel suggestions, perspectives, and judgments.

There are many things that would make their jobs easier. Greater federal funding for restoration programs is certainly at or near the top of that list. But increases in those funds are likely to end up in the bank accounts of large engineering firms. There is currently no institutionalized mechanism for providing the RABs with independent technical assistance. Here and there, a few bases have allocated money for such assistance. We've been lucky to put together a few small grants from the Massachusetts Department of Environmental Protection and some small foundations, combined with the volunteer efforts of college professors and their students. But these are the exceptions rather than the rule, and perhaps nothing would make more of a difference to both the effectiveness of environmental restoration and to greater public trust and democratic involvement than that kind of support for independent technical advice. People need a better realitty that allows them to articulate what substances are there, where those substances are likely to go (and when), what the potential effects on local health might be, and what the best strategies are for remediation or at the very least containment. They need to be able to challenge when necessary—and it often is—the military's own studies and those of their contractors. Or they at least need to hear from someone who is not wearing a uniform, or who is not getting rich off of military contracts, that a restoration plan or health study is okay. They might not get answers or certainties, but they will get judgments they can rely on, and which may even have their own limited uses. And that will definitely make for pursuing sounder sciences, and for better muddling through.

Chapter 6

Defining Disease
Questioning Chemical Sensitivities

[Multiple chemical sensitivities, or MCS, is] characterized by recurrent symptoms, referable to multiple organ systems, occurring in response to demonstrable exposure to many chemically unrelated compounds at doses far below those established in the general population to cause harmful effects. No single widely accepted test of physiologic function can be shown to correlate with symptoms.

Dr. Mark Cullen[1]

The phenomenon of multiple chemical sensitivities is a peculiar manifestation of our technophobic and chemophobic society. . . . The concept of MCS . . . appeals to the widespread fear of chemicals, the distrust of science, medicine, technology, and government; environmental worries; and the modern American mind set of victimization.

Dr. Ronald E. Gots[2]

Multiple chemical sensitivities (MCS) is, as its name suggests, multiply problematic. Even considered separately, the science, clinical treatment, and political and legal sociology of MCS are each a tangled, messy mass of multiple questions and demands. Consider these areas together, as we must, and we face an almost overwhelmingly intricate, difficult, and powerfully charged assemblage that is at once scientific, legal, political, and economic. The assemblage includes the successes and failures of *experimenting* with symptoms and sign-forces to arrive at biomedical realitties; *articulating* these differences with a variety of theories and models, metaphors and names, institutional and cultural structures; of *powering* one's own inquiries with funding, metaphors, and other social resources, trying to learn from and offer assistance to a community of affected people, while being buffeted by other powerful social interests, contrary experiments, and forceful ideas; and *judg-*

ing at every step along the way, trying to arrive at better experimental protocols, more subtly articulated concepts, effective social and political strategies—all these and more combine to make multiple chemical sensitivities a magnificently complex subject, a multiply contentious science, a monumentally conflicted sociology, a messy constellation of signs, requiring a mass of contradictory strategies.

Like all good scientists, we love the challenge.

Belief/Disbelief

As we've shown, a kind of belief in certain theories or concepts as yet unestablished is often an essential part of experimenting productively. Should we place "belief" in MCS, then, at the core of our scientific and social pursuits? To be more precise, should we believe in MCS as an *illness*?

The distinction between disease and illness is important here: If we think of "disease" as a set of clinically and scientifically well-defined (more or less) causal processes, and "illness" as more of a phenomenological indication of individual suffering, then we will be starting in a better place. Because it should at least be clear, based on available evidence, that people are suffering, even if science and medicine have yet to come up with a widely accepted explanation of that suffering, and even if that suffering is often dismissed as psychosomatic. But as we'll see, even this modest beginning encounters a great deal of resistance.

Lynn Lawson opens her book on MCS with a phenomenological account of illness and suffering, describing the occupants of a sauna as they pursue detoxification:

> Several men in the group were firefighters suffering from an acute exposure to polychlorinated biphenyls (PCBs) while fighting a hospital fire. One woman (I'll call her Annemarie), from a small city ringed by mountains, had had multiple exposures to pesticides and more recent exposures to paint and other pollutants at her workplace. Tom, a young, lean, muscular professional golfer, had been incapacitated by pesticides on golf courses. Julia had been poisoned by formaldehyde in her house . . . Valerie, exquisitely sensitive to chemicals, often sat in the sauna holding an ice pack to her jaw to relieve the intense pain. Her problem was thought to be caused in part by old silver/mercury amalgam fillings, removed but, unfortunately, replaced with other toxic substances that her dentist had not adequately tested for compatibility . . . Linda, a grade-school teacher, was apparently made ill by her school's heating system, and Anne, another teacher, was disabled by her school's use of a pesticide containing dioxin . . . Mollie, a social worker, was so chemically sensitive that in her parents' house she could only sit on one chair—an untreated

wooden one. She reacted to the paints, stains, and fabrics on the other chairs. Don, a housepainter, could no longer work with paint. . . .[3]

What was making such a diverse group of people, with a diverse range of symptoms, prompted by a diverse range of substances, so ill that they were submitting themselves to a harrowing regimen of saunas, dieting, exercise, and self- and medical monitoring? We know that PCBs are toxic, we can articulate a convincing scientific narrative for the possible dangers of formaldehyde and dioxin, but paint? Dental fillings? Theron Randolph, the doctor who started the area of treatment and research he called "clinical ecology" in the early 1960s, prefaces Lawson's book with an admission of uncertainty: "We don't know, for instance, why the man who works in a filling station and the woman who sprays synthetic perfume on herself every day don't seem to be bothered by their chemical exposures. We don't know why some workers can clean out tank cars seemingly without ill effect, while others doing the same thing can, ever afterwards, have a serious problem with chemicals. The difference may be genetic, but if so, we don't know the modus operandi."[4]

Instead of taking this uncertainty as an opportunity—indeed, as an urgent demand for—inquiry and action, many members of the scientific community often see it as paralyzing or illogical. One statement that reflects a common perspective of the scientists who don't "believe" in MCS is the following: "If the question cannot be answered as to what MCS is, how can there be approval of research protocols or acceptance of investigative results? In order to appropriately address the controversies surrounding this phenomenon we must know where we're going!"[5]

It's a rather odd attitude for a scientist: wanting to know the answer before deciding what the questions are and how to answer them. It's more than odd: it's totally backwards, and in fact, antiscientific. If there is anything that characterizes the sciences at their best, it's the quality of not knowing where they can or should go, and then experimenting with ways to get there. The sciences pursue questioning precisely when the answer isn't known, and you're not even sure of the right way to ask the question.

Yet more than a few of our colleagues, in both the sciences and the social sciences, have reacted similarly to our involvement with MCS, shaking their heads in disbelief: How could we even be interested in something that is so clearly an example of junk science, let alone try to develop institutional responses to it? Many of these colleagues exhibit what can only be called a kind of allergic reaction, an intense sensitivity to the topic of chemical sensitivities: Having read little if any of the scientific literature on the subject that would allow them to make better judgments on the issue, they remain firm in their belief that it can't be real. MCS must, in their view, be some combination of bad science, irrational hysteria, and self-serving conviction.

So what about us? Are we believers or nonbelievers in MCS? True to our faith in the work of muddling contradictions, we are believers in *and* skeptics toward MCS and thus are neither advocates for *nor* enemies of it. These statements are our two attempts to work in an excluded third term that might be more productive for creative inquiry.

This attitude of belief/disbelief is essential to analyzing and acting on MCS for a number of reasons, and at a number of levels. At the levels of experimenting and articulating, to assume *either* an attitude of total faith in *or* complete rejection of clinical symptoms, experimental leads, technological advances, and theoretical possibilities, will generate *more* error and *more* unproductive, rigid certainties than it will fruitful clues and open inquiries. In the realms of what we've called powering, where institutions, social interests, and broad cultural forces are at work in the sciences, there are already hosts of advocates and enemies, individuals and institutions fully dedicated to either the "truth" or "falsity" of MCS. Our institutional work through ISIS is an attempt to do something else that falls between these camps. But while we may not be advocates or believers, that doesn't mean we think we're being "neutral." We try to be neither neutral *nor* interested, and both partial *and* impartial.

Confusing? It should be. Let's go on.

What Is MCS?

MCS has been defined and derogated, but no one can say what it really is, at least not yet. However, this is not the insurmountable obstacle some scientists think it is. If we can't say what MCS is in reality, we can at least say what different people *say* MCS is in reality. We can describe what assemblages of names and diagnostic categories, immunological or biological articulations (theories), measured biochemical events and neurological patterns, institutions with their fiscal, legal, and cultural concerns, and a dozen other elements go into constructing a (future) realitty for MCS. There will be many assemblages, each complicated enough in itself, and even more so in their interaction. We may witness the "genesis and development" of scientific facts (as Fleck did with syphilis)—or not. And in the end, there won't be an answer, but only a more complex set of articulated elements to be judged.

Let's begin with clinicians and professional groups. There is both agreement and disagreement, areas of overlap and of difference. The definition quoted in the opening of this chapter, posed in 1987 by researcher/physician Mark Cullen of the Yale Occupational and Environmental Medicine program, was followed a few years later by the more phenomenological definition of Nicholas Ashford and Claudia Miller: A person with MCS "can be discovered by removal from the suspected offending agents and by rechallenge, after an appropriate interval, under strictly controlled environmental conditions. Causality is inferred by the clearing of symptoms with removal from the offending environment and recurrence of symptoms with specific challenge." Still later, in 1992, a subsection of the Association of Environmental and Occupational Clinics offered a set of diagnostic criteria that included "symptoms that occur in three or more organ systems" (a more specific formulation than Cullen's multiple organ systems), and "exclusion of patients with other medical conditions" (with the important exception of psychiatric conditions). That same year, a working group at a conference convened by the National Research Council (hardly a "superstitious" federal science agency) laid out less exclusionary criteria that admitted "symptoms and signs in one or more organ systems," and did not exclude people with "preexistent or concurrent conditions, e.g. asthma, arthritis, somatization disorder, or depression."[6]

Thus, different criteria of categorization and diagnosis produce different definitions of MCS and how many people have it. Should researchers therefore agree on one set of criteria that they will all use in future clinical and laboratory studies? That would certainly allow better and easier comparisons among different clinical, epidemiological, and laboratory studies. Our own opinion, however, leans toward encouraging or at least tolerating difference and multiplicity. Especially at this stage of the game, when the ultimate outcome is far from clear, it seems to us that a number of possibilities should be kept open. That would make comparison and compilation harder, and it would seem to work against the ideal of "objectivity." But such a plurality of approaches would be no less objective than an enforced constriction and standardization. Plurality would necessitate a confrontation with what truly constitutes objectivity anyway: By *these* criteria, using *these* tests, on *this* population, with *these* uncertainties, MCS is *this*. And using *those* criteria, *those* tests, on *that* population, with *those* uncertainties, MCS is *that*. What MCS "is" can't be separated from questions of how we conceptualize it, what instruments and research protocols we use, what metaphors are activated, and what social interests are at work.

What's in a Name?

We use the term "multiple chemical sensitivities" (MCS) in this chapter, but among people who are ill there is a lot of disagreement about the best name for their illness. As we've seen, names are always sign-forces, articulated within a dense network of other names, all pushing and pulling with their subtle charges.

The following exchanges are unattributed quotations from an on-line discussion group devoted to the immune system and its disorders:

CHEMICAL INJURY is the best description for the affliction. When you are exposed or poisoned with toxic chemicals, your immune system, brain, liver, nerves etc. can and likely have been INJURED as a consequence to the exposure. . . . I never liked the term MCS, as it just defines one of the many symptoms of Environmental Illness. . . .

Place the causality where it belongs . . . in the chemicals, not in "us." . . . "MCS" takes the cause and locates it in us, who are sick, instead of in the environment where it belongs. . . . If not "chemically induced disorders of biological response," then Environmental Illness.

Calling it "chemical injury" only makes assumptions about how the illness is caused. If a person suffers MCS from genetic, organic brain dysfunction which increased during the person's late 20s, then this isn't "chemical injury." Chemical injury is likely a common cause, but it may not be the only one.

That is an argument the tobacco manufacturers love to use: "It's not the cigarettes making people sick, rather there must be some fluke genetic disorder in such individuals." . . . How about "Chemically Induced Dysregulation"?

[I]f we're going to be scientific we shouldn't disregard all the facts on the table and define the disease based on a single-cause assumption. . . . What about the 20% who can't identify [a single] exposure? What about the people who got MCS after pregnancy or head trauma? What about the patients whose MCS cleared after they cleared up their *Candidiasis*? Why hypothesize a single cause for MCS and then define the disease on that hypothetical cause?

The term "chemical injury" says nothing about the type of injury and strikes me as no more meaningful than other broad categories like physical injury and psychological injury. Another problem is that MCS is a chronic illness that results from toxic exposures. This is quite different from saying just that you have been injured by chemicals, which could mean almost anything. . . .

These few samples—and there are many, many more—show not only how difficult it can be to get agreement on even the most basic issues, but also that disagreement can generate new ideas, sharpened thinking on multiple possible biological explanations, or just fruitful exchange and recognition of difference. Such differences can be difficult to manage scientifically as well: Should researchers start with genes, yeast infections, toxicological mechanisms, or some-where else? And what about the profound distrust of establishment science and the real anger and pain that reinforces the certitude that the chemical industries are the real "cause" of these illnesses?

A recent attempt at renaming MCS brings some of these differences, and their very real links to industry, into sharper relief. In February 1996, a panel convened by the International Programme on Chemical Safety (IPCS), which is cosponsored by the World Health Organization, the United Nations Environmental Pro-gramme, and the International Labor Programme, recommended discontinuing the use of MCS as a name, "because it makes an unsupported judgment on causa-tion"—i.e., chemicals. The panel instead proposed a new name and acronym, Id-iopathic Environmental Intolerance (IEI), a more overarching term that would include other idiopathic or "unexplained" conditions like chronic fatigue syn-drome and Gulf War veterans' illness.

This event prompted MCS Referral & Resources, an organization which does very solid work on behalf of people with MCS, to document that "only 7 of the 17 'experts' convened by the IPCS had ever published papers on MCS, and several testify for chemical companies against MCS patients"; and that other participants included "medical officials from Coca-Cola, Bayer, BASF and Monsanto." The or-ganization further documented how "the IPCS has been strongly criticized in re-cent years for the corporate bias evident in many of its other controversial evaluations of toxic hazards, including chrysotile asbestos, methylene chloride, chlorofluorocarbon refrigerants, and the fungicide benomyl," and how the pro-gram had been publicly taken to task by an official at the National Institute of Oc-cupational Safety and Health for "basing its evaluation of chemicals on work prepared by scientists with close ties to industries involved in the manufacture and sale of the substances under evaluation." All of which led them to conclude that the proposed name change was highly charged:

> The "intolerance(s)" seen in MCS patients are neither to or from the "en-vironment" and neither are they idiopathic: patients report intolerances only to certain specific chemical, physical and other sensory stimuli. Since unpolluted natural environments do not pose problems for most MCS patients, there is no justification for a name like Idiopathic Environ-mental Intolerance that suggests exactly the opposite. For all these rea-sons, MCS Referral & Resources calls on patients and physicians alike to reject the proposed IEI terminology as nothing more than a ploy of the chemical industry to shield its markets and liability.[7]

Names carry a lot of weight, and continue to be invented and debated in this area of inquiry. A new national organization has been formed, for example, under the name of the National Coalition for the Chemically Injured (NCCI). For now, we continue to work with the name MCS, for several reasons. It keeps open the questions of *multiplicity*: multiple potential causes, multiple symptoms, and the reactions to multiple incitants experienced by people (not just the single chemical exposure that might have "caused" their illness). It focuses our inquiries on *chemicals*, which is the best place to concentrate. And the questions and problems of *sensitivities* are the most puzzling and interesting.

Gloveless in Seattle

One of the most prominent and, unfortunately, typical examples of the highly charged assemblages of MCS comes from Washington state, and the production facilities of the aerospace giant Boeing. The case shows how diagnostic categories, industry interests, state regulatory structures, lawyers, and scientific experts who move between the worlds of universities, peer-reviewed journals, industry, and state agencies all come together in the muddiest of situations—and people who are ill, or at least claim to be ill, are left swamped and stuck.[8]

Boeing's plant in Auburn, Washington, includes Building 17–02, built during World War II as cold storage for cadavers. Things have changed since Boeing bought it from the government: With the autoclaves used to bake airplane parts, the temperature inside this huge shed of a structure can exceed 100 degrees Fahrenheit in summer. Some of the workers there (again, many of them women) did "lay-up," taking fiberglass preimpregnated with phenol formaldehyde, layering it with Kevlar and graphite, and baking the result in the autoclaves to make ductwork, paneling, and cockpit parts for commercial aircraft. (Production had shifted in 1987 to use the preimpregnated fiberglass, to comply with more stringent Federal Aviation Administration standards on flammability.) Meanwhile, at Boeing's Developmental Center near Seattle, workers were building planes that we don't all get to fly in: the Advanced Tactical Fighter, or the Stealth fighter, made famous during the Persian Gulf War. The lay-up process was similar there, but these radar-absorbing materials were preimpregnated with a substance called "Avimid K-III." In both places, workers were exposed at their benches through skin and nose, and at the autoclaves, which emitted blasts of hot, chemically "impregnated" air when opened. (Six hundred workers at Lockheed's infamous "Skunk Works" plant in Burbank, another World War II–era building, had previously sued Lockheed over similar health problems from working with another Stealth composite; Lockheed settled out of court for $33 million, most of which it would pass on to the federal government "as a legitimate cost of doing business," as one Lockheed spokesman put it.)

The "chemical situations" faced by workers at both the Auburn and Seattle plants were made worse by a combination of social and institutional factors. The union contract at Auburn allowed for mandatory overtime; people might work seven days a week, ten to twelve hours a day. Workers claimed that they weren't trained to use respirators, and that the gloves they used ($26 per pair) tended to come off, and weren't replaced when they wore out because of the expense. Workers heated food for their Friday potlucks in the curing ovens; when asked about the wisdom of this by a reporter, one woman answered: "Honey, we were too far gone." "I'm lucky," said another, "I can still remember my name, age, and telephone number. There are people down there who are walking vegetables." At the Seattle plant, a month after work started with the new Stealth materials, Boeing's industrial hygiene office relayed a series of changing "tip sheets" on Avimid K-III. The first said no respirators were needed, and workers could use latex gloves. The tip sheets became more stringent over the next ten months, with handwritten addenda. Wearing thick butyl rubber gloves might be advisable, the next one read, and inhaling one of the five toxic components of Avimid K-III could damage eyes. DuPont recommended butyl rubber gloves since Avimid K-III contained a solvent that could eat through latex gloves in about an hour. Boeing appears to have known about these problems, based in part on the Safety Data Sheets from the chemical's manufacturer, DuPont, but ignored or downplayed them because they made the work more difficult and slower. Workers report never seeing the tip sheets, never using respirators, and wearing only the less cumbersome latex gloves.

But enough horror stories from the shop floor, of which there are many more. Let's go up a level. Not in terms of literal space (which would take us to the engineers and secretaries upstairs at the Seattle plant who were reportedly affected by the chemicals emanating from the production floor, but who weren't on a "need to know" basis about the Stealth materials), but in the figural assemblage-space, to the individuals and institutions responsible for testing materials, people, and compensation claims.

In court depositions (often the only place where knowledge gets articulated and accounted for in these legal, adversarial situations), DuPont admitted that it had done toxicological studies on only the individual components of Avimid K-III, and could say nothing about their synergistic effects. Worse, the studies of individual chemicals were done only for inhaling them, not absorbing them through the skin. Boeing monitored only air concentrations and not skin exposure, despite federal rules restricting skin exposure to one of the chemical components. This is the kind of "knowledge gap" that can be found in many situations in which MCS is at issue: only a small percentage of chemicals have even been tested—under far from the most stringent criteria—and only for certain effects and pathways. The effects of complex combinations of chemicals remain almost uncharted territory.

The "lucky" ones at Boeing and in many other cases were the ones who were so sick, and whose sickness could be linked to a known chemical cause, that they fit into a diagnostic category recognized by doctors, the state workers' compensation system, and the insurance industry—something like "occupational hepatitis" or "toxic encephalopathy." But the symptoms of most fell in the cracks or, as is most often the case, were actively placed there. One of the most controversial characteristics of MCS, and the one which presents the greatest challenge to biomedical researchers, is what is called the "spreading effect": While an initial, acute exposure to something like phenol formaldehyde may be the original cause or problem, diverse symptoms can in many cases be provoked by many other compounds, with dissimilar chemical structure or properties. And those compounds—in perfumes, paints, felt-tip pens, carpets, and so on—appear to have enormous effects at very, very low concentrations which go unnoticed by most people, and which defy the conventional dose-response models of immunology or toxicology.

Like anyone claiming to have MCS, Boeing workers had to run a gauntlet of doctors and other examiners; they end up with boxes full of documents, and sometimes not much else. Workers have to appeal to Boeing first, to see if their insurance will cover what medical tests there are. Boeing and the state Labor and Industries department (which runs the workers' compensation system) do not recognize the diagnostic tests relevant to MCS as "objective," and routinely deny claims. MCS proponents will point to studies which purport to be objective, but which they say are funded by industry to prove that MCS is psychological and not physiological.

One such study was headed by the University of Washington's Dr. Gregory Simon, with funding from Boeing's Health and Safety Institute. Other members of Simon's research team included many doctors with direct or indirect links to the unsympathetic Labor and Industries department or to Boeing.

Simon reported that of the thirty-seven workers at Boeing's Auburn plant who had filed compensation claims, a "significant minority . . . developed a full multiple chemical sensitivity (MCS) syndrome. Symptoms persisted even after the workers left work or transferred to new locations. These subjects reported a generalized sensitivity to numerous common exposures so that the symptoms originally precipitated by workplace exposure could now be produced by synthetic fabrics, paint, or gasoline fumes." He stated that testing by the Washington State Division of Occupational Safety and Health and the National Institute for Occupational Safety and Health measured less than 10% of the permissible exposure levels for phenol and formaldehyde in the air at the Auburn plant. (Simon doesn't mention how these tests were performed, or if Boeing plant officials knew the state regulators were coming and "cleaned things up." "Comprehensive medical examination . . . showed no immunologic differences" between the study subjects and healthy laboratory controls. More important to Simon were the tests measuring "social conditions," although he doesn't say what these tests consisted of. They seem to be based on the same kind of "anecdotal evidence" that is often used to dismiss MCS complaints:

> Social conditions at this worksite appeared ripe for what I will call psychological contagion. Many workers described a high level of perceived health threat . . . many believed they had suffered chemical poisoning or immunologic damage. Symptoms often began with highly visible attacks, which publicly emphasized the level of danger. . . . Both labor and management were under high levels of stress. Production pressures had resulted in longer work hours and forced overtime at the time the new manufacturing process was introduced. Success of this new process was essential if management was to meet fast approaching product delivery dates.[9]

Since the only shop-floor detail that seems to have been important to Simon was "stress," it's no surprise that he pays much more attention to psychiatric tests than to toxicological or immunological tests. Simon used a combination of an "abridged version of the Diagnostic Interview Schedule," "selected subscales of the Hopkins Symptom Checklist–90," "the Whitely Index" ("designed to measure hypochondriacal attitudes and beliefs"), and "the Barsky amplification questionnaire . . . designed to measure symptom amplification or the tendency to report a large number of physical symptoms" to construct psychological profiles of affected workers. On the basis of these tests and their results, Simon suggested "several psychologic mechanisms that might contribute to chemical sensitivity." One was a "behavioral sensitization model," a sort of Pavlovian response to an odor associated with a now absent chemical exposure. "In response to a high level of perceived threat," workers at the Auburn plant "became acutely symptomatic. As the level of threat declined, they desensitized themselves and improved spontaneously."

Simon's second model "explains MCS as a form of somatization or symptom amplification." Some workers "appear to have chronically high levels of psychologic distress" and an associated "tendency to amplify symptoms"; "chemical sensitivity develops when a person with chronic symptom amplification develops the new belief that these symptoms are due to chemical exposure. . . . Treatment based on this model . . . [is] reducing symptom sensitivity and learning to attribute physical sensations to benign causes." (Simon doesn't appear too interested in specifying the sources of such "beliefs" in either direct experience, or in things like manufacturer's tip sheets, which suggest that workers should "believe in" wearing thick gloves.) Simon's final model "focuses on the social environment and views MCS as an exaggeration or displacement of legitimate concerns about health risks," which "leads to increased vigilance and a tendency to attribute common symptoms to environmental hazards. . . . The workers examined here would have received much less attention had they only complained that forced overtime was a strain on their families or that the odors involved in the new manufacturing process were irritating and unpleasant."

Boeing used the Simon study and another study headed by Dr. Patricia Sparks in its defense against workers' claims, and more generally against the scientists, doctors, lawyers, and affected people who believe in MCS. The "it's all in your head" argument is the most frequently used and most effective argument against MCS and related conditions. Denied access to certain biomedical tests, or having accepted test results officially deemed irrelevant, Boeing workers were left with the characterization of "psych cases."

Again, MCS proponents trace the connections in these assemblages: In its denial of "objective" medical tests for MCS, the Washington State Medical Association was being advised by Sparks (among others), and Sparks was not only a toxicologist with the Occupational Health Services program at Providence Medical Center, but a consultant and medical examiner for Boeing, as well as a consultant to the Department of Labor and Industries. A former manager of Boeing's occupational medicine program left Boeing and joined Sparks' research group at Providence Medical Center. But in addition to tracing these and other conflicts of interest (including how the editorial boards of peer-review journals are dominated by many of these same, corporate-connected scientists and doctors), MCS "believers" have also dissected the scientific studies of Sparks and Simon, pointing out many flaws in definitions, methodologies, assumptions, statistical analyses, construction of control groups, and other elements of conventional science.[10]

The entire complex assemblage seems to break down into any of a number of related, simple oppositions: sick/healthy, workers/management, poor/rich, nonsense/science, powerless/powerful, the individual/the system. And indeed, a narrative in which the system has a lot to lose, and is using everything in its power to protect its position, seems quite compelling. One estimate has put Boeing's potential workers' compensation bill at $450 million. The journalist who investigated and wrote the Boeing story on which we've drawn here could conclude, without appearing too far out, that "working in concert, Boeing and 'Big Medicine' are using their economic power and political clout to manage the high-stakes debate over multiple chemical sensitivity. . . . Key to their strategy is framing MCS primarily as a *psychological* condition, despite a growing body of medical and scientific evidence to the contrary."

In Washington as well as in the rest of the country both sides appreciate the stakes. As the American Council on Science and Health, which teaches consumers that MCS isn't real, phrased it in a 1994 report: "MCS junk science is costing society millions of dollars. It is restricting people's lives in unnecessary ways and diverting them from effective medical treatment. It can cause enormous problems in the workplace and can cost people their jobs. It may burden the health care system, severely tax the insurance industry and wreak havoc with workers' compensation programs." In sum, they warn that the "economic implications [of MCS] for many industries and insurers are potentially catastrophic. Unless the problem is properly addressed, the millions of dollars now changing hands through claims and lawsuits will become billions, wreaking havoc with many industries and insurance programs and ultimately raising the costs to all consumers."[11]

Who's to say whether that's sober scientific projection or hysteria?

An Us/Them Summary, and Its Breakdown

Although we try to avoid framing problems in terms of oppositions, they can be useful organizers, as long as we don't get attached to them. In the case of MCS, world views tend to divide into those of proponents and opponents, terms we used in the preceding section. Both are, inevitably, caricatures of much subtler and more complex differences.

MCS proponents articulate (as a verb, we prefer this to "believe" or "know") that there is something objectively wrong with their bodies. They can say how they react to a panoply of certain odors that leave other people unfazed; they get headaches, or they become exhausted, or their muscles or joints hurt, or they can't think straight or concentrate, or they develop a rash, or their muscles spasm, or they have trouble breathing, or their vision blurs, or they become nauseated, or their heart races—or some combination of any of these and still other symptoms. These symptoms can sometimes persist for hours, days, and weeks after an exposure. The symptoms are the end result of either an acute exposure to a specific toxin in the past, or the result of longer-term, lower-level exposure to an encompassing atmosphere of chemicals.

Many MCS proponents also show a tremendous faith that the sciences, when properly done, are capable of providing an answer to these problems. They assiduously track developments in immunology, toxicology, biochemistry, neurology, and many other disciplines, looking for the answer that they know is out there, somewhere—causal mechanisms that will result in objective technological tests that will lead in turn to diagnoses, treatments, and in some cases some form of compensation, either from state programs or private litigation. But since mainstream scientific disciplines and professions as yet can offer little or no helpful relief, MCS proponents often turn to nontraditional therapies and "ways of knowing" such as clinical ecology, acupuncture, and homeopathy.

What keeps the sciences from being done properly and objectively, according to many MCS proponents, is an array of powerful social and institutional interests, centralized in the chemical industry. By funding certain scientific institutions, by employing or funding scientists who dominate conferences and regulatory panels, by mounting media campaigns, these interests prevent MCS from becoming "real," and continue to dismiss it as a form of hysteria or other psychological disorder. To the extent that people with MCS *do* have psychological conditions such as depression, MCS proponents argue, it's either as a direct result of their exposures (through some as yet unknown neurological pathway) or as a more indirect result of the social isolation and frustrating encounters to which they are subjected. Finally, MCS proponents say that the relevant interests have to hide the truth about MCS because that truth threatens their economic survival.

How would we describe the world of the MCS opponents? To these doctors and scientists, MCS is at most a psychosomatic phenomenon. It does not fit standard toxicological or immunological models, knowledge, or diagnostic categories. They can find no objective biological measure or sign of the condition, and rather than questioning whether their inquiry infrastructure is incomplete or inadequate, this is taken as conclusive evidence that the illness is "all in their heads." According to MCS opponents, while their studies may be institutionally or financially connected to the chemical or other industries, such linkages do not undermine the objectivity of their sciences. Their objectivity lies safely outside these influences in the world of the laboratory. MCS opponents state that people claiming to suffer from MCS, then, are for the most part experiencing either a hysterical reaction to modern industrial society and its attendant risks (in which case they should be pitied or placed in psychotherapy), or acting out of self-interest (in which case they should be barred from legal and administrative hearings, and put back to work).

MCS proponents characterize the contention that MCS has no biological basis as "cigarette science," a reference to the Tobacco Institute's attempts to cover up the true linkage between smoking and cancer with studies that purport to be "objective" but are in fact funded generously by the tobacco industry. As an editorial in *Our Toxic Times*, the newsletter of an MCS activist organization, put it:

> Like all "cigarette science" the truth will eventually come out. In the meantime, the MCS community needs to prepare for the battles ahead. The battles will be over maintaining the credibility MCS has gained as a public health threat . . . and demanding that MCS not be written off as a psychological illness unless credible, unbiased research into all the possible physiological causes has been done.
>
> With all these orchestrated attacks, the chemical industry must be afraid that science is too close to an explanation for MCS that doesn't in-

volve its sufferers being crazy. Rather than facing up to the problem and
dealing with it, industry is doing what it always does. Asbestos workers,
Vietnam veterans, silicone victims, joint replacement patients, etc. have
all come under industry orchestrated attacks.

It's now our turn—MCS has come of age.[12]

Both MCS proponents and opponents, then, believe in the ability of biomedical
practices and knowledges to come up with a satisfying, and presumably helpful, set
of answers. But note a key difference: MCS proponents make explicit linkages between the sciences and the institutional framework and interests in which they are
carried out, while the MCS opponents deny the relevance of this social context. MCS
proponents make these linkages with a skepticism informed by a history of how
"power" has worked in the twentieth century to deny the "truths" of damaged bodies. Many MCS proponents have worked hard to track the connections between the
sciences and industry, and between scientists and other official bodies. They know
that their articulations, experiences, and demands are being resisted, that resistance
is always a sign of some kind of power or force, and that one of the places in which
resistance and power can almost always be found is in established institutions.

The social setting in which any science is produced and deployed is something
that has to become part of inquiry into the sciences—and in the long run, part of
scientific inquiry itself. In that sense, we "side" with the MCS proponents far more
than with its opponents. But at the same time, we don't want to reduce these stories about MCS to simple power moves, but to keep the complications and muddiness within sight at all times. MCS is controversial, ambiguous, and a challenge
to work on not simply because it is resisted by the powers-that-be wielding cigarette science, but because it is an emergent condition at the intersection of physiology, culture, labor politics, economic forces, toxicological and epidemiological
sciences, and all manner of beliefs. At best MCS is in a stage comparable, in Ludwik Fleck's analysis, to syphilis *before* the Wassermann reaction.

Even those doctors and scientists who are MCS proponents know that even in
the absence of industry influence, the biology and physiology of MCS remain ambiguous and resistant to inquiry. They nevertheless pursue the sciences of MCS
rigorously and carefully. We can't survey here all the ways in which researchers
and clinicians are pursuing the sciences of MCS, but Nicholas Ashford, Claudia
Miller, Iris Bell, Grace Ziem, and William Meggs are among those scientists and
doctors who, in our judgment, are doing well-controlled, well-defined, creative
work in the sciences related to chemical sensitivities. They work with T-cell sub-
sets, porphyrin anomalies, mechanisms of neurogenic inflammation, electroen-
cephalographs, diverse biochemical and immunological assays, and many other
research tools and theories. They can also be quite critical of the study designs and
scientific theories of their colleagues who are also "MCS proponents."

The sciences of MCS are being ably pursued by more than a few practitioners. But while that's very important, it's also not enough for such a complex assemblage. Claudia Miller points to the need for two tracks, a science track and a public health track, with the help of an example from history:

> There's an issue, however, separate from the science. That is the public health questions that we're confronting right now. I'm certainly reminded of John Snow and the cholera epidemic. He saw people getting ill and noticed that the people lived around the Broad Street area and that there was a pump around that area from which they were drawing their water. The people who seemed to be getting ill were the ones who drank that water. So he ordered the pump handle to be broken, and the epidemic stopped. It was not until 30 years later that Koch discovered the bacterium that causes cholera. So, Snow knew of no mechanism at that time, and yet there was a very important public health intervention that could be performed, that saved lives. Still, the science had to proceed. So I think we need to focus on both science and public health.[13]

How Not to Deal with Complexity

Without question, the least helpful way to analyze and manage the complexity of the scientific differences on MCS is to slot everything into two opposed categories: real science and junk science. Peter Huber, the author of *Galileo's Revenge: Junk Science in the Courtroom*, maintains an utterly naive faith in precisely this kind of clear distinction. Inveighing against court decisions around the topic of MCS, Huber puts everything into two neat categories: the "real science" of immunology, based on facts, causes, evidence, and disinterestedness, and the "junk science" of clinical ecology, grounded only in wishful thinking, belief, hysteria, and the desire to punish and/or make money. All the confusion, all the difficulties, all the complexity that might challenge us to think and rethink what's going on between human bodies and the environment, are contained by applying this simple scheme.[14]

But even a very staid report, *Science and Technology in Judicial Decision Making*, drafted by a very staid committee for the Carnegie Commission on Science, Technology, and Government, takes issue with Huber's category of "junk science." The Carnegie Commission committee notes that "judicial decisions that appear to be based on 'bad' science may actually reflect that the law requires a burden of proof, or confidence level, other than the 95 percent confidence level that is often used by scientists. . . ." This committee also argues that so-called "real science" has never been the neat package of truths and methods that purists say it is:

> Although the judicial system's ability to manage and adjudicate science and technology issues can be enhanced, it is wishful thinking to

imagine that all problems will vanish because the "truth" will emerge once the correct procedural approach is adopted. It is equally naive to believe that the scientific world can produce the "right" answer on request. . . . Critics of the judicial system's handling of scientific claims often have an idealized view of science that scientists themselves reject. Although these critics accept the indeterminacy of legal concepts, they speak of scientific "facts" as though they were objectively true . . . even though scientists themselves concede that scientific hypotheses remain open to challenge until the incentives for attacking them disappear.[15]

So, even relatively mainstream commentators now know that the rest of us are well advised to follow (some) scientists' lead in abjuring idealized views of their enterprise, and that even as we passionately cling to the concepts of "truth," "facts," and "right," something scary is happening to them that forces even the most earnest experts to put them in scare quotes.

If scientific hypotheses remain open until the "incentives for attacking them disappear," as the Carnegie Commission report stated, what counts as an incentive? In the case of MCS, such incentives are many and powerful, creating an atmosphere that does no good either for the sciences or for affected individuals. The attacks can come on many fronts: on individual dignity, on professional competence, on the words and numbers produced in laboratories. All rolled up, they are attacks on the entire diagnostic category named MCS.

A SAD Story

That such incentives include powerful economic and social forces is an argument made by more people than MCS proponents. There are scientists who recognize that our articulations of "disease" are produced from more factors than simple, direct access to some physiological reality. Those scientists and doctors most adept at thinking comparatively, at pondering historical examples, at seeing the muddle that exists in the assemblages of the natural, social, and political, are the best hope/allies for people with MCS. Sometimes these people are ensconced within the medical establishment. Dr. Norman Rosenthal, who is Chief of the Section on Outpatient Psychiatry of the Clinical Psychobiology Branch at the National Institutes of Mental Health, has written a remarkably illuminating, thoughtful, and sensitive paper comparing MCS to SAD, seasonal affective disorder. In MCS, he writes:

we see a condition in its early phases of research and treatment. As yet no case definition has been agreed upon by all workers and researchers. It is not codified in any official manual of medical or psychiatric conditions, and some clinicians and researchers question whether it is even a legitimate medical condition. All of these circumstances existed approximately fifteen years ago with respect to another clinical condition, seasonal affective disorder (SAD). Since then, however, the existence of SAD has been widely accepted, criteria for its diagnosis have been developed, and it has been included (albeit in the form of an adjectival modifier, "seasonal pattern") in the standard manual for psychiatric diagnoses, the DSM-III-R . . . and now, the DSM-IV.[16]

SAD and MCS are similar in that diagnosis in both cases has relied extensively (but not exclusively) on patient reports of symptoms; causes for each may be said to lie in environmental stressors rather than in the individual; and both have been recognized informally and, to a lesser extent, formally for quite some time, but only recently become more evident to both the public and professionals.

Yet there are significant differences between SAD and MCS. A specific treatment for SAD, light therapy, was recognized as effective at about the same time the syndrome was first described; effective treatments for MCS are still very much in debate. More importantly, as Rosenthal points out:

> [S]easonal rhythms are part of the natural world and there is a rich literature on naturalistic and experimental studies of seasonal rhythms in animals on which researchers have been able to draw. . . . [I]n the case of SAD, no one can be blamed or held liable; darkness is a feature of our natural landscape. This is not so in the case of MCS, where someone can be held liable for the injury or injuries. The financial implications of such liability could be interpreted as a reason for the victim of such an injury to exaggerate its impact. Alternatively it could be construed as a reason for those with a financial interest at stake to minimize the severity of a patient's syndromes, dismiss them as being "psychological" in nature, or even dismiss the existence of MCS altogether. Given the high stakes involved, it should come as no surprise that the heated controversy about MCS is greater than one might expect to surround a topic of purely academic interest.

On the labeling of people who say they have MCS as hysterics, Rosenthal shows us how to combine a respect for uncertainty with ideals of fairness and kindness. He also shows us how we nevertheless might articulate different, but in any case rigorous and workable, criteria for legal and administrative as opposed to research situations:

> Given that we don't know what causes MCS and that our technology may well be inadequate for determining subtle forms of neurophysiological abnormalities in MCS patients, it would seem not only kinder, but also

fairer, to entertain seriously that the patient may be correct and that the injury, however subtle and elusive at this time, may nonetheless be real. While different standards of proof may be required for special purposes, such as determination of liability, an open-minded approach in our therapeutic dealings with MCS patients would appear to be appropriately respectful of their suffering and our ignorance.

This attitude may come in part from a recognition that there exists, as Rosenthal titles one of the sections of his paper, a "politics of diagnosis." Here again, the ability to denaturalize what were supposed to be natural categories isn't a sign of an antiscientific attitude. Attention to the very real ways in which language operates, to the importance of such articulations to both laboratory and political settings, and to the differences and conflicts that are almost inevitable in these kinds of situations, are all necessary for doing good work in the sciences. But it is all much more complicated and delicate than simply trying to "objectively" describe a reality "out there":

> Diagnoses are not actual entities, like the earth and planets, but rather human artifacts, like countries, the boundaries of which are sometimes drawn by consensus but perhaps more often haggled or fought over. In the case of SAD, the diagnosis was fairly readily accepted into DSM-III-R, but only as "seasonal pattern," an adjectival modifier of any recurrent form of mood disorder. The acceptance of a diagnosis by the medical community confers an important status to a medical condition. It makes it easier to seek third party reimbursement, lends weight to claims of disability or liability, and assists in obtaining research grants, though in itself it doesn't guarantee success for any of these ventures. . . . Thus, it seems clear that acceptance of MCS as a legitimate diagnosis, together with all the benefits that attend such an officially recognized condition, will result not only from scientific endeavors, but also from political efforts.

So do we believe everything we hear about MCS, in the name of "good politics?" If only things were that easy.

TILTed Pursuits

If we had to name what we thought the single most important thing readers of this book could do about MCS, it would be: Exercise judgment. And that is quite the demanding workout. There is no "single most important thing" involved in the processes of judging, especially when it comes to such a multiply riddled topic as MCS. Eventually, you will have to pore over dozens and even hundreds of laboratory and clinical studies, reading not only for the data but the way those sign-forces have been articulated, as well as for what judgments occurred regarding sample size and selection, methods of analysis, margins of error, and so forth. You should know some things about immunology, toxicology, neurology, occupational health, and more than a few other disciplines. You will need to inquire into the so-cial, economic, and political contexts of the research and the researchers. You will have to weigh legal considerations, and you will have to be prepared to acknowl-edge uncertainty and gaps. You should find people who have been diagnosed as having MCS, or are looking for some other articulation of their illness, and listen to their stories with your own imprecise combination of empathy and skepticism.

Don't feel overwhelmed—just start.

One particularly good starting place is a thick book published by the Depart-ment of Health and Human Services (HHS), *Multiple Chemical Sensitivity: A Scien-tific Overview*. It contains the proceedings from three conferences sponsored by HHS, the Public Health Service, the Agency for Toxic Substances and Disease Reg-istry, and the National Research Council. It has articles from MCS proponents, op-ponents, and more than a few people in between, a number of which we've excerpted here. It provides scientific studies, personal testimonies, "outsider" commentary, and extended, rich dialogues and discussions among the participants at the various workshops.

From reading this as well as other sources, one thing is certainly clear to us: The claim that MCS can be reduced to a "psychological disorder" is an injustice at many levels—personal, social, and scientific. It disregards potential explanations from the biological realm, or even just the *potential* for potential explanations as bi-ological investigators pursue the limits of their own uncertain disciplines. It con-fuses cause and effect in the most unhelpful way. The primary use of such psychological "explanations" is to deny people access to further medical tests, or to insurance or other forms of needed compensation. As articulations, in other words, "mass psychogenic illness" and "somatization disorder" are linked with lit-tle more than the desire to marginalize or dismiss. There are many doctors, epi-demiologists, toxicologists, and other researchers who are pursuing a much denser set of articulations for MCS, exploring linkages to multiple physiological causes as well as social factors.

It is indeed possible that such pursuits will come to naught. "Chemical sensi-tivity," writes Claudia Miller, "could be a new paradigm that has the potential to explain many chronic and costly illnesses, including fatigue, depression, headaches, and asthma, or it could be nothing at all. Not understanding MCS, we take an immense gamble." She thinks, and we agree, that gambling *for* its realitty is worth it for "establishing sound environmental policy. If there is a subset of the population that is especially sensitive to low-level chemical exposures, a strategy for protecting this subset must be found. . . . It would make little sense to regulate chemicals at the parts per billion level or lower if what was required was to keep people from becoming sensitized in the first place. Indeed, by understanding the true nature of MCS and who is at risk, we may prevent unnecessary and costly overregulation of environmental exposures in the years to come."[17]

Is someone like Claudia Miller "committed" to articulating better physiological mechanisms for dealing with MCS, or what she calls TILT—toxicant-induced loss of tolerance? Yes. But that commitment also incorporates the tentativeness and openness that characterizes the sciences at their best. What makes Miller an exemplary scientist is not merely her analytic abilities in the lab and clinic and the fact that she describes robust, workable, and replicable criteria for testing her theory. As we saw above, she uses history to hone her thinking and suggest new possibilities for both scientific and social work. She also knows how to play a metaphor. By suggesting that MCS, chronic fatigue syndrome, Gulf War illness, sick building syndrome, and other new and puzzling medical entities may be the tip of the iceberg signaling an as yet submerged, "emergent theory of disease" she names TILT, Miller sets a range of forces and questions in motion—historical, biological, metaphorical, and social.

Miller points out that many of the criticisms made of MCS as a disease category—that too many different kinds of chemicals cause responses; that too many organ systems are involved in these responses; that no single, specific biological mechanism can be proposed; that no common biological marker has been identified—were at one time or another leveled against the germ theory and immunological models of disease. It took decades to address the controversies arising from the fact that different kinds of germs cause a range of responses, involving multiple organ systems, via different specific mechanisms, with no single biomarker common to all bacteria and viruses. The same is true of immunology, where different antigens can provoke different responses in more than one organ at a time, through a range of biochemical mechanisms, marked by different antibodies or immuno-proteins. Each of those histories was marked by bitter professional disputes and the mistakes of ardent researchers on all sides of the issues. And in each of them, some effective preventative measures (antiseptics, sanitation, avoidance, immunization) could have been or were taken before our knowledge of the complex physiological and social mechanisms qualified as noncontroversial science.

Miller's work on TILT parallels that of Iris Bell, the developer of a model of time-dependent sensitization (TDS) to explain the multiple symptoms of MCS. Bell draws on neurophysiological research into how various substances, from alcohol to antidepressants to organophosphate pesticides to nerve gases (Bell works in a Veterans Administration hospital), may travel directly from the olfactory to the limbic system—that is, from the nose deep into the brain—the structures of which can be "kindled," or made permanently hyperreactive to the smallest of stimuli. The limbic system occupies the center of a complex regulatory network involved in eating and drinking disorders, dysfunctions of the immune and reproductive systems, emotional disturbances, and autonomic responses like breathing, heartbeat, and flushing of the skin. Because of their multiple functions and physiological effects, Bell argues, the limbic and central nervous systems may offer more fruitful possibilities for research on MCS than the immune system. (We are not plugging either the TDS or the TILT model. We value them not as answers but as productive means of continuing the pursuit of MCS.)

Both TILT and TDS involve a two-stage physiological process. In the first stage, an acute or chronic exposure to a "toxicant" or toxicants (including but not limited to chemicals) results in a damaging "loss of tolerance" in some people. The precise mechanism is as yet unclear, and could in fact be multiple mechanisms in multiple systems (limbic, central nervous, immune). In the second stage, a range of symptoms can be triggered by extremely small quantities of substances that would leave other people unaffected. Both Bell and Miller, reasoning mainly from clinical studies and patients' own accounts, note that the process resembles addiction. The addict and the chemically sensitive person both experience intense cravings for alcohol, nicotine, caffeine, other drugs, or even certain foods, but while the addict seeks out these substances, the person with MCS just as avidly avoids them. In both cases, the process is biphasic: Any incorporation of the substance results first in a stimulatory phase, and then in a withdrawal phase of fatigue, irritability, headaches, depression, or other symptoms that can last for hours to days. With MCS, multiple exposures to different substances over time can mask the effects of any one, and result in a cacophony of symptoms whose causes and effects are difficult to disentangle.

To test these theories a person must be isolated in what Miller calls an environmental medical unit, a facility as specialized as our current intensive care or coronary care units. As with detoxification from addictive substances, the person diagnosed with MCS has to be isolated from all potential chemical exposures for at least several days, so that the overlapping stimulatory and withdrawal phases from multiple toxicants can settle down and be unmasked. Using double-blind and placebo-controlled procedures, the researcher can then expose the person to particular substances, and measure the responses more precisely through biochemical analyses, electroencephalography, positron emission tomography (PET), and other technologies.

Miller's term, TILT, works in part because it provides a useful analogy for the body that has lost its tolerance: "With a pinball machine, a player has just so much latitude—he can jiggle the machine, nudge it, bump it, rock it, but when he exceeds the limit for that machine, the "TILT" message appears, the lights go out, and the ball cascades to the bottom. The machine's tolerance has been exceeded and no amount of effort will make the bumpers or flippers operate as they did before. The game is over."[18]

A skillful player (robust health, clean environment, or just dumb luck) can keep the game going quite well, within limits. But when the limits are exceeded—prolonged use/abuse, parts wearing out, an overloaded environment, or a manufacturer's conspiracy—play stops. A colleague of Miller's suggested that the analogy extends to the social experience of the MCS patient, how they are bounced around from one specialist to another (averaging, in one study, nearly ten physicians).

TILT also describes the situation faced when researching MCS or working with people who have chemical sensitivities. It's something we've experienced and have had confirmed by other researchers in the field. The position of helper, sympathetic researcher, or service provider becomes very difficult—because people who are TILTed are indeed shaken and de-centered. They are hurt and can reasonably expect to be hurt again, through no fault of their own. But their difficulties and reactions to those difficulties can also make it hard for others to be as sympathetic or open as possible.

The TILT metaphor can be extended to still another level, that of the larger process of defining this new theory of disease. The many legal, economic, personal, and institutional forces that are evident in this controversial area have made the machine of scientific research and medical care extremely touchy and difficult to move within. These forces continue to hit TILT, keeping laboratory scores down, stopping the play of inquiry. We need much more subtle moves, a better awareness of all the bumpers on the table and the forces they exert, and more expert players.

We also need more than a few rolls of quarters. As Miller points out, "knowledge will not come cheaply. Further studies on chemical sensitivity that involve blinded challenges in a controlled environment, that utilize brain-imaging, state-of-the-art immunological testing or other sophisticated tests, and that compare adequate numbers of patients and controls, will be costly. Funding agencies will need to invest adequate sums to acquire answers in this area as they have for other diseases, such as breast cancer and AIDS. Until sufficient research funds become available, chemical sensitivity no doubt will continue to pit physician against physician, perplex policy makers, and impoverish patients and corporations alike."[19]

Chapter 7

Producing Multiplicities
Inquiry Infrastructures for Molecular Genetics

Unstable Equilibria

"It is frequently asked, 'What is the relative importance of nature and nurture?'" the distinguished British geneticist (and socialist) J.B.S. Haldane wrote over fifty years ago in *Heredity and Politics*, a book whose title suggested the inevitable conjunction of questions of science and questions of power. Despite the frequency of the question, or perhaps because of it, Haldane stressed that the most important point was "to realize that the question of the relative importance of nature and nurture has no general answer, but that it has a very large number of particular answers."[1]

True enough—although we would add ". . . a very large number of particular answers *which change over time and place, since neither 'nature' nor 'nurture' are stable over the long term.*" In a very small chapter in a very small book, however, written during a time in which the tone and frequency of such questions becomes extraordinarily high-pitched, we can only sketch out the terms of specificity that Haldane rightfully demanded. But even to hint at the varieties of particular factors and questions—genetic, cellular, developmental, social, political, and others—that come into play under the oppositional categories of nature and nurture is an important undertaking in this era when "it's all in your genes" is such a powerful phrase, so ready-to-hand.

When did that era begin, and how did we get here—and what do we do now that we are here? Those questions, too, resist any general answer. They require instead that we attempt some careful delineation of the swirling influences of practice and theory, tools and ideas, politics and science that together compose this era. But in such terms of composition, we might be better off thinking of the later compositions of John Cage rather than Johann Sebastian Bach: surprising arrangements of odd and disparate elements (nevertheless carefully "prepared," sometimes) that push the limits of comfortable rhythms and tonalities, rather than the intricate and lovely, but ultimately logical harmonies of a classical age. Included in today's genetics score would be agencies like the National Institutes of Health, the theoretical and mechanical inventions of many scientists, multinational pharmaceutical and chemical companies, venture capitalists, vast ignorance, new laws and norms regarding the patenting of organisms and their subcomponents, and changing relationships between university and industry researchers. And of course the inchoate but powerful desire to have an answer to the "really" questions of genetics, especially when they're about things like intelligence, sexual desire, addiction, violence, and other complex human behaviors.

The gene now has the same iconic stature in our culture as the atom did earlier in the twentieth century. The image of the double helix symbolizes the progress and power of the sciences, the genius of scientists—and, more obliquely, the unsettling, unforeseen dangers lurking beneath the sparkling surface of scientific progress—just as the whizzing electrons in the elliptical orbits of the Bohr atom once did. It seems that since we no longer require the atom to enforce bipolar "peace" through threatened global annihilation, we can project a more optimistic future of better health and longer life through genetic manipulation—although we no longer kid ourselves that these things will be "too cheap to meter."

The iconic gene kicks off other cultural crazes and trends. Isolating the gene for Marfan's syndrome, for example, brought out proposals to exhume the remains of Abraham Lincoln just to see if he suffered from some form of the condition. Kary Mullis, who won the Nobel Prize for his work in the invention of the polymerase chain reaction (PCR), one of the most powerful, elegant, and useful tools of molecular biology today, has tried to market jewelry containing the DNA fragments of celebrities. Taxi signs and billboards, some featuring a pregnant Mona Lisa with her mysterious smile, remind people they can call 1–800-DNA-TYPE and for $600 (an extended payment plan can be worked out) definitively settle questions of paternity. Genes and genetic disorders serve as the plot basis, or at least a kind of "maguffin," in everything from The X-Files to the movies of Arnold Schwarzenegger and Michael Keaton.

Even scientists working in these fields are weary of the simplified, triumphalist gene-of-the-week stories that appear with numbing regularity in the media, announcing the discovery, or even just the hoped-for imminent discovery of the "gene for" breast cancer, hemochromatosis, familial polyposis, retinoblastoma, and other relatively "simple" human disorders. Then there are those announcements of "breakthroughs" on conditions that everyone admits are more complex, but to which molecular genetics offers what is often said to be the best approach currently possible: alcoholism, schizophrenia, Alzheimer's disease, bipolar and other forms of affective disorders, attention deficit disorder, and so on. And then there are the minefields which are nevertheless regularly crossed with reports of "solid" or perhaps merely "suggestive" or "tantalizing" findings, available in both peer-review and tabloid format: intelligence, homosexuality, "cuddling behavior," neu-roticism, criminality, and many others.

How are people supposed to think about these kinds of reports—at least one of which you will surely read, watch, or hear about within three days of reading this? As science and discovery, or as myth and doctrine? Who is being "strictly scien-tific" and who is being "merely ideological" when they argue that intelligence or desire is really mostly genetic or really mostly social? Don't we have to depend on molecular genetics now—the best, most accurate, least biased form of genetics that we've ever had—to settle these kinds of uncertainties and questions about ourselves, questions upon which so much depends? On the other hand, if we take the view that genetic explanations are just a social construction, how are we going to avoid being duped by biological or ideological charlatans? You know our writ-ing pattern by now, and know that those are rhetorical questions that we use to muddle through this text, and that we're going to be steering for that complicated, contentious, and far from conclusive zone *in between*.. . .

The first thing to remember is that not all biologists believe that their sciences are best thought of as a simple matter of discovery. The great molecular biologist François Jacob reminds us: "Like other sciences, biology today has lost many of its illusions. It is no longer seeking for truth. It is building its own truth. Reality is seen as an ever-unstable equilibrium."[2] It's vital to understand how biology's truths are built, and how an unstable equilibrium called "biological reality" is, for a short or long time, stabilized. But, as we've repeated almost ad nauseam, that doesn't mean that biological reality is "just" made up, a "social construction." What would it mean to think of things like breast cancer, intelligence, sexual desire, or alco-holism in Jacob's paradoxical term, an "ever-unstable equilibrium?" It would mean learning to think of such objects of knowledge as *multiplicities*.

Calling something a multiplicity is *sort of* like saying it's complex, but it also says something more. A few paragraphs back, we contrasted the musical composi-tions of Bach and Cage. Bach's compositions are complex: intricate combinations of notes, instruments, themes, recursions, and so on. But the complexity can be mastered and appreciated, and the next performance is going to be *almost* identi-cal to the previous ones; a Bach composition is a complex but stable equilibrium. A Cage composition is a multiplicity, an ever-unstable equilibrium. It can't be mas-tered or finalized, but only becomes more complex; new elements and interpreta-tions are added on by chance, by personal choice, by local conditions. The next performance is guaranteed to be different. A multiplicity is inexhaustible in its dif-ferences and possibilities.

Genes, genetic explanations, diseases, and human traits such as intelligence or desire are not simply complex; they are multiplicities. Some of the consequences of such a shift in thinking and doing are described here, along with a few helpful questions and suggestions for engaging creatively and critically with multiplicities. Because multiplicities are "ever-unstable," and because new scientific objects, new technologies, new institutions, and new words will continue to aggregate around them and change them—particularly in a rapidly changing field like molecular genetics—scientific literacy in this area can't be organized around answers and certainties, and involves instead the development of a repertoire of questions and tools for thought.

The historical affinities between biological explanations of human conditions and conservative or even authoritarian politics that we touched on in Chapter 4 casts a heavy shadow today, and can evoke the most knee-jerk reactions to each announced "discovery." It takes extra effort to remember that the division between nature and nurture, biological and social, is neither scientifically nor historically stable, nor are biological explanations for complex traits inherently "dangerous" or "inhuman," and social explanations safer or more humane. Ours is an inquiry-based effort that requires reading the scientific literature and attending conferences, interviewing working geneticists and molecular biologists, spending time in government and corporate laboratories to get a feel for daily realities, listening to what "ordinary people" have to say on these issues, and other pursuits of specificity, complexity, and multiplicity.

That work takes multiple forms: research and writing of books and articles, holding public seminars, and teaching in college classrooms. In each of these pursuits, our goals parallel those of contemporary genetics itself: to keep things moving, to develop new tools for thought, to prevent knowledge from hardening into stable certainty, and to stay aware of what always remains, at one time or another, unknown. We try throughout this chapter to give readers a feel for how this works by referring occasionally to what we do or *would* do in researching a book, in speaking publicly to "nonexpert" audiences, in providing tools for thought to college students.

As multiplicities, things like intelligence, sexual desire, or alcoholism (and breast cancer and hemochromatosis and . . .) have *no proper place* on either the side of "nature" or the side of "nurture." They can't be finally stabilized as either "inside" an organism, in its genes and other biological mechanisms, or on the "outside" of the organism, in the environment or in social and cultural milieus. Moreover, such traits can't even be stabilized as some complex but ultimately determinable interaction between these two, along the lines of "intelligence is 85% nurture and 15% nature." The question for a complex reality is: How do we figure out what the percentages and their interactions *really* are? The questions for a realitty of multiplicities are: How and why do we tell ourselves that these percentages are real? How and why do we decide that the "proper" place for intelligence or breast cancer is in "nature"? If the trend today is to place everything inside, on the side of nature, what combination of experimental tools, thought-styles and languages, institutional forces, and cultural patterns together compose this trend? By what operations of metaphors, technologies, and social practices does a multiplicity like sexual desire get channeled into a stable place, whether that's "the gene" or "the culture"? Where are the gaps in such articulations? What are the assemblages, with all their complexity, specificity, and ambiguity, that make something "really genetic"? What are the different effects of reducing an irreducible multiplicity like alcoholism to the nature or nurture side of the divide, or to some complex but specifiable combination of the two? How should we judge those effects? What kinds of tools can we get from the sciences, the social sciences, and science fictions, for keeping up with these "ever-unstable equilibria"? And what kinds of experiments can we imagine running inside and outside our bodies, our laboratories, our worlds?

Talking About Genes

Viral, bacterial, plant, animal, and human genetics have all been around for at least fifty years, but it's only in the last few decades that they have been studied and practiced at the molecular level. These biological sciences are now part and parcel of the information age and share the genealogy of their contemporaneous scientific endeavors, cybernetics and communications theory. Since World War II, we've used many of the same ideas and terms to conceptualize organisms, computers, and mechanical assemblages alike: information-processing, signal-transmitting, feedback-dependent systems—which exhibit, under certain conditions, the ability to self-organize and evolve. How did we build this particular set of biological truths?

François Jacob can again teach us something, this time about how to pursue the history of biology. One could approach this history as a "succession of ideas" which exhibit "independence," he tells us—facts and theories which, simply because they are "real" and "true," reproduce, spread, invade, and die off. But Jacob prefers another way:

> The alternative approach to the history of biology involves the attempt to discover how objects become accessible to investigation thus permitting new fields of science to be developed. It requires analysis of the nature of these objects, and of the attitude of the investigators, their methods of observation, and the obstacles raised by their cultural background. The importance of a concept is defined operationally in terms of its role in directing observation and experience. There is no longer a more or less linear sequence of ideas, each produced from its predecessor, but instead a domain which thought strives to explore, where it seeks to establish or-

der and attempts to construct a world of abstract relationships in harmony not only with observations and techniques, but also with current
practices, values, and interpretations. . . . Each period is characterized by
a range of possibilities defined not only by current theories and beliefs,
but also by the very nature of the objects accessible to investigation, the
equipment available for studying them and the way of observing and discussing them. It is only within this range that reason can maneuver. It is
within these fixed limits that ideas operate, are tested and come into conflict.[3]

We don't know if Jacob ever read Ludwik Fleck, but the affinities between his
articulations of how the sciences work and those of Fleck should be clear. (We *do*
know that Jacob read Foucault, and the work of anthropologist Claude Lévi-
Strauss, linguist Roman Jakobsen, and other scholars from other disciplines.)

One of the best practitioners of this kind of developmental history of genetics is
Evelyn Fox Keller, who brings to the job the requisite combination of scientific
training and insight, historical acumen, philosophical flair, and adeptness at feminist and literary theory. Given the rediscovery in 1900 of Gregor Mendel's ground-
breaking but unknown work with pea plants over thirty years earlier, the
twentieth century might seem to be best characterized as a century of genetics. But
Keller reminds us that genetics first had to catch up with the more established and
in many ways more sophisticated fields of embryology and physiology. "[E]ven in
the early days of genetics," she writes, "when the gene was still merely an abstract
concept and the necessity of nuclear-cytoplasmic interactions was clearly understood, geneticists of [T.H.] Morgan's school tended to assume that these hypothetical particles, the genes, must somehow lie at the root of development."[4] (Morgan
was perhaps the most important and influential American geneticist in the first
half of this century, who with his colleagues Alfred Sturtevant and Calvin Bridges
pioneered research on the fruit fly, *Drosophila melanogaster*. We're sorry that we
can't stop here to look in more detail at this work, its momentous significance, and
its own fascinating complexities.[5]) As Morgan would write in 1924, "It is clear that
whatever the cytoplasm contributes to development is almost entirely under the
influence of the genes carried by the chromosome, and therefore may in a sense be
said to be indifferent."

In other words, scientists such as Morgan were committed to the primacy of the
gene long before they had good material evidence for such a view, and fully aware
of the complexity of interactions explored in the sciences of embryology and physiology. This view that genes are the basement level of control, the prime mover in
biological systems prodding everything else—proteins, organs, limbs, and character traits—into existence, Keller calls the "discourse of gene action." And this
language of the "entirely influential" gene and "indifferent" cell was alive and oper-

ative long before there was "good material reason" for it. This is a simple fact of history. "Scientists usually assume that only their data and theories matter for scientific progress, that how they talk about these data and theories does not matter, that it is irrelevant to their actual work," Keller contends. "But in introducing this particular way of talking, the first generation of American geneticists provided a conceptual framework that was critically important for the future course of biological research. . . . It enabled geneticists to get on with their work without worrying about the lack of information about the nature of such action—to a considerable degree, it even obscured the need for such information."

What did the "discourse of gene action" do? It provided "powerful rationales and incentives for mobilizing resources, for identifying particular research agendas, for focusing our scientific energies and attention in particular directions. This discourse of gene action has worked in just these ways. And it would be foolhardy to pretend it has not worked well. The history of twentieth-century biology is a history of extraordinary success; genetics—first classical, then molecular—has yielded some of the greatest triumphs of modern science." Hardly an antiscientific perspective, which Keller is sometimes accused of having by science purists. Like Jacob, her point is that in trying to account for the success and power of genetics, we have to attend to the work of language or discourse just as much—not *only*, but just as much—as the work of theorizing, experimenting, and all the other activities which constitute the sciences. Language can be as influential as any gene, any apparatus, any institutional funding source. If "the ways in which we talk about scientific objects are not simply determined by empirical evidence but rather actively influence the kind of evidence we seek (and hence are more likely to find), we must consider other factors if we are to understand the strength and persistence of the discourse of gene action."

(Among the other factors Keller cites as influences contributing to the primacy of the gene in complex and often subtle ways are nationalism—Americans staked out the nucleus as their scientific domain, while Germans established themselves as cytoplasmic investigators and theorizers—and gender—the nucleus was articulated with masculine qualities [sperm is practically all nucleus], while the cytoplasm was articulated with feminine ones.)

We have to skim over much of this history and the series of beautiful experiments that constituted such an important part of it. Running the risk of hewing too closely to the "great men" model of the history of science which we think is so unfair and unhelpful, we'll quickly list a few highlights: the work of George Beadle and Arthur Tatum with the mold *Neurospora*, from which came the "one gene, one enzyme" hypothesis that allowed geneticists and biochemists to muddle through brilliantly, until its limitations and oversimplifications became all too apparent; the work of Oswald Avery and colleagues at the Rockefeller Institute that established

DNA as "the transforming principle," the substance that, without question, changed the functions and products of bacteria; the work of Erwin Chargaff that carefully delineated the precise ratios of the purines and pyrimidines that made up DNA; the Watson and Crick elucidation of DNA's structure (aided by the less-than-collegial appropriation of Rosalind Franklin's crystallographic data); the elegant experiments of Matthew Meselson and Frank Stahl which clarified how DNA replicated; and Jacques Monod's and François Jacob's elaboration of their operon model, the "genetic switch" turned on and off by the presence or absence of certain substances in the cell. Each of these scientists and many, many others designed encounters with the realitty of bacteria, viruses, molecules, molds, and other organisms and their components—experiments that were remarkably productive and, in their own ways, definitive. By the mid-1960s, what was only half-jokingly called the Central Dogma was firmly in place in the heads, hands, tools, techniques, and institutional worlds of most biochemists, molecular biologists, and (at least bacterial and viral) geneticists: DNA codes for RNA codes for protein. Like God, DNA was there at the beginning of the equation, firmly in charge, orchestrating the whole show.

But at the same time, the sciences of genetics consist of much more than just these experimental results. They were influenced and impelled by private and public institutions such as the Rockefeller Foundation and the National Institutes of Health, and they incorporated ideas and techniques from the physical and information sciences. And because of these and other influences, the ways in which scientists talked about genes and gene action were, by the time of the last few experiments mentioned above, starting to mutate.

From Recombinant DNA to "Fantastic Infrastructure"

We'll return to that mutational strain of talking and thinking later in this chapter; it now seems that quite important social and scientific developments were occurring in and around the biological sciences in the 1950s, even though their full import still remains largely a matter of hopeful interpretative analysis and suggestion.

But what Keller calls the discourse of gene action undoubtedly remained the dominant way of articulating and experimenting in molecular biology. A way of thinking and acting—both within the sciences and in society broadly—which drew on talk about genetic blueprints and master codes powered this new hybrid discipline of molecular biology throughout the 1960s and 1970s, and still does today.

It was, as Keller noted, an extremely effective discourse for conceptualizing new problems, guiding new experiments, and attracting funding—a powerful package deal of reinforcing components. One of the most important sources of financial resources to fuel these developments in the late 1960s and early 1970s was in the National Institutes of Health's Viral Oncology Program, which scientists presented as a key component in the War on Cancer. If Congress supported this kind of basic research into the inner workings of viruses and bacteria, they argued, it was precisely the "disinterestedness" of this basic science of molecular mechanisms that stood the best chance of producing a breakthrough knowledge directly applicable to this pressing social problem.

While we may not have gotten a cure for cancer for these hundreds of millions of dollars of public funds, we did get the world-changing sciences and technologies of recombinant DNA. The tools and ideas of recombinant DNA that are so often referred to in the most awe-inspired, almost miraculous terms are indeed useful and powerful in their own ways—the restriction enzymes that can snip DNA at particular points; the ability to splice the resulting fragments together and into other living organisms, where they actually work to make substances like interferon (pioneered by Walter Boyer and Stanley Cohen); the capacity to sequence DNA and "read the code" of an organism, and work out why some of its proteins are misshapen or some of its regulatory or developmental patterns disrupted (developed by Frederick Sanger, Walter Gilbert, and Alan Maxam); the inverse capacity to synthesize DNA from scratch, making new "artificial" strands that can be used as diagnostic probes, or as working inserts into bacterial cells (worked out and later mechanized by H. Gobind Khorana, Leroy Hood, and others). It can easily be said that all these, and now literally hundreds of other new abilities, sprang from the minds and hands (and graduate students) of a few brilliant individuals. But it can just as easily be said that these abilities—which blossomed in the late 1960s through the mid-1970s—exist *only* because of the preceding decades of difficult, collective work on the genetics and biochemistry of bacteria, flies, viruses and other organisms, and *only* because of the enormous public resources that were channeled into them to create new techniques, collective databanks and cell lines, and more skilled scientists to carry on the work.

Our point here is not the one that is sometimes made—that because these knowledges and abilities were achieved with a generous influx of tax dollars, the patenting of genes and gene products that started soon afterward and has now become the widespread, standard practice is somehow wrong or reprehensible. While we think that patenting and other intellectual property issues involving genes and cell lines are in need of extensive democratic review, debate, and rearticulation, we don't think it's particularly helpful, strategic, or even politically astute to simply be opposed to what is usually reductionistically referred to as the "commercialization of life." But we have to skirt that thicket of questions and issues for now, limiting this chapter to the various ways in which scientific thinking about genes and genetic explanations is stylized, in Fleck's terms, and the consequences of those stylizations.

These stylized knowledges and techniques, when hooked up to new flows of venture capital and to the research and economic infrastructure of existing petrochemical and pharmaceutical industries, also became the basis for the new biotechnology industry, upon which we now pin so many of our hopes for health and vitality in both the personal and international economic spheres. The early biotech start-ups of the late 1970s and early 1980s—Cetus, Amgen, Genentech, and Biogen, to name just a few of the biggest—clustered in the San Francisco Bay Area and in Cambridge, Massachusetts, where the research labs of major universities and hospitals provided the intellectual talent and craft skills that the industry needed. The often uneasy but always powerful combination of federally funded university research, private industry research, as well as the scientific work funded by large private foundations such as the Howard Hughes Medical Institute, spurred the rapid development of many new tools, techniques, and databases for producing, organizing, and working with the information and substances of genetics.

Since the mid–1980s, these enormous and widespread changes have been most frequently referred to under the name of the Human Genome Project (HGP), often summed up as "the federally funded, $3 billion effort to map and sequence all the genetic material in the human chromosomes," an organized effort which officially began in 1990 and is scheduled to be completed by 2005. But that popular definition is somewhat misleading: No one really knows how much the entire effort will cost, and the $3 billion figure does not include all the money from private industry that is going into these areas of research. The HGP includes the study of many nonhumans—mice, fruit flies, nematodes, yeast, and other organisms—that offer important comparative lessons for us humans. But most of all, that succinct definition of the HGP places the emphasis, wrongly, on the end point or goal.

Many of the scientists who were key proponents of the HGP in fact emphasized a different set of metaphors: The HGP wasn't about arrival, of finding the holy grail of genetics, but about new tools of pursuit. As Leroy Hood, the main inventor of the automated DNA sequencer and synthesizer, put it in congressional testimony: "Number one, the [HGP] is about developing new technologies. This is the first biological project that has as a major imperative developing new kinds of technologies. . . . [I]t is going to create a fantastic infrastructure for biology and medicine which will be centered on the map and the sequence of the human genome."[6] Another proponent, the mathematician-turned-molecular geneticist Eric Lander called the HGP the Route One of genetics. "To put it simply," he testified, "it is vastly more efficient to build highways than to have each individual clear a path through the wilderness. In the 1950s, Congress showed such wisdom by funding the construction of a comprehensive national highway system. We have reaped the economic benefits of this investment many times over. If Congress had not invested in a timely fashion in highways, the nation would have been the poorer. To-day, biomedicine is saying: it is the critical time to build highways."[7]

In an interview with one of us (Fortun), Lander expressed his preference for this more mundane metaphor:

> I eventually stood up and said, Come on, it's Route One we're talking about. This isn't so fancy. We are talking about building a highway, for God's sake. . . . I guess I found it very hard, very troubling that people were cramming this down people's throats as the holy grail. Yes, it's very important, but it's infrastructure, and I thought Route One conveyed better than anything else, a) how you did the project—that you didn't do it in small pieces, but b) that it wasn't such a grandiose thing either. And so it conveyed both of the things, namely . . . people were proposing building this, by everybody go out and build a mile of highway on either side. And that's crazy, because you're going to have too many gaps and things like that. But the other aspect of it is that it somehow demystified the mystique of the human genome, and put it in what I still think is an appropriate context: It's an infrastructure project. It's a very important infrastructure project. You try to build infrastructure efficiently. And that ought to be our basic metaphor.[8]

That is indeed a better way to think about the HGP: not as a goal, but as what we've called an inquiry infrastructure. It has been an especially productive infrastructure, which can spew out new genes, new substances, and new genetic explanations at an ever-increasing pace. (The culture of speed in genetics and the HGP today, which all these highway metaphors evoke, is something we explore elsewhere.[9]) It has also led to a new name for a new discipline, "genomics," a term which encompasses the tools, techniques, and computer databases for the study, not merely of single genes and their products, but of entire genomes: all the chromosomes in an organism, their structures, functions, and their interactions.

The genomics infrastructure also undergirds new developments in the biotechnology industry, and new alignments of diseases, research strategies, and institutions. In this new world order, genomics-based companies like Millennium Pharmaceuticals strike deals with Hoffmann-La Roche so that the pharmaceutical manufacturer will have rights to Millennium's work on obesity and Type II diabetes (in exchange for $70 million), or with Eli Lilly ($45 million) for Millennium's work on atherosclerosis. Sequana Therapeutics can make alliances with Glaxo, Boehringer Ingelheim, and Boehringer Mannheim for its work on (respectively) Type II diabetes, asthma, and osteoporosis (at an aggregate economic level of $172 million). These linkages represent one typical arrangement in which genomics firms have a "dedicated corporate partner" for specific disease entities.

Incyte Pharmaceutical's agreements with Pharmacia-Upjohn, Pfizer, and other large pharmaceutical firms represents a different type of economic partnership, in which Incyte supplies more general access to its specific technologies, database resources, and software tools for accessing and combining different data sets. All in all, major pharmaceutical companies made agreements with genomics firms totaling almost $400 million in the last half of 1995 alone—approximately twice the annual funding level provided by the National Institutes of Health (NIH) and the Department of Energy (DOE) for genomics research.[10]

There's so much more in this history that demands the literacy skills to think critically about the sciences today: the development of techniques such as the polymerase chain reaction (PCR), for example, which now allows scientists and technicians to take the smallest amounts of DNA from a crime scene, blood sample, or the tissue of mummies, and amplify it into quantities necessary for everything from research into evolutionary origins or the study of rare genetic traits, to paternity suits and murder trials. (It's a fascinating and complex story analyzed through interviews with many of the key players conducted by the anthropologist Paul Rabinow, in *Making PCR: A Story of Biotechnology*.) Or the debates among biomedical researchers and between NIH and DOE concerning if, how, and *where* the HGP should be carried out. (Robert Cook-Deegan provides a thorough insider's account in his *The Gene Wars: Science, Politics, and the Human Genome*.) Or how oncogenes emerged over this period as the complex intersection of new tools, new theories, and new institutional alignments to become the most favored explanation for the cause of cancer. (The sociologist Joan Fujimura examines these reciprocal interactions in *Crafting Science: A Sociohistory of the Quest for the Genetics of Cancer*.[11]) And that would still only scratch the surface of events and issues, leaving unattended many pressing questions of genetic discrimination in health benefits and employment, prenatal diagnosis, gene therapy, privacy rights, patenting, and many others. (Not to mention all the other ways in which the bodies of men and, especially, women are being rearticulated these days: organ transplants, in vitro fertilization and other reproductive technologies, new antidepressants and other drugs, replacement hips and knees, and so forth.)

The point we are extracting from this enormous territory of fast-paced change is modest but broad: We now have the kind of inquiry infrastructure that makes scientists believe that many of today's "really?" questions about our bodies are at least answerable in terms of genes, if not already answered. The fact that those answers often work well enough to be *sold* certainly reinforces the power which genetics has, and the awe in which it is held. It is this complex assemblage of powerful tools, major social institutions, vast public and private resources, and a long-standing "discourse of gene action" that all of us are now learning to live within, ask new questions about, and, we hope, improve upon.

But that, fortunately and unfortunately, is not the whole story.

"A Funny Thing Happened . . . ": Organized Complexity

If this "fantastic infrastructure" of genomics now allows geneticists and other scientists to make an increasing number of stronger claims about the power of genes, it has also had the somewhat paradoxical effect of simultaneously calling those claims into question.

"A funny thing happened on the way to the holy grail," Keller quips. The "extraordinary progress" we've made with our infrastructure for manipulating genes "has become less and less describable within the discourse that fostered it. The dogmatic focus on gene action called forth a dazzling armamentarium of new techniques for analyzing the behavior of distinct gene segments, and the information yielded by those techniques is now radically subverting the doctrine of the gene as sole (or even primary) agent."[12]

It's a subversion that's been going on for some time now, if only among a rather limited group of scientists. At the end of World War II, new disciplines—information theory, operations research, cybernetics, systems theory, and perhaps most importantly, computer science—and their practitioners began to come together, in new government institutions and in private conferences like those sponsored by the Josiah Macy Foundation. Keller aggregates them under the name "cyberscience," which she argues was "developed to deal with the messy complexity of the postmodern world. . . . For cyberscientists, Life (especially corporate life, electronic life, and military life—the modes of life from which these efforts emerged and on which they were focused) had become far too unwieldy to be managed by mere doing, by direct action, or even through the delegation of 'doing' to an army of underlings kept in step by executive order. The problems of the mid-twentieth century were clearly, as [the Rockefeller Foundation's] Warren Weaver put it in 1948, ones of 'organized complexity.'"[13] So are many of the problems of the late twentieth century, as we shall see below.

So, even while genes are sold, literally as well as figuratively, as the code of codes and the blueprint of life, that relatively simple story is beginning to break up under its own momentum. "As we learn more about how genes actually work in complex organisms," Keller continues, "talk about 'gene action' subtly transmutes into talk about 'gene activation,' with the locus of control shifting from genes themselves to the complex biochemical dynamics . . . of cells in constant communication with each other. Current research—drawing on the phenomenal technical successes of molecular biology, and even on the sequence information emerging from the Human Genome Initiative [sic]—invites (ever more insistently) a shift in locution in which the cytoplasm is just as likely as the genome to be cast as the locus of control."

Living in an era in which genes seem to explain everything while holding out the promise of finely tuned control is complicated enough. But we seem to be living in a time of transition, in which we have to deal with several, often contradictory, things at once: Genes are in control; genes are not in control; scientists rely on a simplifying reductionism; scientists recognize greater and greater complexity. We're learning that organisms, their components, their disorders, and indeed all their traits are multiplicities: assemblages so excessive in their agitated complexity that they never permanently settle down into one perspective or another.

Even the venues of popular culture are beginning to exhibit the somewhat befuddling combinations. A recent *U.S. News and World Report*, for example, acknowledges that "more and more experts, including dedicated biologists, sense that the power of genetics has been oversold and that a correction is needed":

> If there's a refrain among geneticists working today, it's this: The harder we work to demonstrate the power of heredity, the harder it is to escape the potency of experience. It's a bit paradoxical, because in a sense we end up once again with the old pre-1950s paradigm, but arrived at with infinitely more sophisticated tools: Yes, the way to intervene in human lives and improve them, to ameliorate mental illness, addictions, and criminal behavior, is to enrich impoverished conditions in the family and society. What's changed is that the argument is coming not from left-leaning sociologists, but from those most intimate with the workings of the human genome. The goal of psychosocial interventions is optimal gene expression.[14]

Things are certainly unnerving in this transitional period: The future promises a return to the past, liberal social analysis reappears as pure natural science, and we all get a new personal best to strive for, "optimal gene expression." Still, the recognition of complexity has its good effects. "Even if scientists were to identify a gene (or genes) that create a susceptibility to alcoholism," *U.S. News* informs its readers, "it's hard to know what this genetic 'loading' would mean. It certainly wouldn't lead to alcoholism in a culture that didn't condone drinking—among the Amish, for example—so it's not deterministic in a strict sense. Even in a culture where drinking is common, there are clearly a lot of complicated choices involved in living an alcoholic life; it's difficult to make the leap from DNA to those choices."

Yet it's exactly that kind of difficult leap between genes and alcoholism that researchers are still trying to make, as we discuss later in this chapter. So while we may, as *U.S. News* puts it, "be witnessing a kind of cultural self-correction, in which after a period of infatuation with neuroscience and genetics the public is becoming disenchanted, or perhaps even anxious about the kinds of social control that critics describe," such a hopeful change doesn't mean there are any fewer questions to be asked. (Why shouldn't the phrases "psychosocial interventions" aimed at "optimal gene expression" make us anxious, for example?) In fact, there are many more. If things like alcoholism are merely complex interactions that vary according to culture as well as biology, then cultural self-correction indeed seems possible. It is an enormous scientific and social challenge, but the belief is that stable, natural equilibria can be reached once we "sort it all out." But if such conditions and traits are multiplicities, or Jacob's unstable equilibria, they won't ever be sorted out once and for all, but always rearticulated and recombined into a new set of practices. Within a "fantastic infrastructure"—that phrase that so nicely captures the paradoxical combination of mundane materialities and extravagant visions—multiplicities are always becoming something else.

Monitoring Complexity

The proliferation of expert panels, advisory boards, and bioethics commissions at every level from the local hospital to the White House are evidence of the complex changes our bodies are undergoing. These social inventions are a partial recognition that knowledge is no longer enough, or at least that scientific expertise is unable on its own to solve the social problems to which it is "applied." Every new technological advance, every new claim about the state of nature only seems to open up a host of social, political, personal, legal, and moral dilemmas that require the attention and direction of still more kinds of experts.

A small percentage of the federal funds devoted to the HGP go to its Ethical, Legal, and Social Implications program (ELSI), which supports research and policy forums on the tangle of issues that comes with this territory. ELSI taskforces and its panels deal with issues such as who should be allowed to administer genetic tests for breast cancer: any family physician? only those who have been certified by a special licensing board? or a new class of specialists altogether? The social issues are complicated by very complex genetics, and still more complex physiology. Going by the popular accounts they most frequently encounter, many people might think that scientists found the "cause" of breast cancer when researchers at the University of Utah "won" a highly competitive race among many scientists and isolated the gene they dubbed BRCA1 in 1994. (The patent for BRCA1 resides with the biotechnology firm Myriad Genetics, for which several of the scientists involved also worked.) But not only can BRCA1 account for just a small percentage of breast cancer cases (less than 15%), it wasn't long before another gene was isolated, BRCA2, and it wasn't long after *that* that hundreds of mutations or variations in both of these genes began turning up. (Such are the complicating and ironic consequences of the productive, fantastic infrastructure of genomics.) That stunning range of differences makes precise genetic screening very difficult. Meanwhile, another panel of ELSI experts reported that even when a precise, positive diagnosis of a "breast cancer gene" can be made for a woman, there really was no clear, positive course of action for her to then take. Still, Myriad Genetics has a test that it wants to sell, that it *needs* to sell to stay competitive in the genomics marketplace. An anthropologist who has worked extensively among cancer patients, Nancy Press, comments on the problem: "The physicians I've talked to are uniformly unaware of the statements made by professional societies on how to use BRCA. It's very worrisome. That's not meant in an accusatory way. They don't have bad hearts or anything. But there's a big gap in knowledge about BRCA, and in the meantime you have these mass mailings by laboratory companies urging doctors to give the tests." And on the immediate future horizon (if not already here): the home-testing kit.[15]

Another relatively "simple" gene-associated disorder, hemochromatosis, recently prompted a similar set of discussions and concerns among ELSI experts. Mercator Genetics isolated and applied for a patent on the hemochromatosis gene which, for the approximately 1 million people in the U.S. who have it, makes their bodies absorb too much iron, which in turn can lead to liver cancer, heart disease, and many other serious conditions. An effective therapy exists for hemochromatosis, a very old therapy in fact: bloodletting. But when a scientist representing Mercator Genetics argued at a recent ELSI meeting for a program of screening large populations for the gene, people immediately raised questions about denial of health insurance benefits, about the 13 percent of the people who suffer from hemochromatosis but who don't carry the gene, and about whether this was any improvement over existing blood tests. There are no easy answers here, no clear delineation of good and bad effects stemming from this commercial gene product. People may indeed be denied health benefits, or insurance companies may decide to cover the cost of screening and bloodletting as a way to avoid paying for liver transplants later. In any case, Mercator Genetics can sell its diagnostic kit.

The complexities get even more tangled when you are dealing not with "simple" disorders which are in fact quite complex in their genetics, their biology, and their social practices, but with those things that even geneticists acknowledge are complex, and which we would say are conditions of multiplicity.

Intelligence: The Bets Are on the Floor

Intelligence is one such multiplicity that recently returned to prominent national attention with the publication of Richard Herrnstein and Charles Murray's *The Bell Curve*. Its primary argument—that there are small but significant differences in intelligence between races (as measured by IQ tests), that such differences are biologically based, explaining and even justifying many social and economic inequalities—has been heard many times before in one variation or another. *The Bell Curve* was not at all based in any new findings or analyses of molecular genetics. But it did appear during a time in which genetic explanations for almost everything were both enormously popular and had at least an air of scientific authority about them. It also appeared at "a historical moment of unprecedented ungenerosity," as Stephen J. Gould points out, "when a mood for slashing social programs can be powerfully abetted by an argument that beneficiaries cannot be helped, owing to inborn cognitive limits expressed as low IQ scores."[16] So was *The Bell Curve* such a social phenomenon and such a source of endless argument and commentary because it was "scientific," or because it was "political"? Isn't there really a simple answer here, like "it was bad science," "it was pseudoscience," or "it was merely ideology"? Why make things more complicated than they really are?

For the simple reason that not only was the "*Bell Curve* affair" *really* complicated, it was a multiplicity. As just one indication, consider the various responses collected in the book *The Bell Curve Wars*.[17] (Bell Curve Wars, Gene Wars, Science Wars, Culture Wars—we thought we were living in peacetime.) The different characterizations of Herrnstein and Murray's book include "a manifesto of conservative ideology" (Gould); "a strange work," although "some of the analysis and a good deal of the tone are reasonable" (Howard Gardner); "all the trappings of a scientific work" but also full of "omissions, misstatements, and eccentric interpretations" (Richard Nisbett); "a chilly synthesis of the work of disreputable race theorists and eccentric eugenicists" (Jeffrey Rosen and Charles Lane); "a very sober, very thorough, and very honest book" (Thomas Sowell); "hate literature with footnotes" (Jacqueline Jones); "a book that deserves reflection and respect" and which "urge[s] us to create a world in which all human beings can achieve their fullest potential" (Andrew Hacker); and "the sudden and astonishing legitimation . . . of a body of racialist pseudoscience" (Michael Lind). Whatever was in *The Bell Curve* could be further articulated in a number of strikingly dissimilar ways, elaborated into different theories and practices.

Isn't this exactly why we need better, harder distinctions between "pseudoscience" and "real science" more than ever, to prevent these kinds of debates? Well, we have people who can make such distinctions forcefully and cogently, but that hasn't prevented the disagreements and debates from happening anyway. Our point in drawing up this list is not that all of these views are "right" in some relativist sense, but only that in a democratic society characterized by marked political and intellectual differences, this book and others like it will work in complex ways. *The Bell Curve* is a "power/knowledge object": whatever its fallacies and omissions and eccentric interpretations, whatever combination of sciences, pseudoscience, and hate literature it might be, it operates with *enough* apparent "reasonableness" in *enough* contexts to count as real. It reinforces what some people "believe" or "know," it has effects on people's lives, it prompts actions. Recognizing *The Bell Curve* as this kind of power/knowledge object is admittedly very unsettling—it means that these kinds of arguments aren't subject to proof or disproof in the senses of these words to which we are accustomed. That's why books like this have appeared regularly over the past century and more, and immediately stirred controversy: the articulations are *both* already "proven" *and* already "disproven."

Unsettling as such a reconceptualization of *The Bell Curve* may be, it also prompts recognition of a number of important lessons. Paradoxically, it means that since the "scientificity" of *The Bell Curve* is only one part of the problem, it is all the more imperative that geneticists and others respond to it. "There's no science in *The Bell Curve*, so why respond to it?" remarked one geneticist at an ELSI meeting during a debate over whether the panel should write an official response to the book. But another scientist, Stanford University's David Cox, knew exactly why: "When *The Bell Curve* came out, you couldn't pick up a magazine without seeing essays about it. The genetics societies were the only ones who said nothing. Their approach was, 'It's not in our job description.' Well, that's a very narrow view of life. The rest of the world was commenting." Cox knows that the kind of simplifications offered in the book can be challenged by new work in genetics. "From the molecular genetics point of view, it's absolutely clear that we're demolishing the arguments of genetic determinism. But the facts have never gotten in the way of people who wanted to use genetics in a deterministic way in our society. And I think this is where my colleagues miss the boat. Just because something is scientifically right and proven, it isn't the end of the story. You have to consider how that fact is going to be parlayed into people's lives."[18]

While we wouldn't necessarily agree with Cox's articulations of "facts" and "proofs," he's right in recognizing that this isn't the end of the story. The inverse in this case is also true: Just because something is "scientifically wrong and disproven" isn't the end of the story either. The story is the assemblage, and all the cultural, political, and even scientific elements of it that make *The Bell Curve*, however much we might regret it, a working realitty.

At the same time, Cox's language can't contain his own story. Whether by accident or intention (or some muddle of these), his curious use of the word "parlay" here suggests a situation of multiplicity rather than complexity. A parlay is a bet, of an original wager plus one's winnings, placed on a subsequent event in the future. Cox is right to question how such a bet is parlayed in the social realm, where it can be "cashed out" in a number of very real ways, as we saw above in the excerpted responses to *The Bell Curve*. "Subsequent events" and articulations in the social world of democratic differences are undoubtedly part of the situation that geneticists need to address. But Cox is gambling on the "fact" that molecular genetics will always, in all subsequent futures, demolish the arguments of genetic determinism. Moreover, Cox doesn't think that's a gamble at all—it's a sure thing.

That may indeed be a good gamble, but it's far from "absolutely clear." The expert, scientific view on intelligence and other multiplicities has shifted before, and you can certainly bet that it will change again. Herrnstein and Murray's bet on linking intelligence and race is parlayed not only in the social realm, but in the realm of the sciences as well. The fantastic infrastructure of contemporary molecular genetics is designed to produce "subsequent events"—isolate new genes, articulate new biochemical and neurological pathways, correlate new definitions of populations with new genetic markers—any one of which *may* articulate *some* working relationship between "genes" and "intelligence"—which will then be quickly correlated with that other unstable equilibrium, our social and scientific articulations of "race." Moreover, it can be argued (and Herrnstein and Murray have argued) that *The Bell Curve* was not about genetic determinism, but about the more complex interactions between "nature" and "nurture" at the level of populations rather than individuals, a view which in many ways is hardly demolished by molecular genetics, but supported. As a multiplicity, "intelligence" isn't so much settled and determined, but is replayed again and again with a combination of old, new, and yet-to-be-imagined things and ideas.

As a democratic society our bets are better placed on the practices by which those replays will happen, rather than on some "truths" that are supposed to underlie them. What kinds of scientific and social practices are we going to gamble on? What are the stakes? What are the parameters of the risks? Who gets to play? Who has the most to lose?

In other terms, the questions for a democratic society are: What are the rules and practices for articulating "intelligence"? What effects do specific practices have? Scientists like Stephen J. Gould and Howard Gardner can already provide us with insights into what such questions entail. Gould's incisive critiques of the biological claims and statistical analyses in *The Bell Curve* build on his brilliant social and scientific analyses of the history of intelligence testing and its articulation with racial, ethnic, and gender differences, in his book *The Mismeasure of Man*.[19] By focusing on the practices by which intelligence comes to be seen as an object of scientific and social knowledge, and the effects of those practices, Gould demonstrates how IQ is not some real, natural cognitive capacity that is represented or measured by performance on an IQ test. IQ is not a reality, but a realitty: it *is* a person's performance on a certain test; the IQ test "performs" something named "intelligence;" the practice and the "truth" are inseparable. The cultural and scientific limitations of these tests can be analyzed and described in detail, as can their effects and uses: creating hierarchical ranks in the military, tracking students into certain curricular programs, predicting certain kinds of job performance. As a performance, "IQ" works in certain highly delimited ways. Howard Gardner and other psychologists have shown how different practices can perform different truths, different "intelligences" articulated with criteria such as creativity, "emotional intelligence," musical talents, or an ability to adapt to new situations. "In short," Gardner writes, "the closed world of intelligence is being opened up."[20] It's a multiplicity, and in its combinations with new scientific and social practices, "intelligence" is becoming something else.

And it's becoming something better. Gardner's expanded categorizations of multiple intelligences are better practices than the narrow, singular category used by Murray and Herrnstein. Such multiple practices perform differences, and respect those differences; they are difficult if not impossible to articulate in terms of hierarchies of ethnicities, gender, or simply individuals. They are a much better bet for building a dynamic democracy than the rigid and closed practices of Herrnstein and Murray. Meanwhile, molecular geneticists will continue their own attempts to make intelligence become something else, something involving a complex interaction of genes and gene products. These attempts within the biological and neurological sciences will go on. But more multiple, subtle performances of "intelligence" should give molecular geneticists a more complex set of tools and concepts for replaying their own sciences.

A Science Named Desire

Research into the biology and genetics of homosexuality has sparked debate and controversy similar to that of research into the biology of intelligence. We look here at the work of Dean Hamer, a scientist at the National Institutes of Health who became an overnight sensation when he published what has become known as the "gay gene" study. Our emphasis again is on developing a set of questions and judgments about the practices most appropriate to the "ever-unstable equilibrium" of a multiplicity.

In his 1994 book *The Science of Desire: The Search for the Gay Gene and the Biology of Behavior* (coauthored by Peter Copeland), Hamer relates some of the background to the story. In it, we can see the way Hamer's own scientific desires might be mapped, and what their limits are. Hamer claims that he and his colleagues "didn't invent a new idea, we just showed it was true." What we want to show here is that it's better to reverse this statement: Hamer and colleagues didn't show that it was true that homosexuality is genetic, they invented new ideas, new articulations regarding homosexuality that tried to link up "genes" and "desire." Now we're being asked to judge what kind of invention it is: how well it works, and what it works for. That process of judging, as we've said, has to be a very complex and subtle one. The kind of politically correct claim of "it's bad to think of gay people that way" won't be good enough. Judging involves that whole slew of things we've called an assemblage: biology, biostatistics, definitions and their lacunae, social context, and so on.

Hamer begins with a discussion of the headlines resulting from the announcement of his "discovery," and his appearance on *Nightline*. The reactions were of course mixed: His "mailbox filled with letters from people thanking me for doing the study. Other letters promised I would burn in hell." The study appeared at a time and in a social and political milieu that was far from conducive to objective reception. "Our results were published during the midst of the great debate over gays in the military. In fact, the date set by President Clinton for the Pentagon to have a new policy on homosexuality was 15 July 1993, just one day before our paper appeared. When the study was published, my phone rang off the hook with questions about how the results might affect the new Pentagon policy." Suddenly his previously boring life in the laboratory was over: "No one outside of my immediate scientific circle ever had paid much attention to my work before, but now strangers were sending me letters, reporters wanted interviews, lawyers subpoenaed me to testify in court, and members of Congress wanted to know just what in the world was going on. All because I wondered what makes people gay."[21]

Well, in fact probably thousands of people wonder every day what makes people gay, but they usually don't make headlines or have their phones ring off the hook. Hamer was more than curious; he produced some new sign-forces, some new combinations of material and symbolic elements which carried the cultural authority of the sciences. At the very least, Hamer's opening story should make it clear to any scientist that the worst way to think about his or her work in such an area is in terms of some disinterested space, where one "wonders." There is an immediately present, highly charged context; isolation is impossible.

And indeed, it was in part the energy generated by these issues that elicited Hamer's own scientific desires. For ten years he had worked on the regulation of metallothionein (MT) gene transcription by heavy metal ions. "People often ask why I switched from a field as obscure as metallothionein research to one as controversial as homosexuality. The answer is the same that most scientists give for why they do what they do: a combination of curiosity, altruism, and ambition (especially curiosity, both personal and scientific), combined with one more factor— boredom. After twenty years of doing science, I had learned quite a bit about how genes work in individual cells, but I knew little about what makes people tick."[22]

That's quite a combination, and one with which any attempts to "control" this area of research would *have* to reckon. That kind of curiosity or desire is an important and ineradicable part of doing science (as is altruism and ambition). As Hamer himself recognized, this was in his terms "an emotionally and politically charged topic. I was pretty sure that if I had attended a scientific meeting about sexual orientation instead of gene regulation I would not have been bored." So we can't, and we shouldn't even try or want to, stop scientists from "wondering" about homosexuality and its biological causes. But what kind of assemblages can we build in which that "wondering" can proceed with the best chances of respecting complexity, contingency, finitude, and difference?

We have to ask many questions about the articulations made by Hamer and his colleagues in their scientific work, what powers those articulations, and what judgments about them might be necessary. Let's begin at the level of the biological articulations, not because these are necessarily more fundamental, but simply because it's a convenient place to start.

At the most general level, the paper by Hamer and colleagues that was peer reviewed and published in *Science* claimed that there was a region on the X chromosome in males (in the mapping terminology used in genomics, this region is called Xq28) that was linked to, or associated with, homosexuality.[23] It's important to note that despite the numerous headlines about the "gay gene" which soon appeared, no specific gene was identified in this study—let alone a protein produced by a gene, or a biochemical pathway by which that gene and gene product could be said to be linked to "desire." The group simply had a genetic marker that showed up in the Xq28 region of a "gay" sibling's genome more often than it showed up in the genome of a "nongay" sibling (or uncle or cousin) subjected to the same testing procedures. "More often" means more than can be accounted for by chance, and is defined by certain very complicated statistical algorithms and procedures, over which even experts in biostatistics can differ.

As of this writing, another lab had tried to replicate Hamer's results and been unsuccessful, and Hamer's original study is under NIH scrutiny after questions were raised about data being properly produced, reported, and analyzed. But for now, we give Hamer every benefit of the doubt—that there wasn't some experimental artifact that messed up his results, or that he didn't transgress "normal" (as defined both formally and informally) standards of judging the relevance of data and their necessary ambiguity.

Some geneticists and biostatisticians can still argue (and are arguing) that Hamer's sample size was much too small for the study to be anything but "suggestive." They know from experience that linking such complex traits to a region of a chromosome is an extraordinarily difficult task, involving a number of assumptions that can crumble under another examination. The history of chromosomal regions like Xq28 being linked to complex conditions in this statistically dependent manner is littered with claims advanced, hyped, and abandoned. The genetics of manic depression, or bipolar disorder, offers a good comparison: As Stanford University geneticists David Botstein and Neil Risch recently pointed out in the journal *Nature Genetics*, sixteen different research groups have published and publicized studies over the last twelve years which claimed linkage between manic depression and markers at fifteen different locations on eleven different chromosomes.[24] Of the many judgments that have to be made in such studies, one is particularly important: that one has a solid, stable, workable diagnostic category. In the case of the sixteen manic depression studies, this was a frequent issue plaguing the research since, given the often relatively small sample sizes and populations involved, the odds of linkage could change quite significantly when a person who was not diagnosed as manic depressive suddenly was diagnosed to be so, or vice versa. The *heterogeneity* masked by such seemingly clear and unified categories as "manic depressive" or "bipolar" is something that has to be carefully accounted for by geneticists, and it is difficult if not impossible, in some cases, to do well. Indeed, these kinds of genetic linkage studies require such monolithic categories, so that people can be put in an either/or fashion into one group or another. As currently practiced, there isn't any room for in-between, muddled, or middled categories in the sciences of molecular genetics. You're either manic depressive or you're not, as determined by qualified physicians or psychiatrists using their own standardized array of diagnostic tests, disciplinary thought-styles, professional training, and other practices.

All of which means that Hamer had to classify his test subjects as either homosexual or heterosexual. Whatever "desire" really is, it had to fall into one of these two categories for Hamer and colleagues. This prompted many people to raise questions not only about problems of self-identification (especially in the "emotionally and politically charged" context that Hamer himself recognized), but also about the overall relevance or adequacy of a study so obviously and necessarily built out of binary oppositions (heterosexual/homosexual, straight/gay) in an area in which there was clearly a wide spectrum of possible positions. (In addition, the original research in Hamer's laboratory looked only at gay men, although a member of his team is now trying to conduct similar research with lesbian subjects.)

The science of the "gay gene" is riddled with uncertainty, ambiguity, and assumptions which make its evaluation just on those terms quite complex. And if the context in which Hamer worked was indeed a charged one, as we've seen, it must be remembered that the affinities between the sciences and politics or culture in these charged assemblages can also be quite complex and contingent. Traditional categories of "progressive" or "conservative" got confused when gay rights groups and activist organizations disagreed on the social implications of grounding homosexuality in genes: Some said it would normalize homosexuality by naturalizing it, and thus making it more socially acceptable and irrelevant for public policy. Others argued that it would further stigmatize and objectify an identity of abnormality, and invite schemes for its cure.

Judging the social implications of a "gay gene" is, then, a difficult demand that should be undertaken with all our caveats in mind. There is a certain amount of contingency involved, but in our judgment, the charged assemblage that is American society is much more likely to pull a "gay gene" into closer alignment with the other kinds of forces, stigmatization and abnormalization. "Normalizing through naturalizing" is indeed possible, but it would take an awful lot of effort by a lot of different people, some of whom are clearly inclined to go another way.

We should at least wonder why "homosexual desire" becomes an object of "wondering" on a genetic level in late twentieth-century America, while "heterosexual desire" does not. Indeed, it's remarkable that nowhere in this book titled *The Science of Desire* does Hamer discuss either his own definition of desire, or the multiple efforts that have been made by other scholars in many other fields to articulate this concept, and its gaps. Can his science be criticized for relying too heavily on these binary categories? Only to the extent that *all* genetic studies have to rely on such hard-and-fast oppositions. Certainly the categories "homosexual" and "heterosexual" are too rigid and imprecise, but we also have to admit that they are, in our own terms, a good-enough kludge job that allows the work of genetics to proceed.

And some work in neurobiology as well. The neuroscientist Simon LeVay made headlines a few years before Hamer with his 1991 study, also published in *Science*, "A difference in hypothalamic structure between homosexual and heterosexual men." A scientist at the prestigious Salk Institute at the time, LeVay has now gone on to found the Institute of Gay and Lesbian Education in West Hollywood. His popular book, *The Sexual Brain*, describes numerous scientific studies which examine the hypothalamus, corpus callosum, the amygdala, and other structures of the brain, comparing differences in size and function between men and women, and between homosexuals and heterosexuals. You can read all about studies on the interstitial nuclei of the anterior hypothalamus, the roles of messenger chemicals like oxytocin as well as the more familiar testosterone and estrogen in sexual behavior, or why 5-alpha-reductase may be the key to understanding transsexualism. We're not going to delve into either LeVay's research, which can and has been extensively criticized, or into the other studies which he discusses. But we were taken by one of his concluding remarks: "Given the explosive rate at which the fields of molecular genetics and neurobiology are expanding, it is inevitable that the perception of our own nature, in the field of sex as in all attributes of our physical and mental lives, will be increasingly dominated by concepts derived from the biological sciences."[25]

As we are so fond of saying: yes and no. It *is* inevitable, given the "fantastic infrastructure" for inquiry into these areas of genetics and neurobiology—an infrastructure that becomes more fantastic daily and produces new things and articulations at an "explosive rate"—that new substances will be isolated; brain structures imaged, compared, and dissected in new ways; new genes marked; new articulations offered as explanations; and new social and political charges generated. Interesting and provocative articulations from biology on the notion of "desire" will continue to pile up, and link up with each other. But will they dominate our perceptions of desire? *Should* they?

Furthermore, if we are going to be deriving our concepts about ourselves from the biological sciences, we should be asking from where the biological sciences "derive" their concepts. This we judge to be the most serious hole in the efforts of Hamer and LeVay (although they are far from alone in making this assumption): that the sciences are the ground from which everything else will be derived, and are themselves underived, merely empirical. As LeVay puts it: "[S]cience can only describe what's out there and attempt to discover how it got that way." Science for Hamer and LeVay will offer the final, underived, certain truth on sexuality; this is both a serious scientific error and a serious social error as well. While each admits to some extent to the interaction between genes and environment, brain and mind, nature and nurture, they won't or can't admit to *even just the possibility* of the analogous interaction between science and culture.

They aren't able to ask the question that now begs to be asked: What kind of culture pursues desire and sexuality through the biological sciences? Why? What are the consequences of those kinds of practices? What if "desire" is something so fundamentally complex that to think in terms of one component or another "dominating" it is a serious mistake? Of course, that question applies across the board, and not just to genetics or the neurosciences—to think of desire as dominated by the sociocultural may also be a serious mistake. A better question may be: What kind of culture can pursue multiple approaches to desire, using each approach to both inform and critique the others, questioning what they derive from each other, respecting both uncertainty and the differences between perspectives, and between people?

The medical doctor and historian of science Vernon Rosario submits that such genetic theories of desire have multiple causes and effects: They add both "a new level of scientific realism to . . . homosexual bio-histories, as well as a good dose of hereditarian chauvinism. Converting gay identity into genetic currency (coining "our" own gene) boosts the trait's value in the genetic economy of the United States, where vast funds go to molecular biology research, and where genes are valuable commodities." We can't avoid such complex and risky gambles, Rosario suggests, because "the political urgency . . . is to recognize science as an instrument of our own creation—a tool for writing alternative, liberatory bio-histories of the future."[26]

Mice in the (Plexiglas) Gutter

As one last example of the ways in which the categories of the social and the natural are being reconfigured in today's genetics, we look at another multiplicity—

alcoholism. We were prompted by an exchange with a social scientist who argued that a genetic explanation for alcoholism was demeaning to individuals, since it robbed them of free will, and that such a "biologization" of a "social problem" was simplistic, reductionist science. He had made, in our view, a kind of snap judgment: He already knew that any genetically based explanation must be either a scientific error of simplification or the social error of an "ideology of scientific reductionism." He didn't have to do any more work if and when someone announced the discovery of a "gene for alcoholism"; he already knew the answer.

Our approach entails a different and more demanding set of tasks. By starting from the position that alcoholism is a multiplicity, we don't start with a "social problem" that is then, wrongly or rightly, "biologized." We inquire into the practices by which an unstable multiplicity is stabilized into the biological or the social, or some combination of these. It means we have to examine new experimental practices and discover what kinds of activities are going on in this field, what molecules and machines are being mobilized, and what's being linked together in the laboratories, and between laboratories and hospitals, and other spaces. In addition to these articulating activities, we have to look at rhetorical practices—exactly what's being articulated, or said, about alcoholism? In what terms? What other interpretations and meanings can be seen in those terms? What kinds of social and political charges does this work hold? How strong are those charges, or how contingent? Only after all that can we *start* to ask how to judge this research. How robust is it? How dense and finely fitted are the articulations? What are the effects of and implications for power, and can we do anything to avoid the nastier stuff? And what other ways are there to think about and act on alcoholism?

Snap judgments are no help here. This is going to take some time, with more than a few detours. And the first detour takes us not to alcoholics *per se*, but to mice. Once we lived in a culture in which we had to go through temperance societies to think about and reform our drinking practices; we now live in a culture in which those discussions and practices have to go through mice.

The paper we focus on is from just one of the many scientific journals devoted to human molecular genetics: "Confirmation of quantitative trait loci for ethanol sensitivity in long-sleep and short-sleep mice," which appeared in the February 1997 issue of *Genome Research*. The authors state that by using "stringent, genome-wide mapping criteria," they have "identified seven quantitative trait loci (QTLs)" that "together account for ~60 percent of the total genetic variance for [alcohol sensitivity]."[27] This has all the appearances of the classical reporting of a "matter of fact": They've used "stringent," controlled methods which they appear to describe in full, the paper has been reviewed by peers and published in one of the more prestigious journals in the field, and it concludes that sensitivity to alcohol—or at least a little more than half of it—*really* is genetic. (As in the example of Xq28 and homosexuality above, "loci" here refers to regions of a chromosome, and not to specific genes. A locus is just an area of a chromosome that is correlated with the trait in question, marked so that researchers may try to pursue any of the hundred of particular genes that might lie within it.)

What would we have to do if we wanted to understand this "matter of fact" according to the criteria described in Section I? What kinds of questions would we ask? What would we want people to be literate about so that they could make some kind of informed judgment on this question of alcoholism and biology?

More importantly, what would we suggest to someone wishing to study this question—whether officially enrolled, engaged as a citizen, or otherwise motivated to become a student of mouse alcohol sensitivity genes?

What kinds of questions would we encourage such students to ask? First: What's the experimental mechanism that produces these effects? They might want to look at how the two different strains of mice that were used for these experiments were bred—how, in other words, very special and very popular kinds of nonhuman cyborgs were created to generate scientific knowledge: long-sleep (LS) and short-sleep (SS) mice. Then they would want to ask about how the eighteenfold difference in ethanol sensitivity between these two strains was determined—which in turn would entail studying the experiments that described the "duration of loss of righting reflex (LORR) after a hypnotic dose of ethanol" was administered to the mice. The ethanol was not administered in a mouse-sized martini glass, but

> injected intraperitoneally between 9:00 a.m. and 1:00 p.m., 2–6 hr after the start of the light cycle. . . . Duration of LORR was determined by placing mice on their backs in a Plexiglas trough after injection. The start of LORR was considered as the minute when a mouse, after being placed on its back, was no longer able to right itself at least three times within 1 min, and the duration of LORR was the time until spontaneous righting occurred at least three times within 1 min. . . . Mice were excluded for leaky or subcutaneous injections. Mice were returned to cages after testing and were subsequently sacrificed for molecular analyses.

The training and perceptual skills required for these microprocesses of judging times and "spontaneous righting" would be a nice study in themselves. (That aspect of scientific work, as well as the formulaic use of the word "sacrificing" in the scientific literature, putting this operation in the symbolic order of ritual rather than that of mere killing, has been examined by the sociologist Michael Lynch.[28])

At the level of theoretical articulations, our student would have to ask about the connections between LORR and BEC, or blood ethanol concentration, which is supposed to "measure CNS [central nervous system] sensitivity to ethanol and . . . rule out pharmacokinetic (metabolic) aspects of alcohol action." If she then followed this work as it tracked down which genes at these loci might be responsible for these phenomena, and hence candidates for further research, she would have to find out about such things as "high-affinity neurotensin receptors," "∝-adrenergic ß 2 receptors" which, "via G proteins, activate adenylate cyclase (AC), which has been studied intensively as a biochemical marker for vulnerability to alcoholism," "prodynorphin," "inducible nitric oxide synthase," and a host of other substances and the genes that give rise to them.

This astounding specificity and apparent precision is accompanied, however, by a degree of looseness at another level of the alcohol-mouse-gene assemblage. Consider the opening paragraph of the paper:

> At least 10 million Americans are reported to have drinking problems. Alcoholism is a complex trait with a significant genetic component and heterogenous architecture in family studies, suggesting that different genes may contribute to the segregation of alcoholism in different families. Alcohol sensitivity is an important predictor of alcoholism. Sons of alcoholic fathers are at an elevated risk for alcoholism and are less sensitive to the effects of ethanol than are sons of nonalcoholic fathers. . . . It is possible that initial insensitivity to ethanol is a predisposing factor for alcoholism. Thus, a first step toward the understanding of alcoholism may involve the discovery of those genes that influence initial alcohol insensitivity.

There is some slippage between terms being articulated here: What happens in the translations between "alcoholism," "alcohol sensitivity," and the most encompassing term of all that opens the paper, "drinking problems"? True, the authors state that they're only considering "possible" links between these categories, and as we pointed out above, molecular geneticists need a well-defined, stable diagnostic category like "alcohol sensitivity" with which to work. They need something that will exhibit differences that show up on their meters. But what if there is not only a gap between loss of righting reflex in mice and the ten million Americans with drinking problems, but an abyss? How many, and what kind of, articulations would be involved in trying to span that abyss?

Despite its lack of creativity, the standard critique of reductionism is still a viable and even necessary pursuit. The drinking problems of ten million Americans cannot be approached solely by observing how long mice flail around on their backs in a Plexiglas trough. Exactly what problems are we talking about—spousal abuse? cirrhosis? chronic unemployment? showing up for work late? How does the economic class of these Americans figure into the problem? Sociogeographical location—urban? suburban? small Midwest town? rural? And how does the way a person lights up the census form—African American, Native American, Asian American, Hispanic, Non-Hispanic Caucasian, etc.—get thrown into our equations?

The language of the paragraph quoted above carries a subtle assumption— "Thus, a first step toward the understanding of alcoholism . . ."—that causes some problems. There is no "first step" toward understanding alcoholism. We've already walked many, many miles down that road; millions of people, including thousands of experts from various professions, already know what alcoholism is; we have deeply ingrained cultural habits of understanding. Moreover, as multiplicities, alcoholism, mice, and humans *really* don't have "first steps," whether that point of origin is located in the gene, in the society, or wherever. Biology is not the ground on which the social gets overlaid: We and our laboratory kin are not organisms "first" and social, cultural, or symbolic "second." Just as the sciences are all these things at once, so are we.

And that includes political. We started to look at *articulating* above, and how seven different areas of some mouse chromosomes were linked to possible biochemical mechanisms, as well as varying mouse behavior, and the problems of 10 million Americans. The articulating of these experimental-theoretical nodes always occurs within other sets of articulations, including institutional ones. The paper's authors crossed spheres, with affiliations from both the academic world— the Institute for Behavioral Genetics and the Department of Psychology at the University of Colorado—and with the state more directly, in this case the Department of Veterans Affairs Alcohol Research Center in Denver. The lead author went on to a job at Millennium Pharmaceuticals. Our students would have to ask about these and other social institutions, and what specific practices each deems to be most effective in dealing with alcoholism.

We can at least say that the question of whether isolating "quantitative trait loci" is pure or applied science is no longer a very helpful or productive one. "Sacrificing" mice and creating correlations between their behavior and the statistical frequency of markers on their chromosomes occur within a dense tangle of articulations, a highly charged field, involving institutional resources, the needs of state institutions, drug and diagnostic markets, and so on. That doesn't mean that the study is necessarily a kind of "cigarette science." But it does mean that we have to be prepared to ask more kinds of questions, and ask them earlier in the whole process. How would the Department of Veterans Affairs use a diagnostic test for alcohol insensitivity, or perhaps a drug counteracting the effects of alcohol? If Millennium develops a screening kit that identifies some or all of the genes said to account for 60% of the difference in reactions to alcohol, who should be enrolled in a screening program—the sons of alcoholic fathers, who are at "elevated risk"? At what age? With what social and personal consequences?

We know that alcohol sensitivity *can* be "practiced" via the tools and concepts of biology and an explication of the biochemistry of ethanol and high-affinity neurotensin and ∝–adrenergic ß 2 receptors. If it couldn't, there would be little point in pouring that glass of wine or bourbon tonight, because it wouldn't have the expected, desired effect. Everyone also knows that the effect can be different in different people. We'd be fools to think there wasn't something interesting here, something to spark the passionate inquiry of any number of life scientists. We are—exquisitely, thankfully, disturbingly—responsive to molecular effects.

But researchers working in the various institutions where molecular genetics is currently practiced know, in some way, that the question is no longer one of determining the molecular basis of alcohol sensitivity once and for all, and for everybody. They know that organisms are too complex at the biological or genetic level, and they even know that alcoholism is frightfully complicated at the social level as well, and that no easy reduction is possible. Remember François Jacob's pronouncement: *The biologists have lost their illusions; they are not seeking truth, they are building it.* These researchers are not pursuing a final understanding or representation of alcohol sensitivity so much as they are building new tools and new practices by which we will reconstruct what alcoholism *is*. It doesn't matter to Millennium Pharmaceuticals that they can't fully explain everything about alcohol uptake in every population. They can, however, work to develop new substances, new diagnostic categories and techniques, and new institutional practices that will have specific but limited effects (and, undoubtedly, "side effects") on certain groups of people. And that's enough "truth" to make a market, and to make a difference in people's lives.

The questions for our student to deliberate and judge, then: What combination of those differences are we likely to get? Which social and laboratory practices are particularly robust and dense when it comes to alcoholism? What effects does a particular construction have on our thinking, our institutions, our attitudes toward others? What are the trajectories it *might* set in motion for imagination, future research, future risk, future social order? How can different constructions be best combined?

Of Sheep and Earthquakes

In the midst of this book's creation, the story of Dolly, the cloned sheep, was all over television and the popular press. Shortly thereafter, we held a public seminar to discuss the event in terms of many of the issues and questions raised above. By that time, people had read so many articles and opinion pieces, and watched so many television pundits pronounce on the meaning of the event, that we knew we had to take a different approach if we didn't want to get caught in the predictable positions. We turned to Philip K. Dick's *Do Androids Dream of Electric Sheep?*, the novel on which the movie *Blade Runner* was based, and asked what a science fiction novel could teach us about contemporary genetics.

The novel, more so than the movie, is about the anxieties we have when we can't decide the difference between a living organism and a machine—an irresolvable, ineradicable anxiety. Usually, anxiety about unstable difference begets violence, which is why the central figure of the story is a cop, Rick Deckard, hired to hunt down androids since the social order considers them machines and not human, not alive. In this future most of the "real" animals have died from radiation poisoning, so the less fortunate citizens own electric pets, a fact they try to hide from the more well-off citizens who can still afford "real" ones. Rick the cop happens to have a pet sheep. In the opening pages of the novel he puts on his Ajax model Mountibank Lead Codpiece, and before he leaves for work goes up to

> the covered roof pasture whereon his electric sheep "grazed." Whereon it, sophisticated piece of hardware that it was, chomped away in simulated contentment, bamboozling the other tenants of the building. Of course, some of their animals undoubtedly consisted of electronic circuitry fakes, too; he had of course never nosed into the matter, any more than they, his neighbors, had pried into the real workings of his sheep. Nothing could be more impolite. To say, "Is your sheep genuine" would be a worse breach of manners than to inquire whether a citizen's teeth, hair, or internal organs would test out authentic.

Rick runs into Barbour on the roof, a neighbor who has a real horse, and Rick asks if he ever thought of selling it. Barbour is shocked, and can't understand why Rick isn't happy with his sheep, which Barbour thinks is real:

> Going over to his sheep, Rick bent down, searching in the thick white wool—the fleece at least was genuine—until he found what he was looking for: the concealed control panel of the mechanism. As the neighbor watched he snapped open the panel covering, revealing it. "See?" he said to Barbour. "You understand now why I want your colt so badly?"
> After an interval, Barbour said "You poor guy. Has it always been this way?"

In fact, it hadn't. Rick had once owned a real sheep, named Groucho, but he died, and so he bought the only thing he could afford at the time, an electric one:

"It's a premium job. And I've put as much time and attention into caring for it as I did when it was real. But—" He shrugged.

"It's not the same," Barbour finished.

"But almost. You feel the same doing it; you have to keep your eye on it exactly as you did when it was really live. Because they break down and then everyone in the building knows. I've had it at the repair shop six times, mostly little malfunctions, but if anyone saw them—for instance, one time the voice tape broke or anyhow got fouled and it wouldn't stop baaing—they'd recognize it as a *mechanical* breakdown ." He added, "The repair outfit's truck is of course marked 'animal hospital something.' And the driver dresses like a vet, completely in white." He glanced suddenly at his watch, remembering the time. "I have to get to work," he said to Barbour. "I'll see you this evening."

As he started toward his car Barbour called after him hurriedly, "Um, I won't say anything to anybody here in the building."

Pausing, Rick started to say thanks. But then something of the despair that [his wife] had been talking about [that morning] tapped him on the shoulder and he said, "I don't know; maybe it doesn't make any difference."

This may be the best place to begin any inquiry into any multiplicity—intelligence, desire, cloned sheep, hemochromatosis, or any of the other objects that are temporarily stabilized in contemporary molecular genetics: "I don't know; maybe it doesn't make any difference." Or maybe it does. Asking whether something is really genetic or really social is akin to Rick asking himself about the difference between real sheep and electric sheep: Does it make a difference or not? What specific kinds of differences does it make? To whom? What cultural conventions and social institutions emerge to manage those differences or make meaning out of them? (Repairmen reassuringly posing as veterinarians, social status accruing to the "authentic" and "natural.") How are differences created and enforced? How will those differences in turn break down, and become something else?

New developments in genetics, and in dozens of closely linked scientific and technological fields, continue to gallop toward us. Old certainties are challenged and often scrapped, new uncertainties arise with every opening of the morning newspaper. The response cannot be to cry "whoa!" or even just "slow down!" There are some instances in which slowing or even stopping may be possible, and for which the regulatory, legal, and policy structures exist to make such a response possible. Research using human fetal tissue, for example, was a field of biomedical research in which lack of funding as well as serious policy restrictions during most of the 1980s kept development at a snail's pace. But for the most part, the multiple fields of biomedical research (of which human molecular genetics is but one part) are expanding through all areas of our world. This chaotic and rapid growth is powered by decades of scientific momentum, fed by a large and steadily growing number of practitioners who encounter exquisitely challenging questions which can't help but spark their imaginations, their desire to know. Our own frail bodies and endless search for optimal health and well-being fuel the expansion as well. Much as we might want to get out in front of this Banzai Pipeline of the biomedical sciences and "control" it—via some yet-to-be-imagined democratic social and political mechanisms—we will in fact always be in its wake, after the fact, struggling to keep up with its changes.

Certainly we will need more people thinking about how to "apply" this or that new scientific discovery or technological innovation. We will need new intellectual property laws which recognize the social and collective foundation of scientific invention rather than the "individual creative genius," and which address growing economic inequalities. (Read James Boyle's thoughtful *Shamans, Software, and Spleens*.)[29] We will need new state and federal legislation regarding "genetic privacy," new genetic screening protocols and guidelines for an array of bewilderingly different disorders, new regulations for the insurance industry, new educational initiatives for doctors and other health care workers whose expertise in questions of genetics leaves much to be desired. On the social and political level, there is a lot of work to be done.

But we're going to need something more: tools for thought, for asking questions about the biological sciences as the kinds of muddling operations we described in Section I. We need more of the public, and particularly more scientists, equipped to ask how genetic explanations and even in some sense genes themselves might be thought about not as matters of fact and discovery, but always as some dense and intricate combination of experimenting, articulating, powering, and judging. All of us need to be able to ask these kinds of questions and to see them incorporated into the way scientists—and perhaps especially now biologists—think about their fields and go about doing their sciences.

Think, for a moment, of an earthquake. It can be called a "natural disaster," but that doesn't respect its reality. An earthquake in Los Angeles is an entirely different event from one in Armenia, precisely because the "natural" cannot help but be muddled with such political, social, and cultural factors as building codes, construction practices riddled with bribery and corruption, the capacities of the state to provide relief, transportation and public health infrastructure, and so on. The challenge for sciences in the twenty-first century is to think of genes—and indeed, atoms, health, the environment, and every other "natural" or "social" object or concept mentioned in this book—in the same way: as events at a complex crossroads of articulated elements, rather than as solid, stable things that operate as primary causes.

We need more people literate in questioning how natural/social selves are built, and how a society committed to democracy and equality will employ those fragile human constructs of inquiry to address questions of social order. If we are not prepared to evacuate Los Angeles *now* because of the inevitability of earthquakes, but instead commit ourselves to strengthening material and social infrastructures, a democratic society is equally obliged to invent analogous strategies for the quakes that rattle human bodies—strategies of support, assistance, and *further becoming* rather than stigmatization, exclusion, or stasis.

Weird Interactions and Entangled Events
Quantum Teleportation

Oh Boy . . .

The recent television series "Quantum Leap" featured a physicist from the future, Sam, who had invented/would invent (the appropriate tenses are unstable) a computer-cum-teleporter named Ziggy that would function, he hoped/would hope, as a time machine. But of course things don't quite work out that way. Instead of being able to travel freely between future and past, Sam ends up trapped, always in some past that occurred within his own life span, the one narrative constraint. Each week he would be serially wrenched into various time zones and geographical coordinates by Ziggy. Taking on personae that reflected the entire post–World War II spectrum of ethnicities, social positions, and genders, Sam had to set right some personal or social wrong in each episode; he was, in effect, responsible for rewriting history.

Sam was aided in this modest task by another telepresence from the future, the cigar-smoking Al, appearing holographically to Sam and relaying various messages from the future Ziggy: personal and historical information relevant to the situation, and the ever-changing probabilities of various outcomes based on Sam's interventions. Those messages were always subject to persistent disruption, statistical uncertainty, and technical breakdown (Al was always banging and coaxing the handheld module that conveyed Ziggy's info), but somehow Sam muddled through, acting responsibly and getting things right, hoping for that final leap that would return him "home," his stable future. But it was always at that moment of success, when past and future were once again configured in a more satisfying pattern, that special effects would light up Sam's body, and the indeterminate disruptions of the quantum world would send him, not home, but to another space and time in the past, another scenario in which to do it all over again next week. This final minute of the show, previewing next week's dilemma, always ended with Sam disorientedly looking around and uttering the simple phrase, "Oh boy. . . ."

It's uncanny how well the fictional portrayal of disorientation, indeterminacy, and sense of responsibility maps onto the nonfictional world of the quantum physicist who coauthored this book. And it's exactly this kind of uncanny correlation that is under scrutiny in this chapter, which deals with current and past episodes in theoretical and experimental quantum physics.

Readers of books like John Gribbin's *In Search of Schrödinger's Cat*, or of the frequent articles that appear in *The New York Times* science section or in the weekly news magazines, will be familiar with the characterization of the quantum world that we are about to present: a world of so-called "spooky action at a distance," in which distant particles and events seem to act telepathically on each other. These books and articles are good guides to the peculiar interactions that take place between theory and observation, between experimental instruments and subatomic particles, and maybe even between consciousness and reality. But these texts limit the strangeness and spooky correlations to a restricted domain of science conventionally defined, and we end up with a "gee whiz" depiction of quantum physics, a window into the strange and awesome theories of the quantum world, which may also be a window into a strange and awesome future.

In a sense, all scientists are emissaries to a strange and creative place in the world that might be called the future of everyday life. You could also just call that place "theory," the word rooted in the Greek *theorein*, to view, usually with wonder. But there's more meaning buried in this ancient etymology—*theorein* was a special form of viewing. *Theoros* was the viewing of ritual games and also the state ambassador sent to see the games or to consult an oracle. As Greek society sent forth some of its members to the games that let them view the gods or glimpse things yet to come, our society sends scientists into the future, or into those imaginative constructions of it that their collective effort produces. And then, if the machine is working properly, they come back and make that world happen.

But as we know, the machine is always a kludge job, and works both properly and improperly. This is the story of how quantum teleportation has gone from future to present, from theory to practice, and from science fiction to (almost) everyday life. These journeys are never complete, of course, and like Sam, quantum teleportation is trapped somewhere between all these opposed terms. Being caught between future and present or theory and practice is improper enough, but the case of quantum teleportation is more improper still. To understand it we have to expand the "gee-whiz" view and explore the weird and indeterminate interactions between the quantum world and the thought-styles of physicists, the social worlds of funding agencies, national security interests, and perhaps even culture writ large. We'll see in the process that how we think about the responsibility of the physicist also becomes less than straightforward, entangled with probabilities and uncontrollable forces. Like Sam's, our information from the future—which should serve as a guide to responsible action here in the present—gets scrambled and disrupted. Lacking such reliable intelligence, we're forced to experiment with other mechanisms that come with no guarantee of success.

We begin, again, with the past.

Reconstructing a History

Our orienting question in this review of past and present episodes in theoretical and experimental quantum physics is this: Is the microworld really fundamental and solid, or is it really created by observation itself?

In this century quantum physics has been the field of "pure science" in which the dictates of logic, reason, causality, and reality itself have broken down most spectacularly and disconcertingly. The central feature of an indivisible quantum of action threw two new elements into the muddle. First, not every conceivable outcome of experimenting is possible. Those that would split a quantum never occur; ordinary particle spins along any given axis, for example, can have only two discrete values. Second, the completely random behavior of a single quantum determines which outcome, from among all possible results, actually occurs. Chance is inseparable from indivisibility, dashing the classical physicist's hope of perfect predictability. In general, quantum theory predicts only the probability of each outcome, not which one will occur in a given run of an experiment. Together with relativity theory, quantum mechanics upset many of the most fundamental preconceptions of nineteenth-century sciences.

The weird relationships between observer and observed; the muddling of waves and particles, matter and energy; the apparent indefiniteness and unvisualizable, counterintuitive results of physical measurements and theory—all have been at the heart of inquiries conducted by physicists for most of the twentieth century. These inquiries are at once a matter of *experimenting*, the tuning and inventing of very exacting devices; of *articulating* the meaning of the results of those actions via physical theory and metatheoretical (or philosophical) concepts; of *powering/knowing* those processes by conjoining them with funding sources and other less obvious energy sources, such as unstated assumptions and cultural values; and of *judging* the subtle differences, contradictions, and demands that arise in all these areas—laboratory, mind, and society.

Quantum mechanics is the field of physics in which questions of *interpretation*, that most literary of activities, have long held center stage. The most inexplicable experimental results, the most exquisitely imaginary theories are more often than not left dangling over the edge of the ontological cliff, while the mountaineers confine themselves to tweaking the ropes supporting the unseen object, trying to interpret carefully the subtle motions of the cords before they fray and snap altogether.[1] The key figures and events may be familiar to those who have read some of the authors mentioned previously: the fervid discussions between Albert Einstein and Niels Bohr that had begun by 1927 over the adequacy of the Copenhagen interpretation for quantum physics; Erwin Schrödinger's thought-experiment involving a kind of zombie cat, both dead and alive (or neither dead nor alive) in some quantum half-world; and the subsequent theories of physicists such as David Bohm and J.S. Bell which suggested that "nonlocality" (among other things)—by which some kind of correlations operated instantaneously, faster than light—might be a fundamental feature of our universe.

Much of that history revolves around theoretical debates, thought-experiments, and questions of interpretation, cobbled about a limited range of experimental possibilities. In the early 1980s, however, experiments in quantum mechanics were no longer so restricted to the realm of thought. Through the work of physicists like Alain Aspect, it began to appear that experiments showed the world, or at least the microworld of physics, *really* to be indeterminate, spooky, random, and nonlocal. And by the 1990s, more and more new experimental arrangements were confirming this view, yielding not only more challenging fodder for interpretation, but also new devices which could actually *operate off of* rather than simply interpret the quantum world.

Historical accounts begin most often with a 1935 paper coauthored by Albert Einstein, Boris Podolsky, and Nathan Rosen (referred to in shorthand as "the EPR paper")—and for good reason. While most scientific papers have a very brief half-life, the possibilities and problems articulated in this paper remained relevant for both theorists and experimentalists for over sixty years; physicists, philosophers, and combinations of the two returned to it over and over again to reconfirm, redirect, or simply sharpen their questioning. When first published the paper served as a focal point for conflicting articulations of what would count as adequate physical theories. It became the centerpiece in a lengthy, spirited, and productive disagreement between the two giants of physics at the time, Einstein and Niels Bohr. To understand what was at stake, consider the first two sentences from the EPR paper. They provide a clear statement of Einstein's working philosophy of science, and were a clear challenge thrown down to Bohr and his colleagues:

> Any serious consideration of a physical theory must take into account the distinction between the objective reality, which is independent of any theory, and the physical concepts with which the theory operates. These concepts are intended to correspond with the objective reality, and by means of these concepts we picture this reality to ourselves.[2]

You don't just publish a series of equations in the *Physical Review*, the prestigious journal in which this paper appeared. Neither, of course, do you publish without them. But the rhetoric with which the equations *must* be articulated does a great deal of work, both for the coherence of the physical theory itself (as if such a thing could be easily separated out from everything else going on) and in its social effect. Here Einstein, Podolsky, and Rosen have articulated the classical conception of scientific theories, which still reigns so powerfully today: Reality is objective and independent, *distinct* from the theoretical representations which "correspond" to it. There's reality, and there's the way we picture it to ourselves, it and us; what God has put asunder, Einstein might have said in a reversal of the traditional wedding vows, let no one join together. ("No muddling!") And there's a subtle jab at the beginning: such theories are the only kind to be taken as "serious." Surely you're joking, Dr. Bohr! is the subtext.

Such sober theories have to fulfill two criteria: they have to be correct, and they have to be complete. "Whatever the meaning assigned to the term complete," Einstein, Podolsky, and Rosen continue, "the following requirement for a complete theory seems to be a necessary one: *every element of the physical reality must have a counterpart in the physical theory.*" The enterprise is from the start marked by a profound contradiction: theories must be complete, yet Einstein possesses only an incomplete definition of "complete;" its meaning is still in the realm of "whatever." Einstein and his colleagues thus suggest what "seem" to be necessary criteria for any future definition. Later in the paper another deficiency shows up: They state that "a comprehensive definition of reality is . . . unnecessary for our purposes." But again they suggest a criterion to begin filling this void, a criterion based jointly on an unarticulated notion of the "reasonable," and the "satisfaction" of an inarticulate desire:

> We shall be satisfied with the following criterion, which we regard as reasonable. *If, without in any way disturbing a system, we can predict with certainty (i.e., with probability equal to unity) the value of a physical quantity, then there exists an element of physical reality corresponding to this physical quantity.*[3]

As we'll see, other physicists could be satisfied with a different conception of "reasonable." Moreover, a recent work under the intriguing name "GHZ theorem" shows that what seemed a reasonable definition for two particle states actually fails completely for simple states of three particles.[4] For now, let's follow the demands of *this* definition of reasonable, requiring a one-to-one correspondence between a physical property and a representative value given by theory. Einstein, Podolsky, and Rosen elaborate a thought-experiment involving two particles, originally correlated but separated by a sufficiently large distance, in which the physical values or properties designated as P and Q are measured. Saying that this pair of EPR particles (as we now call them) is correlated means that if one of the pair has the quality "P," then the other *must* have the quality "not-P," or if one is "Q," then the other *has to be* "not-Q." You can be completely assured of that—the physicist *knows* it—without having to do the other measurement, without in any way looking at or disturbing the second particle. That at least is the situation in Einstein and company's serious, reasonable, one-to-one universe in which things and theories are distinct.

But according to the (nonserious, muddy, and unreasonable) logic of quantum mechanics, EPR argued, that process of measuring P or Q results in a paradox: "On this point of view, since either one or the other, but not both simultaneously, of the quantities P and Q can be predicted, they are not simultaneously real. This makes the reality of P and Q depend upon the process of measurement carried out on the first system, which does not disturb the second system in any way. No reasonable definition of reality could be expected to permit this."[5]

There's that vague but powerful word again, "reasonable." Unfortunately for the trio, their definition of "reasonable" turned out to be incomplete, insufficient for a changed world. Eventually it has even proven incorrect. E, P, and R attempted to take quantum mechanics as true, to concede its correctness (consistency and agreement with experiment), and to show that it is incomplete by "reasonable" standards. Not only by reasonable standards, but using only cases of definite causal connections, cases where the probabilities are all zero or one. That avoids any loophole created from the randomness that indivisible quanta sometimes entail.

Since probabilities of one or zero imply definite results, the predictions will be correct under all possible theories—that is, under deterministic classical theory (whenever it matches the quantum results), quantum theory itself, and any new, improved, more complete theory that may finally emerge. EPR couldn't have known that only a slight increase in complexity, from two particles to three, provides cases in which Einsteinian elements of local reality contradict one another. What is "real" about a particle (because it can be predicted with certainty) definitely *does* depend on the process of measurement carried out on a remote system. But more about that later . . .

Before going more fully into Bohr's response, we would like to point out some of this man's most remarkable and admirable characteristics. First is his extraordinary attention to language, characterized by physicist David Bohm as "a highly implicit and carefully balanced mode of saying things, which makes reading his work rather arduous but which is in harmony with the very subtle content of quantum theory."[6] Those qualities of balance and subtlety characterized not only his thought and use of language, but ran through his working style as well. One of Bohr's chief amanuenses, Léon Rosenfeld, recalled how the publication of the EPR paper

came down upon us as a bolt from the blue. Its effect on Bohr was remarkable . . . everything else was abandoned. . . . In great excitement,

Bohr immediately started dictating to me the outline of . . . a reply. Very soon, however, he became hesitant: "No, this won't do, we must try all over again . . . we must make it quite clear. . . ." So it went on for a while, with growing wonder at the unexpected subtlety of the argument. . . . The next morning he at once took up the dictation again, and I was struck by a change in the tone of the sentences: there was no trace in them of the previous day's sharp expressions of dissent. As I pointed out to him that he seemed to take a milder view of the case, he smiled: "That's a sign," he said, "that we are beginning to understand the problem."[7]

Bohr soon published his own understanding of the problem in the next volume of the *Physical Review*, bearing the same title as the EPR paper: "Can quantum-mechanical description of physical reality be considered complete?" His answer pointed to the root cause—ideas that seem reasonable at the scale of everyday classical phenomena may need changing *at their roots* to fit microscopic, quantum-dominated experiments:

The apparent contradiction in fact discloses only an essential inadequacy of the customary viewpoint of natural philosophy for a rational account of physical phenomena of the type with which we are concerned in quantum mechanics. Indeed the *finite interaction between object and measuring agencies* conditioned by the very existence of the quantum of action entails . . . the necessity of a final renunciation of the classical ideal of causality and a radical revision of our attitude toward the problem of physical reality.[8]

(Despite the sweeping statement about "the problem of physical reality," Bohr also tried to delimit precisely where "customary viewpoints" broke down: with "phenomena of the type we are concerned with at the quantum level." So while the Copenhagen interpretation of quantum mechanics and the famous Heisenberg indeterminacy principle are often used to point out the fundamental irrationality, indefiniteness, or constructed nature of the world writ large, these principles and theories should technically be restricted to the quantum level—a caveat not always heeded by otherwise science-literate commentators, especially some from science studies.)

Bohr and his associates in Copenhagen provided, in short, new definitions of "reasonable." New definitions that were *better* not so much because they represented an independent reality better—Bohr didn't find such a metaphysical statement either coherent or useful—but better because they were *more productive*. Physicists could do more with less, with a restriction on how far they would push ideas derived from everyday large-scale life in exploring the microworld that can only be reached through extraordinary effort and specialized equipment. The subsequent history has been remarkably kind to Bohr's approach, and rather ironically *un*-Einsteinian in several interesting ways.

Although he could never produce a final, knockdown, watertight, *scientific* argument against the Copenhagen interpretation, Einstein remained in that hard-to-pin-down but very palpable and forceful state best summed up with that concept from the EPR paper: "unsatisfied."[9]

The same might also be said of Erwin Schrödinger, who in the same year of 1935 proposed another thought experiment which would become known as "Schrödinger's Cat." In this thought experiment as well, the crucial component was the conjunction of the microworld of quantum physics, where information according to quantum mechanics was always incomplete, statistical, and shifty, with the macroworld of cats, humans, and other bodies, where information is visibly and grossly final. Like Einstein, Schrödinger was committed to a physics that aimed for a perfect and complete representation of the real world. In this thought-experiment, the physicist places a purring, pacing cat into a box which also contains a radioactive atom, a hammer, and a cyanide capsule. The box is closed, so that the physicist has no further information about what goes on inside. If the radioactive atom decays, it triggers a chain of events: the hammer falls, the cyanide capsule breaks, the cat dies; if the radioactive atom doesn't decay, the cat continues to purr and pace. When a time corresponding to the half-life of the radioactive atom has elapsed (a time which can be established in a very solidly factual manner, by the way), the radioactive atom has a fifty-fifty chance of having decayed or of remaining intact. According to quantum mechanics, however, the physicist, without opening the box, can only speak about the state of the wave function that corresponds to the radioactive atom: in theory, the atom is not *either* decayed or intact, but *neither* decayed *nor* intact, as we prefer to put it. Or one could say that it was *both* decayed *and* intact, language getting kind of tricky at this level.

All well and good (if rather bizarre) for radioactive atoms, maybe—or for their wave functions—but not very acceptable in the feline world. The problem that Schrödinger was pointing to arises when the quantum world is explicitly *articulated* with the macroworld of everyday reality. If quantum logic carried the day, the cat would end up being a kind of zombie: neither dead nor alive, but both dead and alive, suspended in a twilight zone. The true fate-state of the cat is decided only when the box is opened, whereupon (this is just one way of putting it) the wave function of the quantum world collapses, probabilities instantaneously become certainty, and the real situation is revealed—or is it created?

Today most physicists would agree that a cat is so large and complicated that its interactions within the box would be enough to decide its fate-state. It could never stay isolated for long, and its quantum state would quickly change to "dead" or "alive" in concert with the state of the environment. But the fact that it's possible to say that the physicist "creates" quantum reality in this act of observation makes for some interesting and haunting effects. As we'll see in more detail a bit later, it can be said (and has been said, by some very eminent physicists) that what the physicist does with this apparatus confers responsibility for whether the cat, in the final analysis, lives or dies.[10] But is that just a way of speaking? Or does the fact that we have arrived at a point in history where we can and must speak about the quantum world entail that we have to rethink what it means to say "just a way of speaking?" Are our ways of speaking, of articulating words and concepts into that thing we call "thought" or "theory," now inextricably tied up with the thing about which we were speaking? This problem of finding a language adequate to these kinds of scenarios in which reality itself (at least at the quantum level) is—is what? Created via consciousness and the brain? Some physicists have spoken in this way. Interacted with creatively, in a participatory manner? That's been said, too. That reality itself (whatever that means) bifurcates and multiplies, perhaps infinitely into many worlds? Other physicists have spoken in that manner. Those questions, and the way a language of responsibility continually arises within them, recur again and again in the remainder of this chapter.

How a Proof Proves: Bloodhounds and Sighthounds

You'll notice that we've left unspoken the outcomes of these debates about particles, cats, probabilities, and realities. That's because, in a certain sense, there were no outcomes. The questions persisted, yet in such a way that they did not impede but rather impelled creative work in physical theory and experimentation—a situation which continues even today, over sixty years later. But even though these questions were in principle open, physicists did not exactly choose openly and freely among approaches; certain theoretical assumptions and even more ambiguous (but no less real) cultural forces exerted powerful effects. We continue here with our incomplete history of these inquiries, hitting only on some interesting highlights.

David Bohm's textbook, *Quantum Theory*, came out in 1951 and soon became a standard text in university courses.[11] It expounded the Copenhagen interpretation. Bohm had sent copies of his book to both Einstein and Bohr, among other physicists; no response came from Bohr, but Einstein asked to meet Bohm. "This encounter with Einstein had a strong effect on the direction of my research," Bohm later wrote, pointing to the nonlocal forces that correlated these two physicists, "because I then became seriously interested in whether a deterministic extension of the quantum theory could be found."[12] Bohm, too, wanted the "real world" and causality back in physics, in some form or another. He proposed a new kind of force, derived from the quantum potential which, although "hidden" to current practices and theorizations, had "an independent actuality that existed on its own, rather than merely being a function from which the statistical properties of phenomena could be derived."[13] In place of the "fragmentariness" of standard quantum mechanics, Bohm tried to provide the "wholeness" of an "implicate order" which was quite real. One small price of this wholeness, which Einstein thought too steep for his restrained pocketbook, was the introduction of "nonlocality" into quantum physics: "[O]ne could at least in principle have a strong and direct (nonlocal) connection between particles that are quite distant from each other." These kinds of connections are indeed so strong and direct that the word "connection" is a misnomer. The concept of nonlocality completely disrupts our conventional notions of particles in different spaces with connections between them; in the quantum world, properties of correlated particles are not in any one place at all.

As it has turned out, we are living in Bohm's future, a future described below, in which nonlocality has exquisite experimental proof in the form of real-world uses such as quantum cryptography, quantum computing, and, in the most science-fictionish scenario that recalls our allegory from TV, quantum teleportation.

These proposals concerning nonlocality and hidden variables, as Bohm admitted later in a retrospective article, "did not actually 'catch on' among physicists. The reasons are quite complex and difficult to assess."[14] Which makes them that much more interesting and important to take up.

The short answer to why Bohm's theories didn't catch on is that they had been proven, decades previously, to be impossible by the mathematician John von Neumann. But this lesser-known episode in the history of quantum mechanics proves very illustrative of the complex, muddled ways in which the sciences progress— which means, as this episode makes clear, along some (complexly determined) paths and not others.

In his 1932 book on the mathematical foundations of quantum physics, von Neumann proved that "hidden-variable theories" were impossible, and moreover, the *only* possible theory was a statistical view such as that of the Copenhagen interpretation. Or as he put it in his book: "It is therefore not, as is often assumed, a question of reinterpretation of quantum mechanics—the present system of quantum mechanics would have to be objectively false, in order that another description of the elementary processes than the statistical one be possible."

How does one prove such a thing? The proof was a combination of a great deal of formidable mathematics, and five axiomatic statements on the requirements and consequences of physical theory, particularly a hidden-variable theory. The most important of these was the last, axiom **E**, which specified the mathematical conditions under which measurements on "dispersion-free ensembles" could be combined. (Dispersion-free ensembles are states of a system which *don't* have any uncertainty in any variable. They are states with exact values of every possible measurement; states for which the hidden variables are complete, and specify what result will emerge.) Since von Neumann showed that one could not derive any dispersion-free ensembles that met these conditions and still reproduced the statistical results given by the Copenhagen interpretation of quantum theory, he claimed he had proven the impossibility of hidden-variable theories.

Everyone regarded the mathematics as impeccable. And for over thirty years, everyone regarded the overall proof as having done its job of proving.

Well, almost everyone. Jump ahead to the end of the story: In 1964 J.S. Bell— perhaps the best foundational quantum theorist of the 1960s and 1970s—wrote a paper (published in 1966) that showed how the proof was not only wrong, but that it was . . . well, let's let Bell speak for himself. As he rather sharply pointed out in an interview twenty years later with *Omni* magazine: "The von Neumann proof, if you actually come to grips with it, falls apart in your hands! There is *nothing* to it. It's not just flawed, it's *silly!* . . . When you translate [his assumptions] into terms of physical description, they're nonsense. You may quote me on that: The proof of von Neumann is not merely false but *foolish!*"[15] It was based on assumptions that should *never* have been made about the "hidden" states.

So how did error, silliness, and foolishness manage to operate quite forcefully for thirty years in quantum physics? How did nonsense work just like sense? How was it that *nothing* behaved just like *something*, and a very big something indeed? How was it that no one "actually came to grips with it," that they handled it in such a way that it not only didn't fall apart, but served as a very resilient and powerful tool?

As Bohm said about the chilly reception given to his theories more generally: the reasons are complex and difficult to assess. It's especially difficult, because the style of the seminars and discussions at Bohr's institute in Copenhagen was such that one would expect this kind of error or misstep to have been doggedly pursued and hunted down. Bohr was a master bloodhound, and he taught his students and colleagues well. Following his lead, the physicists who gathered at Copenhagen positively took each other's work apart line by line, assumption by assumption, unforeseen consequence by unforeseen consequence. They kept their noses to the ground and were rigorous as hell—and yet something slipped by them.

But something is *always* going to slip by. No one—no *group* of people, even— can control every nuance, detail, philosophical point, experimental constraint, or theoretical requirement. There is always some gap or hole springing up, or rather *plunging down*, somewhere in the process. Even the best and brightest will occasionally lose the trail.

Another part of the answer can be teased out of the same metaphor. Like any pack of bloodhounds, the Copenhagen physicists always started from, and circled back to, the main house. It's hard to appreciate the force with which the statistical/empirical/ phenomenological interpretation of quantum mechanics operated then. In part, that force had to do with the fact that it was the first workable interpretation offered, establishing a strong scent to follow. Also contributing to that force, as the historian Paul Forman has argued, may have been a "Weimar culture" fascinated by acausality and indeterminacy.[16] Physical theory which privileged a certain kind of mathematics (the kind practiced by von Neumann, and wielded forcefully by Heisenberg, Dirac, and others) was also part of the trail. And we should certainly add considerations of the social authority that had accrued to Bohr, Heisenberg, Pauli, Dirac and others committed to the Copenhagen interpretation. They had been compared by other physicists at the time to a church spreading a particular gospel, which made these other physicists uneasy and suspicious, but without entirely knowing why. The remarks of the American physicist P.W. Bridgman made in 1960, a few years before Bell's disproof of the proof, are indicative:

> Now the mere mention of concealed variables is sufficient to automatically elicit from the elect the remark that John von Neumann gave absolute proof that this way out is not possible. To me it is a curious spectacle to see the unanimity with which the members of a certain circle accept the rigor of von Neumann's proof.[17]

The situation was indeed quite complex. The social and intellectual authority of an elite circle, commitments to a style of mathematical and physical reasoning or to a cultural sensibility, proven ability to explain experimental results and generate new verifiable predictions—all these and more were in forceful operation, and were a social fact. It would be unhelpful to draw any simple moral lesson about how authority or commitment to certain approaches are bad for science and need to be eradicated. Because they're also good for science: If they kept these bloodhounds from looking up from the trail very often, they also turned up more than a few rabbits and foxes.

But what you can do—and what *could* and *should* have been done in this particular historical example—is add a few sighthounds to the pack.[18] A friend in Texas who knows about hunting (unlike us) supplied this part of the metaphor: Where a bloodhound will take up a scent and pursue it almost obsessively, a sighthound will stand out in a field and shift its gaze around rapidly. Seeing something, or seeing just a movement, it will focus, and take off in pursuit. And whether it finds something or not, it will enthusiastically do it all over again.[19]

We said above that almost everyone believed von Neumann's proof. As it was, there were a few sighthounds who were able to take a different view of the proof and see its flaws years before Bell's antiproof. It shouldn't be surprising that these figures came from a (slightly) different field, philosophy. Only two years after von Neumann first published his proof in German, the philosopher Grete Hermann claimed to have shown that axiom **E**, which hidden-variable theories were supposed to obey, was in fact *impossible* for a hidden-variable theory to adhere to. The only thing the proof proved, she argued, was that if your axioms are self-contradictory, your results will be too. Then in 1944, the positivist philosopher Hans Reichenbach also challenged one of the proof's basic assumptions, showing how von Neumann presumed the eternal validity of the Copenhagen interpretation of quantum mechanics, and only the Copenhagen interpretation—a very big assumption indeed.[20]

By looking at that which the physicists had overlooked (for whatever complex reasons) these two historical figures offer a number of lessons. First, doing physics is a complicated affair involving, at the very least, doing philosophy. There should be institutionalized mechanisms to bring such "interlopers" into the discussion and process. Second, doing so would not overcome, but would disrupt the patterns of social and intellectual authority that constantly threaten to ossify thought and cover up error. As we'll see, the situation in quantum theory today is similar (and also different). There are currently a number of plausible interpretations of quantum theory on the table, and we now have an excellent opportunity to apply some lessons from these early years.

As most physicists today learned it, the Copenhagen interpretation was basically justified pragmatically, with the slogan "It works. Use it." Not "It's true"—or "simple," or even "well constructed." And working meant that the mathematical relations always predicted experimental results correctly, if you learned your mathematical manipulations well. But such a slogan begs further questions: What works? And for whom? How does it work in the world? Questions pointing directly at the acts of experimenting and articulation. Perhaps even pointing to the remarkable effectiveness of quantum mechanics "in life": modern electronics, optics, and nuclear physics—all understood by quantum mechanics—appear in forms as diverse as supermarket checkout scanners, power plants, computers, and bombs. The slogan also made it hard to think about the questions raised by Einstein.

The Spin Doctor

David Bohm made the EPR paper a little easier to think about. Instead of particles moving along a line and bearing the properties Q and P, Bohm worked with spinning particles, and focused on measurements of the way the axis of spin is oriented in space. His spin version of EPR applies to any of the particles in ordinary matter—electron, neutron, or proton. And it also works for the polarization of the photon, Planck's original quantum of light.

All of these particles have spin, which—like a planet's rotation—is along a definite axis in space. That axis defines the direction we assign to the spin. For example, as the earth rotates its spin points along the south pole-north pole axis toward Polaris, the pole star. Similarly, any spinning particle will point toward *some* direction on the celestial sphere.

A good experimenter with a good budget can easily make a device to distinguish hundreds of thousands of different spin directions. That's because a massive particle's spin makes it magnetic, just as the earth is. So the physicist builds a magnet to make the measurement. And the direction of the magnetic field, at the location of the particle under test, defines the measuring axis. The axis direction of objects as big as a typical laboratory magnet can be oriented at any desired angle to within a small fraction of a degree.

So, instead of measuring P or Q (which were actually momentum and position) with their continuous values, Bohm suggested measuring particle spin, which has only two possible outcomes. Point the magnetic field west, and the only measured values that ever occur are spin fully along or fully opposite that direction: we have a WEST state or an EAST state. The spin in a perpendicular direction (say with measuring field pointing north) remains completely undetermined. The WEST state will have a fifty-fifty chance of being measured as NORTH- or SOUTH-spinning in such a field. This is the Complementarity Principle for spin: values in perpendicular directions cannot be measured at the same time, because the experimenter's magnetic field can only lie in one direction and because measurements using that field must entail a change of a quantum of spin. In a state in which one of the directions has a definite spin value, the other spin-direction values are totally indefinite.

The Bohm spin-state idea was much easier to think about making into an experiment than the original EPR state. Instead of a complicated two-particle "translational" motion (which resisted attempts to construct it for over half a century), the necessary EPR-Bohm state was simply a state with total spin zero, made out of two particles spinning in opposite directions. The photons coming from a suitably excited atom would be in exactly such a singlet state.

And with a spin-zero state made from spinning particles, it's easier to see why the experimental results seem to depend on one another's reality. Whatever is done to each particle separately cannot change the fact that as a whole system, the two are spinning in opposite directions. So when the strangeness about being able to measure only one of three perpendicular directions is translated into measuring way across the room, at least we have a feeling for what the answer must be over here near us. It will be as if the particle here were suddenly thrown into a state exactly opposite whatever the outcome at the other particle's experimental station (see Appendix for a fuller description of the EPR-Bohm spin-state).

We emphasize "as if" because what is actually happening with the particles is something that John Bell placed in the domain of the "unspeakable." "The 'Problem' then," as he once put it, "is this: how exactly is the world to be divided into speakable apparatus . . . that we can talk about . . . and unspeakable quantum system that we can not talk about?"[21] When physicists try to speak about the unspeakable quantum entanglements between particles, things start getting strange: They say that forces are exerted "instantaneously," which either means they're the spooky effect of a "ghostly pilot wave," or they violate the theory of special relativity and travel faster than light, or some signal is traveling backward in time. Such attempts at articulation are fine, and even important to thinking, as long as you remember that you're trying to speak the unspeakable, and so have already made your first mistake.

Our own manner of speaking about these weird entanglements is in terms of doing things—making measurements, building detectors, and so on. The interactions between particles are not actions at all, but just the effect of sharing indivisible quanta of spin in a state of zero total spin.[22] This guarantees that whatever you subsequently measure about one particle's spin, the same measurement on the other's will give the exact opposite result. The correlation is complete; any of the spin variables you want to know can be determined without direct interaction on the particle in question.

Such a situation of perfect correlation is summarized by the idea of "entanglement" of two particles—the translation given to Schrödinger's term *Verschränkung* for the property exhibited by EPR pairs. (*Verschränkung* actually deserves a more entangled translation, one that would intertwine "entanglement," "crosslinked," [as in crisscrossed arms] and "being tightly mixed up with something.") And we note that because they are not actions these correlations cannot be used to send signals faster than light. Entanglement gives correlations between events at remote locations, but doesn't change the overall pattern or probabilities of outcomes at one location.

EPR to GHZ and the Alterity of Realitty

In the 1960s and 1970s, as we've mentioned, it was Bell who did some of the most important theoretical work in quantum physics. He had a surprisingly stern moralist's suspicion of quantum theory and approached his work wanting to disprove the Copenhagen "nonsense." In later years Bell spoke scornfully of its "unworkmanlike" character and the poor craftsmanship of something that *starts* its description of the fundamental microscopic world based on a conceptually complex and huge-number-of-particle object like a measurement tool, which "ought to be mentioned only at the end of the last book, not the beginning of the first." And once he'd disproved von Neumann he sought to find a test in which hidden variables could reveal themselves. Yet today he is most known for "Bell's inequality," which worked out certain theoretical conditions that any *local* hidden-variable theory—i.e., one without the spooky correlations between distant particles—must obey. It applied to the probabilistic cases of quantum theory, rather than just to the perfectly determined outcomes.[23] The inequality was indeed subject to experimental intervention, and it was immediately clear that outcomes obeying quantum mechanics would offer empirical evidence that could violate these simple relations, and thus could be said to confirm the model of the physical world articulated by quantum mechanics.

It wasn't until the early 1980s that such experiments became possible, among them those of the French physicist Alain Aspect, who took different measurements of the polarizations of two photons whose creation from a single atom's decay assured that they were in the singlet spin-zero state on which Bohm had focused. During the photons' flight to opposite ends of the laboratory, switches randomly chose different measurements of the polarization. The correlations between the measurement outcomes could be explained by quantum mechanics perfectly, but since they violated the Bell inequality, there was no local theory that could explain the results.[24]

In 1985, physicists Michael Horne and Anton Zeilinger proposed a new form of EPR particle experiments involving "down-conversion photon pairs" produced via nonlinear crystals.[25] This down-conversion process produces a wonderful, prolific kind of entanglement. Several groups of physicists began taking advantage of these new experimental possibilities. In an important review article in *Physics Today*, Daniel Greenberger, Horne, and Zeilinger describe this "revolution in the laboratory preparation of new types of two-particle entanglements," and how new possibilities for experimental articulation of beam-splitters, particle detectors, polarizers, and other devices were leading to all kinds of new theoretical work.

While Bell had already killed his own dreams of a hidden theory with Aspect's—and others'—experimenting help, the real surprises were about to come from considerations involving states of three particles, instead of only two. The first surprise was a fatal flaw in the very definition of Einsteinian elements of reality: When there are three spinning particles, many different elements of reality exist. Enough of them can be found that some actually contradict each other.

The second was the possibility of quantum teleportation. First theorized in 1993, by Charles Bennett of IBM's Watson Laboratory and his collaborators, quantum teleportation's first experimental realization came only four years later in December 1997, at the Innsbruck laboratory of Bernstein's collaborator Anton Zeilinger. The Austrian team next produced laboratory examples of the three-particle states that predict results that completely contradict Einstein's "reasonable" definition of reality.

Quantum teleportation requires quantum states of three particles, as does the disproof of Einstein's definition. The disproof uses a special, fully entangled state of three particles. David Mermin worked out this crucial state and called it GHZ after Greenberger, Horne, and Zeilinger. A paper under their names had appeared based on Greenberger's idea of applying Bohm's trick to a singlet state made up of four particles, arranged in two pairs—each of which had their own spins parallel instead of opposite. Mermin showed that the contradiction of the EPR definition of an element of reality needed only three particles. (For a detailed analysis of both the GHZ theorem and quantum teleportation, see the Appendix.)

Einstein would no longer be pleased, and definitely not satisfied. There is no preexisting, independent local reality for microscopic variables, just as the Copenhagen interpretation had argued all along. These developments in quantum theory remind us of the double reading of reality: Quantum reality is inextricable from our acts of measuring and knowing and at the same time surprises us with its absolute alterity, behaving in utterly unexpected ways. But what's truly remarkable about this nondeterministic, radical alterity is that it can be used in perfectly reliable ways—say, to "teleport" the properties of a particle from one place to another.

Quantum Leaps

In late twentieth-century America familiarity with teleportation is a given. Kids, grown-ups, nonagenarians alike know it through movies, comics, TV shows, short stories, and novels. The original coining of the word "teleportation" lies somewhere in the history of science fiction, probably in its golden age of the late 1940s to mid-1950s. *Star Trek* never uses the word; there (or *then,* in the twenty-fourth century) teleportation is so commonplace that the particular technology used to achieve it is called, simply, the "transporter."

If scenes from movies and TV are supposed to be in real time, then distance appears to present no difficulty, as long as you are within range of the transporter. The object being transported begins to appear almost immediately at the new location, going from planet surface to orbiting starship instantaneously, no matter how much time light would take to travel between them. Those scenes contradict special relativity, which dictates that no signal, and no object, can ever go faster than the speed of light.

Quantum teleportation is slightly different: No object is transported from here to there (or anywhere). Only information is transmitted, and not even that travels faster than light. In all the correlations and "passions at a distance" of quantum strangeness, this prohibition on the speed of information transfer is always respected. And the information cannot be "beamed" or directed to a particular location.

The classic 1993 paper by Charles Bennett and colleagues on quantum teleportation has as its central feature the entanglement of quantum particles. In their most general form, these entanglements "assist in the 'teleportation' of an intact quantum state from one place to another, by a sender who knows neither the state to be teleported nor the location of the intended receiver."[26] (See the Appendix for a detailed explanation.)

Quite simply, then, two experimenters conventionally known as Alice and Bob, are trying to communicate clandestinely and accurately. Quantum teleportation can be seen as one effect in a larger field of operations that Bennett has called "quantum information processing." This prompts another of our questions to stay close to: How is quantum teleportation itself entangled with another technique, quantum cryptography, a method of sending coded messages that are impervious to eavesdropping and impossible to break?

When we first heard of teleportation, you can imagine our delight and excitement: finally a place where entanglement, the choices involved in experimenting, and the construction of realitty combine to make something possible. And what a something—the long-held dream of teleportation. But the entanglement of quantum teleportation with quantum cryptography—that's an invitation to the willies.

The Willies

Quantum teleportation is physics at its exciting, challenging, mind-bending, re-alitty-creating best. That's what's so alluring about doing it, and also what makes it such wonderful grist for popular science writers. But you'd never know from those popular accounts, which are by and large stripped of all social context, that this amazing world of realitty-invention also gives some of us the willies.

Getting the willies is *sort of* like having an ethical or moral dilemma: you worry about what the right thing to do is. But to understand why getting the willies is at the same time quite different from the more garden-variety kind of ethical dilemmas that physicists may encounter—the willies are more haunting because they are less clear, more muddled, *spookier*—it might help to return briefly to some historical figures and examples.

There's no question that any history of physics in the twentieth century will have the Manhattan Project and the work of physicists on nuclear weapons at its temporal and moral midpoint. That seems as reliable a prediction as there can be: whether it's a history written in the twenty-first, twenty-fifth, or twenty-ninth century (call us crazy optimists), those scientific and social events that resulted in the first atomic weapons being built and used will be found in the middle of all subsequent historical inquiry into our era.[27] And in the middle of the Manhattan Project, those future historians will always find J. Robert Oppenheimer. And in the middle of Oppenheimer's life, they will always find two interesting utterances of his. If you've read Richard Rhodes's bestselling *The Making of the Atomic Bomb*, or seen one of the many dramatic, documentary, or more muddled docudramatic films depicting these events, Oppenheimer's reflections on physics and what it had become responsible for will be familiar. One of them he attributed to the *Bhagavad Gita*: "I am become death, the destroyer of worlds." The other is less poetic: "We physicists have known sin."

The magnitude of these sayings! To try to make some sense or meaning out of what had happened in the discipline of physics, to try to put physicists into some proper perspective, Oppenheimer reached for the biggest, grandest, most cosmotheological metaphors he could find. Only these ancient words of epic and eschatological proportions, it must have seemed to him, could encompass the moral and social responsibilities incurred and, simultaneously, transgressed by many physicists at the century's core. The bomb exploded. The cosmos trembled.

Meanwhile, back on earth, lines of power simply reconfigured. In the postwar period in the U.S., new federal institutions were created to channel immense new levels of support to the physical (and other) sciences: the National Science Foundation (NSF), the Office of Naval Research (ONR), the Office of Scientific Research and Development (OSRD), and a host of others. The national laboratories at Los Alamos, Brookhaven, Livermore, Oak Ridge, and other sites took in ever larger amounts of money, personnel, territory, and equipment. It became more and more difficult to tell academic, pure, or basic research from applied or military: by 1951 a Department of Defense survey had found that an average of 70 percent of the research time of physicists at 750 colleges and universities was "devoted to defense research." In 1954, according to the NSF, 98 percent of the federal government's $22 million given to academic physicists came from the Defense Department and the Atomic Energy Commission. "In the fifteen years following the war," as historian Paul Forman has documented, "the central fact of scientific life in physics was unprecedented growth based upon military funding."[28]

A question arises from this rather crude juxtaposition of two key features of the postwar landscape in physics: What are the connections between moral and social responsibility, funding, the creative impulses of physicists, and the military? How can or should a physicist think about these linkages, between the work that he or she does at the blackboard and what happens on the other side of that blackboard? These questions seem particularly troubling in the areas of inquiry which focus on the quantum world, where, as we've seen, the much-publicized and popularized theories of quantum physics suggest that the physicist may be "responsible" for the outcomes of experiments—or even more strikingly (and somewhat megalomaniacally) is said to participate in the "creation" of the quantum world.

What once might have been a relatively straightforward ethical choice on the part of the scientist confronting the national security state—"I'm not going to work on nuclear weapons," or "I'm not going to accept Defense Department money"—is now a more tangled affair. This is not to belittle the complexity or weight of the former kind of ethical decisions; many physicists have made hard and even heroic choices to refuse to work on particular problems or in particular contexts where they knew their work would be used for ends which they judged to be illegitimate, unethical, or immoral.

The key word here is *know*. What happens when you *don't* know what your physics is going to become, or what things it will lead to? What happens when the grant money you've accepted is "clean" but the meaning of that word has been muddied beyond easy recognition, because the worlds of the military, the national security establishment, and "pure" physics are now so thoroughly entangled after decades of work, billions of dollars, and constant traffic of people, ideas, and equipment between these worlds? What happens when you're not only worried about the future use or application of your science, but about what possibilities are being built into it now at the supposedly abstract, theoretical level?

What happens is, you get the willies. You sense that something is wrong . . . but you're not quite sure what. You hear a call to responsibility . . . but you're not at all certain what responsibility in this situation involves. You think there's something strange and spooky happening on this dimly lit road . . . but your companions tell you you're imagining things. Your flesh crawls . . . but your mind can't find a reason for it.

And there's no way to reconcile these contradictions.

Worlds Entangling

We got our first, rather mild case of the willies relatively early in the process, reading the scientific literature on quantum teleportation closely. The language of coded information, "unknown" quantum states, tightly controlled protocols for information exchange, "unspeakability"—a language which is part and parcel of the recent history of quantum theory, suddenly suggested the qualities of a spy narrative to us. But that's absurd, right? Everybody knows that spy stories don't get published in the leading physics journals.

But in earlier papers on quantum cryptography, we had seen the same characters "Alice" and "Bob" that were teleporting particle states involved in the theories of creating perfectly secure codes, and of message transmission that unequivocally ruled out any form of eavesdropping. And there it was clear that those names had come from the literature on "classical" cryptography. At least one physicist that we know of has decided not to work on quantum cryptography because of its entanglement with institutions like the National Security Agency. It's an ethical decision which we respect, but at the same time, the boundaries here are simply not that easy to draw. As a scientist, the minute you get interested (if you are any good at all in your field) that interest itself provides a valuable commodity: listeners and questioners who make a researcher think—whether the researcher is you, your student, or some innocent third scientist, you are already contributing to the field. Your interest begins to articulate an effort, to power an approach, to stimulate new experimenting and new thinking. Working on quantum teleportation means contributing directly and indirectly to work in the areas of quantum cryptography and quantum computing.

It's a lesson in disciplinary entanglement that was confirmed in our next, more forceful experience of the willies. Bernstein was a coauthor of a highly theoretical paper concerning the properties of a certain kind of mathematical operator called a unitary matrix. True, the paper concerned a "real" device, a sort of multiple-beam generalization of a simple half-silvered mirror. But it was essentially mathematical. The fundamental work answered a long-standing question: could any mathematically real-valued operator be made into a physical measurement? The answer was yes, at least for a general class of interesting cases. At a conference soon afterward, Bernstein met a very nice, very engaging physicist-cum-engineer who expressed interest in the proof, and the physical device which might realize it. He seemed very taken by the work, and was eager to get the reference, which had been in the previous month's *Physical Review Letters*. It hardly seemed noteworthy, the kind of collegial interest and exchange of fundamental ideas that goes on all the time.

But this fellow was a *lot* more practical with his physics than we are, more daily-life practical. In a telephone conversation after the conference, he told Bernstein of a proposal that he'd wanted to submit to the space-based Ballistic Missile Defense (BMD) system, popularly called "Star Wars." He was going to try to construct a communications system for the electronic battlefields of the future that would take advantage of quantum cryptographic techniques to securely code and transmit multiple messages over fiberoptic cables.

We received a copy of the grant proposal in the mail a short while later. And lo and behold, something we'd never seen or heard of before: attached behind the proposal is the "pure physics" paper in the leading theory journal. Apparently, the reviewers at the BMD research office didn't even have access to *Physical Review Letters*, and so had requested that the article from this seemingly distant and abstract field be attached to the proposal. So here on our desk at ISIS was this multimillion-dollar communication idea being sold to the military, with the simplicity of our beautiful theoretical device at its heart.

Now we are not simply antimilitary. One of us (Bernstein) worked for the Navy in summers to get through college and has been supported by Air Force research contracts, consultancies with the Armed Forces Colleges, the Navy, and Army Office of Scientific Research–sponsored conferences, and once held top-secret clearance. And together we work quite cooperatively with the military on the cleanup work of the Restoration Advisory Boards, as we described in Chapter 5. There are all kinds of people in the services, motivated by everything from peacekeeping to earning a living, people who love the technology and people who love the uniforms and the weapons, people who are learning new skills, people who work diligently toward disarmament, people who just want to survive until honorable discharge, and everything in between as well. But the almost immediate uptake and attempted application of an abstract mathematical-physical formalism to the world of high-tech battlefields was more than a little eye-opening and willies-inducing. We thought we were exploring, at a fundamental theoretical level, how the experiments and ideas of the quantum physicist help create realit*t*y; we didn't think someone would right away go out and try to make a realit*t*y entirely different from the one we would have liked.

While that one encounter at a conference mushroomed so quickly into something beyond our reckoning, another scientific conference that Bernstein attended somewhat later suggested deeper, further entanglements in a changing social context. For some time, Bernstein had been attending small conferences and workshops on quantum information theory, comprised mostly of the initial cadre of theoreticians and a few experimentalists. Like many scientific conferences, these were exciting events, charged with heightened feelings, the thrill of discovering how to work something out that has been elusive, the joy and passion of colleagues finding new insights in each others' articulations. At a certain point, however, the size of the meeting rooms suddenly grew, and while these conferences never lost their cutting-edge feel, something was changing.

One such conference on quantum computation was at the National Institute of Standards and Technology (NIST). Arriving there, Bernstein peered in the doorway to the conference room. Someone had opened the rear wall to double the size of the seminar room. There were about twice as many people as expected: over seventy attendees, whereas the e-mail announcement went to a list of approximately forty people almost all known to Bernstein from earlier meetings in Italy. Who were these extra people? What was the difference that made attendance swell? Most were mathematicians and physicists from the National Security Agency, some were from the Department of Defense; all were now paying close, "personal" attention to an area of research they had been monitoring from some small distance. The conference was quite spooky, literally and figuratively.

Other changes were noticeable later, at one of the annual meetings in Italy that have helped launch the subject of quantum computing into the forefront of physics. Here were new European researchers, many of them women, who again filled a room twice the size expected. But this time, there was a professional explanation: the realm of quantum information and computing had become a field unto itself. The American government had privatized or normalized its program. University researchers from Princeton, MIT, and CalTech were now taking Air Force and National Science Foundation money—millions of dollars—for the scientific work on questions the NSA was interested in: Can you build a real-life quantum computer? Which proposed system(s) is most promising? What else can you do with all the wonderful new theoretical tricks—fault tolerance, quantum state protection, error correction, entanglement production, entanglement for protection from decoherence by the environment? Indeed which of these can be implemented and for what purposes, even before—if ever—quantum computers themselves are built? And the Europeans were not about to be left behind. Several individual countries, the European Union itself, and international training grants had all been deployed.

These annual workshops are sponsored by the Elsag Bailey company, and held at the Institute for Scientific Interchange in Turin, Italy. They started because a prescient Italian industrialist, whose multinational company has much stake in high technology and its human capital, decided to bet on quantum computation research before there was hardly any advantage to it, even in theory. Certainly one could hope for something from the strange mixture of correlation and connection, of being able to put many different configurations together by "wavelike" superposition. These give a massively parallel aspect to the quantum computer: in fact *all* the possible values to a given problem can be input simultaneously. But those workshops before the NIST meeting had also taken place before Peter Shor of Bell Laboratories had published his proof that a quantum computer could solve certain mathematics problems with relative ease.

Ah, and what problems! The movie *Sneakers* gives you some idea of what a quantum computer could accomplish. In this film, a curious young private security operative decides to try out an apparently important little black box which his boss had just captured from an Eastern European scientist's lab. His pal says, "Try the Federal Reserve transfer node—$900 billion a day goes through there. . . ." A little fiddling, a bit of tooling around with the digital probe, and bingo! the incredible scramble of indecipherable symbols and characters on the screen begins to clear, the data start appearing: figures, names, cities and dates of all kinds of global financial exchanges replace the hashed-up mixture that had covered the display a second before. "Try the National Power Grid" is the next suggestion. Soon, "We're in!" and the locations, status, and map symbols for the northeastern United States fill the monitor. *Every* code would yield to the little gadget.

That black box is what the hypothetical quantum computer (suitably connected and interfaced) could hypothetically do if programmed with Shor's algorithm. Because factoring a product of two large prime numbers is a one-way problem. It's in a class of problems thought (but *not proven*) to be so hard to solve, but obviously so easy to check (just multiply the two factors and see if it gives the original number) that it can serve as a public key code system. One-way problems allow the sending of coded messages between people who have never met (or otherwise exchanged information). You state the product but it doesn't tell anyone the key (which is one of the huge primes) unless they already have the other prime in their possession. One-way problems are the basis for the encryption of all truly secure messages now sent. Quantum computation promises a technology capable of breaking *any* public key code.

Oh Boy . . .

We've told a number of open-ended stories in this book, but perhaps none more so than this story of quantum teleportation. We've stayed close to questions throughout, shunning simple morals or straight-line messages, but the lessons of this chapter are particularly entangled and, indeed, questionable. We've seen how physicists have held different definitions of what is "reasonable" to expect from nature or from physical theory. We've seen how the most reasonable and definitive proofs can, after years of operating as perfect truth, turn out to be a kind of nonsense. We've seen how language can mislead even as it points the way to new articulations of quantum theories, and how some things may always remain unspeakable. We've seen how the real spins of particles are made through our acts of experimenting, at the same time seeing that microphysical reality not only isn't what we thought it was, it *can't* be what we thought it was. And we've seen how the coolest, most abstract quantum theoretical work can be a source of wonder, fun, and delight for the physicist, *and* a source of the willies as that work becomes entangled with some pretty scary social forces and problems.

If we were to draw one of our lobsterlike diagrams for today's quantum physicist, as we did earlier for Darwin and Galileo, it would have to be a kind of quantum lobster. But a quantum lobster can't be depicted with the ink-on-paper technology you hold in your hand. The articulated linkages between the disparate elements of realities, theories, experiments, languages, and social forces would have to shiver and shimmer, fade into and out of the realm of the real, evoke probabilities rather than certainties. And if the questions of experimenting and articulating become particularly complex and fluid in this nonclassical realm, our questions of judging the power/knowledge assemblages involved are even harder to approach by the classical paradigms of responsibility and its direct moral linkages of cause and effect. Responding to those questions of judging and responsibility requires us to shift metaphors once again, from a quantum lobster to a bucking bronco.

The mathematician and pioneer of cybernetics Norbert Wiener also found himself and his work fully entangled in the complex dynamics of postwar American science, where the military-industrial complex seemed to dominate and control many fields of scientific inquiry. Wiener, too, worried about how his "pure" research might be entangled, in not always explicit or direct ways, with the needs and desires of the military or the corporate order. He too questioned what scientific responsibility could mean in such a densely woven social web:

> I tried to see where my duties led me, and if by any chance I ought to exercise a right of personal secrecy parallel to the right of governmental secrecy assumed in high quarters, suppressing my ideas and the work I had done. After toying with the notion for some time, I came to the conclusion that this was impossible, for the ideas which I possessed belonged to the times rather than to myself. If I had been able to suppress every word of what I had done, they were bound to reappear in the work of other people, very possibly in a form in which the philosophic significance and the social dangers would be stressed less. I could not get off the back of this bronco, so there was nothing for me to do but to ride it.[29]

Wiener decided that the best he could do was "to turn from a position of the greatest secrecy to a position of the greatest publicity, and bring to the attention of all the possibilities and the dangers of the new developments" in his field of cybernetics.

It seems a grossly inadequate response, but also an honest one, and we've adopted something like it for our pursuits of quantum physics. There's no getting off the back of the quantum bronco. Even if you personally decided to get off, the whole rodeo is by this point so fabulously resilient and well-stocked with talent that the show will go on. The most responsible action may indeed be to ride the writhing beast, perhaps gaining a better feel for its contradictory, bucking movements in the process. Or, like those physicists who for decades thought that a hidden-variable theory of quantum mechanics had been proven impossible, we may yet be fooling ourselves. In any case, we will at least have observed Wiener's advice, to call greater attention to the possibilities and dangers (and the inevitable crisscrosses of these terms) entangled in the pursuits of quantum information.

To return to our opening allegory: For today's (or tomorrow's) quantum physicist, there's no final leap home, where scientific, political, or moral knowledges—and the entanglements between them—are reassuringly secure. Instead, it's always once more into the breach, wrenched into a space of new questions, surprising events, incomplete information, and suggestive but suspiciously elusive correlations. Responsibility here, for us, means *not* getting off the quantum bronco, and taking the risky ride instead.

PART III

"... a tolerance for contradictions ..."

I have become persuaded that, in scientific research, neither the degree of one's intelligence nor the ability to carry out one's tasks with thoroughness and precision are factors essential to personal success and fulfillment. More important for the attaining of both ends are total dedication and a tendency to underestimate difficulties, which cause one to tackle problems that other, more critical and acute persons instead opt to avoid. Without pre-established plan and guided at every turn rather by my inclinations and by chance, I have tried . . . to reconcile two aspirations that the Irish poet William Butler Yeats deemed to be irreconcilable: perfection of the life and perfection of the work. By so doing, and in accordance with his predictions, I have achieved what might be termed "imperfection of the life and of the work." The fact that the activities that I have carried out in such imperfect ways have been and still are for me a source of inexhaustible joy, leads me to believe that imperfection, rather than perfection, in the execution of our assigned or elected tasks is more in keeping with human nature.

—Rita Levi-Montalcini, *In Praise of Imperfection*

In perceiving conflicting information and points of view, [la mestiza] is subjected to a swamping of her psychological borders. She has discovered that she can't hold concepts or ideas in rigid boundaries. . . . La mestiza constantly has to shift . . . from convergent thinking, analytical reasoning that tends to use rationality to move toward a single goal . . . to divergent thinking, characterized by movement away from set patterns and goals. . . . The new mestiza copes by developing a tolerance for contradictions, a tolerance for ambiguity. . . .

—Gloria Anzaldua, *La Frontera* (Borderlands)

Chapter 9

Muddling Through

At the Interface

Not a day went by during the writing of this book that didn't include an announcement of some event which challenged how we think about the relationship between government, the sciences, and the democratic society in which we all live—or at least hope for and work toward. New drug therapies of great promise and uncertain risk, new environmental hazards postulated or refuted, more data on and more elaborate theories of climate change, new ways to conceive and bear children, enormous anxieties raised by mammalian cloning, the reorganization of work through new information technologies, another reorienting "big picture" of the cosmos—the scope and pace of social change accompanying the sciences is overwhelming. We live with an implacable need for more, and better, social policies bearing on scientific and technological change.

The fact that we have written only tangentially about the governmental policy aspects of the sciences doesn't mean we think they're unimportant. But better science policies will absolutely not be enough. Policy has to come, but it will inevitably come too late. To demarcate the pursuit of the sciences from the promulgation of policy, as our society does, is to forever play a catchup game. Because the sciences make realitty, they exercise their power in many ways other than the conventionally political ones. When it comes to the sciences, politics—in the sense of social power being created and directed—happens in ways other than government action. The power of science happens before policy, and policy—as necessary as it is—is always after the fact.

The need for more, and more critical, readings of the sciences is not a small demand, and it is one which it is difficult to remain optimistic about in the current cultural climate. In 1996 a fiscally and politically conservative Congress dealt a quick and relatively quiet death to its own Office of Technology Assessment (OTA), one of the most important sources of inquiry into the effects of science and technology on democratic society. Since its establishment in the early 1970s, the

OTA had been one of the few institutions where social problems and promises associated with the sciences and technologies could be analyzed. Even if Congress didn't always know what to do with them, OTA reports provided a great deal of information on hazardous waste and its cleanup, nuclear weapons production and proliferation, the economic impacts of computer technologies, the patenting of biological materials, and hundreds of other topics.

At the same time this source of critical readings of the sciences was being dissolved, a disquieting report came from the National Science Foundation (NSF), the government agency so crucial to the funding and directing of the sciences in the post–World War II era. The NSF had convened a panel of experts for a workshop on "Science, Technology, and Democracy: Research on Issues of Governance and Change" in 1994; their subsequent report opened with the following sentences:

> [I]t is hard to imagine a modern industrial democracy without close coordination between scientists, engineers, and public officials. Yet the practices and processes that link science and engineering research to government in industrial democracies are largely implicit. The scope and pace of changes based on science and technology challenge democratic governance. Attempts to translate the social structures for research and governance from one society to another, or to modify it in existing ones, go awry because the human dimensions of that interface are poorly understood.[1]

Buried in the bland bureaucratese is a rather stunning statement: Science and technology raise fundamental questions of social governance, but the ways in which they do so, and the ways in which we might respond, exceed our grasp. In other words, this thing called modern industrial democracy happens in the absence of our full understanding. Immense, rapid changes occur "implicitly," with practically no foreknowledge or planning. We don't know how we got here, let alone what we might do. The fantasy of a social evolution that can be mastered, understood, and guided may very well be just that: a fantasy.

And yet we harbor a persistent hope and desire to make the implicit explicit, to understand the interfaces between science, technology, and government more fully, in hope that this understanding will lead to better planning and governing. Where does this desire come from? What makes us think that we can get even a little bit more control of this monstrously complex event—a modern industrial democracy that is practically, economically, and culturally dependent on the sciences—that has so obviously been taking place without our full willingness for so long?

And what if we can't? What if being at the interface between the sciences, technologies, policies, and something apparently separate from these called "human dimensions," means not being in control? What if we're in the middle, and can't get to a safe perspective and effective leverage point outside the system?

Where Were We, Now?

There is another response to the intellectual, political, and ethical demands the sciences place on democratic society besides "better policies and more control." As much as a vital democratic society needs new legislative and regulatory frameworks for the sciences, it needs something else equally, if not more. There's something indirect, subtle, accidental, and perhaps even unknowable about the incredible power of the sciences. They change the world, through processes which we can only dimly understand. Social or political mastery seems out of the question. Yet we need *something* to help us live better with the sciences, in the middle.

We may not know, now, what that something is. But we do know that to have a better shot at making that something, we will need scientists, engineers, and all other citizens capable of reimagining the sciences along the lines we've outlined here: as experimental conglomerations of things, thoughts, and so much else, which literally make our world. We need a diverse polity outfitted with critical literacy skills, able to continually question and reread the sciences and their world-making powers again and again.

Such a recognition of the need for new science literacies is visible throughout our scientific culture, although expressed in ways orthogonal to ours. The Smithsonian—that venerable nineteenth-century institution symbolizing (and carrying out) the promise of state support for the sciences and technologies – recently drew fire for its exhibit "Science in American Life." The exhibit provoked outraged responses from many scientists for showing the "other side" of the social promise science has come to signify in our culture: the environmental havoc caused by exuberant use of pesticides, the growing influence of private industry on scientific careers and research agendas, the clustering links between the sciences and the military, the social and ethical dilemmas of genetic engineering, and other less than pure situations.

The exhibit is hardly an exposé, and the dominant tone (often expressed in the words of working scientists represented by cardboard cutouts) is one of respect for the achievements of science and scientists. In an activities room children, and their parents, can undertake hands-on explorations of various scientific phenomena. But, like the war veterans who demanded changes in the Smithsonian exhibit on the *Enola Gay* and the American use of atomic weapons at the end of World War II, some scientists felt that the historical record should be a separate matter from the public display of that record in edifying social institutions (especially ones supported with federal funds). The president of the American Physical Society complained to the Smithsonian that the $5.5 million exhibit was "a portrayal of science that trivializes its accomplishments and exaggerates any negative consequences. We are concerned that this presentation is seriously misleading, and will inhibit the American public's ability to make informed decisions on the future uses of science and technology."[2] The American Chemical Society, which contributed a good portion of that $5.5 million, also protested the exhibit's "built-in tendency to revise and rewrite history in a 'politically correct' fashion"; one chemist involved in organizing the exhibit complained that it was largely the work of "social scientists and pseudoscientists who had no idea of how science worked."[3] These sentiments echo a long-held concept of scientific literacy: "Scientists will tell you what you need to know. You need only accept and believe that they have the ability to muster control, of nature and society alike." Or, "We'll teach you the sounds—you don't need to learn the alphabet."

But the sciences are less about the ability to control than they are about the unleashing of new forces, new capacities for changing the world. Social or political mastery of the sciences is out of the question because the natures of the sciences, technology, democracy—as well as the relationships between these questionable terms—themselves remain a question. To paraphrase Foucault once more: What are these sciences that we use? What are their limits? What effects have they exerted, or had exerted on them, in the past? Now? In a time still to come? Where do their promises unavoidably crisscross with their dangers, and how can we live better, more justly, at those confounding and unmasterable intersections?

Still other questions: Can the sciences really answer "really?" questions? And what kind of answer could we expect for that last question, which is a "really?" question about "really?" questions? The answer to that may in fact be another question to stay close to: what would it mean to stay close to questions without resolving them? What kinds of intellectual and institutional practices can we experiment with that would allow us to give good, workable answers to "really?" questions, but without the word *really*?

New science literacies have to begin with a serious engagement with histories of the sciences. In Section I, we questioned what it has meant, in different times and in a variety of situations, to be "practicing a rationality" called the sciences. We showed how the sciences have always been exercises in muddling, in ways which have furthered the generation of new ideas, new useful artifacts, new ways to imagine ourselves and our world—new realitties. We developed a set of four concepts/actions—experimenting, articulating, powering/knowing, and judging—to serve as an alternative theoretical frame for thinking about the sciences, distinct from the empirically ill-fitted and conceptually feeble one of facts, hypothesis, theory, testing, representing, and so on. To that extent, Section I was not only an engagement with the history of science, but a statement of theory, our own ideal of what the sciences are.

In Section II we gave accounts of some specific pursuits of the sciences to which we've stayed particularly close in the present. We tried to show how complicated and demanding specific questions can get on all levels—scientific, social, practical, ethical—and how those situations can still be worked even in the face of enormous complexity. The contemporary issues examined in Section II show how muddling through is particularly important in times of dramatic change, when things have stopped making sense and there is an urgent need for perspectives sensitive to new social realities. So in addition to being the present to the pasts of Section I, Section II could also be thought of as the practice of the theory of Section I—stories about what happens when the ideal hits the real.

But the sciences are still more complicated than that. This structure needs a third element, a third articulation that will punctuate many of the same points of the first two sections, but with a slightly different twist. In terms of the conceptual scheme which we borrowed from Charles Sanders Peirce and kludged into Chapter 2: Section I of this book was our "Firstness": ideality, mind-stuff, our reimagining of what the sciences are, based on how they've been practiced in the past. Section II was "Secondness": actuality, rude awakenings, tales of hard encounters with one world or another in the present. They can't really be separated from each other, and indeed, Sections I and II were only different articulations of many of the same points, with past and present mixed. This Section III, then, should correspond to Peirce's "Thirdness" and should reiterate the sciences, again, from a subtly different vantage point.

Peirce's Thirdness was more subtle and elusive than either Firstness or Secondness, and gave his commentators the hardest time. Thirdness is a "gentle force" that somehow binds the other two together, and keeps them from simply colliding and bouncing off each other. Thirdness has to do with an interpretation that interpenetrates the web of ideas and things; Thirdness contextualizes, extends the articulated webs ever outward. Thirdness stabilizes and weaves, but Thirdness is also the fount of difference, instability, and change. Even as it mediates the stark opposition of Firstness and Secondness, Thirdness generates further contradictions. Thirdness works behind the scenes and indirectly, but there is no scene without it, and no direction. Thirdness may have something to do with those qualities we call imagination, creativity, and even genius in the sciences. At the same time, Thirdness is what calls for muddling through.

Gentle, subtle, ambiguous, muddled, indirect Thirdness eludes full and concise description in a language tailored for directness. If Firstness and Secondness correspond to the pasts and presents of Sections I and II, then the Thirdness of Section III may correspond to the sciences of an unknowable future. Thirdness is Feynman's "half-assedly thought out pictorial semivision thing," or Bateson's "sort of secret" that "one *cannot* tell" and in which one has "no direct control over the matter." Thirdness has been the indirect subject of this entire book.

One and One Is . . . ?

Demographics and marketing are not our expertise. We don't know how many of the people who will buy this book are working scientists, students of the sciences, or simply people interested in the sciences. You may or may not spend long hours in the laboratory, oscillating between frustration and rapture, and then relax by reading about someone else's oscillations in the sciences. You may or may not spend your days toiling to solve arduous equations or poring through mounds of scientific articles, and then extending your passion for the sciences in your nighttime reading. But in any case, chances are you bought this book because you're in the habit of buying science books, and there's a lot of material to feed your habit. You're interested in and enjoy reading about *The Double Helix, The Coming Plague, The Search for the God Particle, A Brief History of Time, The Structure of Scientific Revolutions, The Mismeasure of Man, The Emperor's New Mind, The Selfish Gene, The Inflationary Universe, The Life of the Cosmos,* and maybe even *The End of Science.* You watch *Nova* programs on the big riddles of cosmology, *National Geographic* specials on threatened biodiversity, and the weekly mix of *Scientific American Frontiers* with Alan Alda. You listen to National Public Radio's Science Fridays program. Maybe your kids watch Bill Nye the Science Guy every day after school. You're fully immersed in our culture.

Even if only some of these marketers' generalizations apply to you, your book shelf may hold John Brockman's edited volume of interviews with scientists, *The Third Culture,* or one of the other "Reality Club" books edited by Brockman, literary agent to the science stars of today. *The Third Culture's* title plays on C.P. Snow's 1959 definition of *The Two Cultures,* which divided the world neatly into scientists and literary intellectuals. Brockman's interviewees come from the fields of evolutionary, molecular, and theoretical biology (Stephen Jay Gould, Lynn Margulis, Richard Dawkins, Francisco Varela, Stuart Kauffman), theoretical physics (Alan Guth, Murray Gell-Mann, Lee Smolin, Roger Penrose, Paul Davies), and computer, cognitive, and "chaos" sciences (Marvin Minsky, Christopher Langton, W. Daniel Hillis)—all hot areas in the sciences. These scientists expound on the extraordinary ferment that each of their fields—and the cross-fertilizations between them—has undergone toward the end of the twentieth century. Read at face value, *The Third Culture* is an enthralling collection of the achievements and trajectories of thought "beyond the scientific revolution," as the subtitle reads.

But there is much more than that to read in these pages. First, a detectable if slight tone of resentment. By and large, these scientists feel that their fellow citizens (mostly in the U.S. and U.K.) are scientifically illiterate. The scientists of *The Third Culture* want to be taken seriously as intellectuals—as they should want, and as they should be. The sciences are indeed a major intellectual, cultural, and social force in our world today, and public understanding and appreciation of ideas from the sciences leaves much to be desired. Like these scientists, we too want people to be able to think better about the sciences, to know what's going on in these fields, to be literate. Few things would please us more than if several million people went out, bought all the books by all the scientists interviewed in *The Third Culture*, and read them.

But read them *critically*. And for all its emphasis on literacy, *The Third Culture* doesn't seem especially concerned with building a critical literacy. Rather than truly pursuing a third culture, this volume seems more a case of wanting the "first" culture of science to simply colonize, territorialize, or subsume the "second" culture of literary intellectuals. Brockman writes that these figures from the "empirical" world of the sciences are "taking the place of the traditional intellectual," whose culture "dismisses science, is often nonempirical," and "is chiefly characterized by comment on comments" in which "the real world gets lost."[4]

> *Stephen Jay Gould:* . . . There's something of a conspiracy among literary intellectuals to think they own the intellectual landscape. . . . Peter Medawar, a very humanistically and classically educated scientist, said it was unfair that a scientist who didn't know art and music pretty well was, among literary people, considered a dolt and a philistine, whereas literary people don't think they need to know any science in order to be considered educated. . . .

> *Richard Dawkins:* I do feel somewhat paranoid about what I think of as a hijacking by literary people of the intellectual media. . . . The very word "theory" has been hijacked for some extremely narrow parochial literary purpose—as though Einstein didn't have theories; as though Darwin didn't have theories.

> *Paul Davies:* . . . The fact that scientists are starting to be heard, capturing not only the minds but the hearts of the population—as evidenced by the phenomenal success of science books—is provoking what seems to be a territorial squeal from the literary side. The backlash has taken the form of hysterical ranting in newspapers and periodicals, and a spate of books denouncing scientists as arrogant and self-serving frauds. . . . For years and years scientists were ignored because they were not heard; now that they're starting to be heard, they're being stamped on by an intellectual mafia.

Nicholas Humphrey: . . . Since they don't understand science, their only
defense is to say that it doesn't matter. But they're fighting a losing battle.
People are voting with their feet. Who listens to what nowadays? Who
watches what on TV? Who's buying what books?

Terms like "hysteria," "hijacking," and "intellectual mafia" lack the precision
and detachment that we usually associate with the sciences, so there may be something more at work here than the noble desire to make sure that people are literate
in the sciences. There's a sense of urgency, perhaps even emergency, expressed
along with a more or less subtle triumphalism and sense of self-importance.

But another motif of the book deserves more attention. In the comments which
these scientists direct at each other and each other's work, there is, beyond the
dominant tone of mutual admiration, an unmistakable overtone of profound difference and disagreement. This one thinks that one is crazy, "mystical," or "roman-
tic"; X *really* understands Darwin and evolutionary theory where Y has got it all
wrong. And yet these disagreements and differences are often grounded in quite
subtle expressions. To quote just a few: "I don't intersect with his mode of thought
that strongly"; "there's just too much of a gap in our approach to things for there to
be much useful dialog [sic] between us"; "he really says nothing particularly interesting"; "I don't understand it at all"; "although the smell is the right one, I don't
think I can buy the actual theory he's trying to stitch together"; "I have gut misgiv-
ings about theories of that kind."

What should we make of such differences? Certainly not too much; they're the
kind of differences that crop up all the time in the sciences, the kind of differences
that make pursuing science an exciting, challenging, arduous, and rewarding experience. But neither should we make too little of them. It's all too easy to say that
scientists disagree with each other all the time, but eventually the truth will out
when theory (Firstness) confronts and is tested against empirical reality (Secondness). For one thing, even when the truth outs, another set of differences emerges;
settle one question and two others immediately pop up.

The real problem with this conventional characterization of the sciences is that
it's much too diluted to match how the sciences are actually practiced. As we've
shown in numerous examples, pursuing the sciences has always involved those
complex, heterogenous processes we've gathered under the rubric of muddling.
So what happens when you start from the assumption that subtle as well as radical
differences among scientists are not only possible, but ineradicable? What happens when you remember that the scientific method not only includes, but depends on these somewhat unmethodical elements for imbuing the sciences with
power? That Firstness and Secondness, theories and facts, concept and thing, are

never sufficient in pursuing sciences and never encountered on their own, but always come packaged with Thirdness?

That, we think, would be the way to begin seeing the emerging outlines of a truly "third" culture, a third culture that in fact has always been with us, emerging, despite simple habits of dividing culture into (1) science and (2) literature. Look back at those quotes we selected: what underlies these differences between scientists is not "they got the data all wrong," or "the man couldn't make an objective observation to save his life" (as is often the case in such volumes, there are few women represented, and it's often Lynn Margulis), or "the logic is shoddy and the calculations are way off." Those things would provide a relatively easy avenue of resolution. And indeed, differences within the sciences can often be resolved by such means. But as we see here, what's frequently involved is something much more murky: a mode of thought, an approach, a smell, a stitching pattern, a gut feeling. The sciences have always, and will always, involve such opaque terms for correspondingly opaque processes. To deal with any "third"—Third Culture, Thirdness, third term—is to find yourself in the muddled middle, where you don't always have good analytical tools for working your way through the opaqueness into an open clearing.

We need a kind of "third culture," but the way to get it is not to simply invert the status of the two terms, to simply elevate science over literature. Moreover, in some sense a third culture has always been present in the sciences or, more accurately, *between* the sciences and its various Others: literature, society, culture, religion, and still more. The fact that the sciences have never been pure, have never been thoroughly rid of the influences of Others, have always been a kind of "third" hybrid term—this is what has made them so powerful, so alluring, so productive, so interesting, and, frankly, so much fun. And such a problem.

It's especially because of that last characteristic that all of us, scientists and citizens alike, need new critical literacy skills for reading the third culture that's already in our midst. To put it in other terms: What we really have to understand and cultivate is not a third culture, but a culture of the third, where *one and one is at least three.*

Cytoplasm and History

Especially in this digital age, there always appears to be an either/or choice from among the oppositions of this and that, on and off, yes and no, true and false. The *real* third culture will be a culture of the third, a world where people know that "this and that" isn't a question of two terms, but a combination of three: this, that, and *and.* And *and* is always in the middle; it's the sign of the dangerous crisscross, the messy collusion, the maddeningly wonderful muddle. This book has described some of the intellectual and political tools for engaging, not with the either/or of this or that—theory or experiment, thought or thing, power or knowledge, social or natural, science or nonscience—but with the both/and of the 3 that lies *between* 1 and 2. We've argued that the sciences have always been both/and, have always had their Other—literature, philosophy, the historical, the social—incorporated within them, a kind of parasite to their host.

That last set of metaphors belongs in an allegory of symbiogenesis, a concept found in Lynn Margulis's work in evolutionary biology. Margulis showed—no, we have to be more precise: Margulis argued in writing, and after a decade or so of derision and dismissal, other scientists began to appreciate what was being shown—that the organelles bearing pieces of DNA in the cytoplasm of the eukaryotic cell, outside the nuclear "control center" where scientific interest and effort have mostly concentrated, are the evolutionary traces of microorganisms that symbiotically entered bacterial cells roughly two billion years ago. That merger made for new survival strategies, new reproductive possibilities, new capacities—in short, a new form of life.

In our allegory, the sciences live, the sciences evolve, the sciences are Science because they're not just a coded expression of some Platonic nucleus of ideas and logic, but because of other working, generative structures *fused* within the same cell. Make no mistake: the nucleus does operate, vitally so. Practicing a rationality and pursuing the sciences can require the most exacting obedience to exact logic, sharply defined concepts and their relationships, and various other signals that resist. But it's just as important not to mistake the nucleus for the cell. Out there in the messy, intricate mix of chemical soups, diaphanous membranes, and writhing proteins, other structures are at work, other messages created and exchanged, other substances and interactions contributing to the overall stability and vital power. Consider this the cellular variation on the metaphorical image of an articulated lobster that we set loose in Chapter 3: the sciences as amoebae, active congeries of disparate substances, structures, and motions, extending itself here, retracting there, continually changing shape.

As it happens, around the time Margulis was writing up her first paper on endosymbiosis, Thomas Kuhn's *The Structure of Scientific Revolutions* was on its way to becoming the most widely read work on the sciences. Kuhn's famous book employed something like this kind of symbiotic cellular perspective. At a time when most philosophers of science were intent on shoring up the nuclear membrane that protected the exclusively coding, logical purity of the sciences, Kuhn started mucking around in the cytoplasm of history, where messy influences abound. The process of revolutionary change in the sciences, he argued via a number of historical examples, neither obeyed simple scientific logics, nor exhibited the simple progress of better approach to an eternal and universal Truth. While there's much in Kuhn that we disagree with, and while he eventually disassociated himself from some of the more radical (and interesting) extensions of his inquiries, his enduring legacy has allowed us to remain close to the question of the interaction between the inside and the outside of the sciences.

It's a legacy that had its own symbiogenesis. Kuhn's work alone didn't lead to the expansion of critical intellectual work in the historical, philosophical, socio-logical, anthropological, literary, and other social analyses of the sciences that has occurred in the academy over several decades. It combined with the social ferment of the civil rights, feminist, and antiwar movements of the 1960s and beyond, all of which challenged the received notions of truth and progress in one way or an-other. The result has been an extensive body of scholarship that examines the ex-tranuclear structures of the sciences in much greater detail and nuance than the heroic narratives often penned by scientists themselves. It's a body of scholarship that has been indispensable for the writing of this book, allowing us to show how the sciences have always been a kind of emergent feature of the *and* between sci-ence and culture. It's shown how we can think about the sciences as a culture of the third—a culture which practices rationality in different ways, in different situ-ations, at different times. These detailed studies of the sciences are an essential part of the critical literacy skills needed by both practicing scientists and everyone else in this culture of Thirdness.

The Third Without Qualities

Slippery and protoplasmic Thirdness still remains somewhat elusive. Some rather famous scientists claim to want something called a third culture, but still tend to get bogged down in the old one-two. Perhaps another detour through history, this time of the more literary kind, can guide us closer to the question of Thirdness without, once again, fully resolving it. As something that can't be fully resolved, Thirdness requires continual revisiting and requestioning. As something that can't be fully resolved, Thirdness may have something to do with contradictions, irre-solvable oppositions which, in one form and another, have always been with us.

In Chapter 1 we saw the physicist Max Planck struggling to get his bearings, while trying to give some direction to what he saw as a directionless culture, in his 1932 book *Where Is Science Going?* As Planck was penning his version of this seem-ingly perennial question, the novelist Robert Musil was at work on his inter-minable *The Man Without Qualities*. Although the events of this consummately modernist novel were cast in the year before World War I, Musil was responding to the same questions—about the sciences, their historical effects, and their limits—as Planck. Musil's novel has reappeared recently in a new English translation, sug-gesting that its questions and characterizations of the age remain pertinent today—good stones on which to sharpen our thinking about the sciences, the work of contradictions, and strategies for dealing with them.

Somewhere around the middle of Volume 1, a dialogue occurs between the man without qualities, Ulrich, who, if he could have qualities, might be a mathematician, and Walter, a man with definite qualities, principally those of being an artist (albeit one fully clothed in the rising bourgeois culture). Begin with the end of their argument:

> Walter continued in a low voice: "You're right when you say there's nothing serious, rational, or even intelligible left; but why can't you see that it is precisely this growing rationality, infecting everything like a disease, that is to blame? Everyone's brain is seized with this craving to become more and more rational, to rationalize and compartmentalize life more than ever, but unable to imagine what's to become of us when we know everything and have it all analyzed, classified, mechanized, standardized. It can't go on like this."
>
> "Well," Ulrich said with composure, "when the monks were in charge, a Christian had to be a believer, even though the only heaven he could conceive of, with its clouds and harps, was rather boring; and now we are confronted with the Heaven of Reason, which reminds us of our schooldays with its rulers, hard benches, and horrible chalk figures."
>
> "I have the feeling there will be a reaction of an unbridled excess of fantasy," Walter added thoughtfully.[5]

As a man with qualities, Walter sees things in terms of definite opposites: an excess of reason can lead only to a reaction of "unbridled fantasy." He thinks he *really* knows exactly what science is—mechanical, analytical, compartmentalized—and therefore what it must inevitably become, and he doesn't want anything to do with it. Planck's worst nightmare, his is a familiar figure today: the technophobic critic who can only fear the sciences in the broadest, almost caricatured terms. But Ulrich is unsure of clear definitions and limits, unsure about what science or reason will become over time once it overflows the confines of childhood memories and stunted imaginations. Having experienced them from the inside, Ulrich can read the popular critiques of the sciences that Walter depends on for what they are: sketchy chalk figures drawn more for moralistic effect than for faithful characterization.

So with Planck, Musil, Ulrich, and Walter, we are very close to the question of the two cultures as it appeared around 1932—so close that maybe we can glimpse a third. A machine might help bring it into better focus—not a machine made out of metal, but one made out of words. It doesn't have to be perfect; it just has to work. Its job will be to make opposed terms more visible, so we can see how they work and how we might get in between. We begin a chart of oppositions and their thirds, the first terms of which are:

science	literature	
Planck	Musil/Ulrich	Walter
fact	fiction	

We characterized Musil above as a novelist, which wasn't quite right. He could also be called an engineer (among many other things), which is why we put him in between science and literature. And the hybrid qualities of the factual Musil can certainly be identified in the fictional Ulrich, which is why they get joined (and separated) with that slash. Ulrich is a mathematician (among many other things) and has experienced the power and allure of science from within:

> "Scientific man is an entirely inescapable thing these days; we can't not want to know! And at no time has the difference between the expert's experience and that of the layman been as great as it is now. Everyone can see this in the ability of a masseur or a pianist. No one would send a horse to the races these days without special preparation. . . . But I'll grant you something quite different," Ulrich went on after some thought. "The experts never finish anything. Not only are they not finished today, but they are incapable of conceiving an end to their activities. Even incapable, perhaps, of wishing for one. Can you imagine that man will have a soul, for instance, once he has learned to understand it and control it biologically and psychologically? Yet this is precisely the condition we are aiming for! That's the trouble. Knowledge is a mode of conduct, a passion. At bottom, an impermissible mode of conduct: like dipsomania, sex mania, homicidal mania, the compulsion to know forms its own character that is off-balance. It is simply not so that the researcher pursues the truth; it pursues him. He suffers it. What is true is true, and a fact is real, without concerning itself about him: he's the one who has a passion for it, a dipsomania for the factual, which marks his character, and he doesn't give a damn whether his findings will lead to something human, perfect, or anything at all. Such a man is full of contradictions and misery, and yet he is a monster of energy!"

We'll add these terms to our machine:

science	compulsion to know	passion
professional expertise	incomplete accomplishment	lay incompetence
perfect	contradictions/misery/energy	human

. . . and hope that it will hold together, helping us operate in what is rapidly becoming a complicated situation. Musil/Ulrich reminds us that there is no escaping the sciences; it cannot be simply a question of "let's do something else." The sciences are, in effect, not "ours" to do with what we like. In the psychoanalytic terms that were then beginning to percolate through Musil's culture, the sciences are akin to the drives and urges of the unconscious: an uncontrollable Outside incorporated within our most interior selves, keeping us off balance. The sciences are forever ahead of us, keeping us always "after the fact," playing catch-up with little hope of winning.

And even as we pursue the sciences, they pursue us. They pursue us, somewhat paradoxically, by resisting us: The natural world seems to force our hand with no regard for what we might prefer to be the case. A fact is a fact: The perception that the sciences supply us with self-evident facts and unassailable truths is almost inescapable, no matter how skeptical one might want to be, or knows one should be. Our character (if you'll temporarily forgive the generalization) is indelibly marked with a compulsion to know, a habit somewhere between science and passion. And that in-betweenness is a space of both painful contradictions and exuberant energies. The acute need for expertise permeates every nook and cranny of our world, from the sciences to sports—even as the limits of that expertise become all too apparent.

And there's one more question, perhaps the most important but at the same time most destabilizing of all:

> "And—?" Walter asked.
> "What do you mean, 'And—'?"
> "Surely you're not suggesting that we can leave it at that?"
> "I would like to leave it at that," Ulrich said calmly. "Our conception of our environment, and also of ourselves changes every day. We live in a time of passage. It may go on like this until the end of the planet if we don't learn to tackle our deepest problems better than we have so far. Even so, when one is placed in the dark, one should not begin to sing out of fear, like a child. And it is mere singing in the dark to act as though we knew how we are supposed to conduct ourselves down here; you can shout your head off, it's still nothing but terror. All I know for sure is: we're galloping! We're still a long way from our goals, they're not getting any closer, we can't even see them, we're likely to go on taking wrong turns, and we'll have to change horses; but one day—the day after tomorrow, or two thousand years from now—the horizon will begin to flow and come roaring toward us!"
> Dusk had fallen. "No one can see my face now," Ulrich thought. "I don't even know myself whether I'm lying. . . ."

Things are getting more complicated and confusing, perpetually in motion, and the one certainty is *speed*. Time moves faster and faster, and no one knows when it will stop this incessant changeless movement through changes: 1933? 2000? We never seem to get to the promised millennial end, but find ourselves once again in the middle:

truth-telling	pursuing	lying
light	dusk	dark
certainty	terror	
either	and . . . ?	or

But no matter how you would like to leave things there in the middle, something that you might call responsibility (issuing, in this case, from the artist's mouth) calls you to act: And? There is always the question of the *and?*, and it is actually two questions at once. It is the question of complication and addition: *And? Haven't you forgotten something? What about throwing this into the mix?* And *and?* is the question of time and ethics: *And? So? What now? What are you going to do for your next act?* And even if you are incapable—because of the constitution of your soul or because of where you are geographically located in this strange time and place of passage—incapable of whistling a happy tune in the dark ("Purifying Science Will Save Us," "Adding Soul to Science Will Save Us," "Democratizing Science Will Save Us"—all catchy tunes that we can't get out of our heads), one still hears the question reverberating in the dusky twilight: What to do, then?

> "Do you realize what you're talking about?" [Walter] shouted. "Muddling through! You're simply an Austrian, and you're expounding the Austrian national philosophy of muddling through!"
>
> "That may not be as bad as you think," Ulrich replied. "A passionate longing for keenness and precision, or beauty, may very well bring one to prefer muddling through to all those exertions in the modern spirit. I congratulate you on having discovered Austria's world mission."

Muddling: the word itself looks and sounds unattractive and pallid. Who would want to use such a gray word to advocate a presumably murky and messy set of activities, and ways of thinking that must surely be indistinguishable from muddleheadedness? Given a choice between pure science and a muddy muddling through, who wouldn't go for the former? And how, finally, can it be said that a desire for precision leads one to prefer the muddiness of muddling through over the keen appeals of modernist exertions?

When Worlds Collude

Many of the questions and characterizations of the world of *The Man Without Qualities* persist in the world of today. The excerpts above should indicate how the ideas contained in the book's narrative are still struggled over now, and its questions remain relevant. But we hope you haven't read only for content, but for the method we applied as well. Our charting of oppositional terms and their thirds, or middles, was a kind of prototype machine for generating meanings in a fictional world marked by irresolvable contradictions stemming from the sciences and the forces they set in motion. By occupying a middle position, Musil/Ulrich was able to develop questions, experimental propositions, and even goals that were somewhat orthogonal to the expectations and demands of "the system."

The accompanying chart is a more elaborate version of that middling-machine, and should serve as a handle on where we've been in this book. In a word, all our meanderings have been designed to lead back to the same place to which the sciences are drawn, and from which they draw their power and beauty: the middle. The terms which run down the middle of our chart—experimenting, articulating, kludging, judging, contingent affinities, powering/knowing, and especially reality—were all invented or borrowed for their capacity to evade and confuse the usual conventions about the sciences.

Thirdness, then, is akin to the third or middle term. We need to break a few habits of thinking about the middle. As logic's Law of the Excluded Middle suggests (either A or not-A, in its most abstract and simple form), the middle is usually considered a restricted area—indeed, so restricted as to be nonexistent. But when you look at the practices and concepts of the sciences in real life and real time, you find the mixings of the middle all over the place. The middle, the mess, is real—the realest of the real. Furthermore, the middle doesn't have to entail a golden mean as an ideal, nor a happy medium, nor the blandness of being middle of the road. Our middle is mean, all right—it's the harsh, stern, and demanding place where apparent opposites mix, gold and other precious metals lose some of their luster, and your limited capacities are there for everyone to see. Our middle is an unhappy medium: forever restless, questioning, insatiable, looking for trouble and almost always finding it. And while compromise is often a necessity in our middle, we remain a little too eccentric to be anywhere near the political center. For the sciences and for so much else in our world, the middle is the space of change and creativity; it's where the interesting problems and questions are; it's where things are unsettled, calling for experimentation; it's where the action is.

How to Muddle Through

The oppositional pattern of thought tends to persist because each side receives a different valuation. Epistemological or ontological propositions about the sciences almost always harbor a moral element or two. They almost always have one exalted term, while the other is at best tolerated, at worst something to be eradicated. As a generalization, the left-hand side of our chart lists the things, qualities, and values associated with the sciences in our culture, and which we mostly think of as

Chart of Oppositions and Their Muddled Middles

A	Excluded Middle	not-A
A	Excluded Middle	not-A
either	both/and	or
either	neither/nor	or
yes	yes and no	no
science		superstition
science		anti-science
science		religion
science	pursuing sciences	literature
science		politics
science		history
science		science studies
nucleus	symbiogenesis	cytoplasm
First Culture	Culture of the Third	Second Culture
Secondness	Thirdness	Firstness
presence	realitty	absence
reality		imagination
transparent	muddy	opaque
pure		applied
rigid		loose
elegant	crafted	messy
clever	kludged	klutzy
real		invented
neutral	charged	interested
free	contingent	driven
objectivity	robust	subjectivity
science		politics
certainty	ambiguity	uncertainty
discovering	experimenting	constructing
theory	signal that resists	practice
theory	sign-force	fact
seen		spoken
literal	articulating	metaphorical
plain language		- jargon
sense		nonsense
science		values
science		politics
knowledge	powering/knowing	power
reason		force
right		might
real	judging	constructed
determined		chosen
functional logics	assemblage	chance
science		society
experts	pluralism	communities
serious		playful
playful	responsibility	serious
safety	criss-cross	danger
complete	and?	empty

"good," while those things and qualities on the right-hand side are associated with the sciences' Others and are at least "not as good as . . ." their partners on the left, if not simply "bad."

The middle terms with which we've worked and played are meant to suggest other ways of thinking and doing. They oscillate between the extremes without ever escaping them. Sometimes they hold closer to what have been the best qualities or ideals of the sciences themselves—qualities or ideals that are often forgotten, misconstrued, obscured, or twisted into unusual patterns. The encouragement of public witnessing of truth-making and fact-building, for example, the modest recognition of multiple possible explanations for a phenomenon, the idea of an "excessive nature" which constantly eludes our comprehension—there are many things about the sciences that we can't give up, and shouldn't give up. At other times, our middle or third terms have drawn more from the Others of the sciences, the "not-A" side of our charting machine. The historical examples and our present-day stories exemplify how scientists have always occupied this middle space, and how it is simultaneously productive, problematic, generative, and open-ended.

The chart maps out a world in which anxieties, desires, and prescriptions about and for the sciences have been with us for most of the century, if not longer, organized by the powerful oppositions between reason and irrationality, science and antiscience, the "two cultures" of scientists and literary intellectuals. Many people in this world want the two sides of the chart to be kept as far away from each other as possible, in which any mixing in the middle is considered a dangerous threat—where more and more, in Foucault's words, it is thought that "any critical questioning of . . . rationality risks sending us into irrationality." It is a world in which many people believe there to be an end (however far in the future) where inquiry will halt in the stillness of objective certainty—and it is science that will take us to that end. And it is a world in which others believe that the sciences are fundamentally misguided and misconceptualized, a species of lying, a mere mask for power, an out-of-control force which must be contained politically and socially if we are to survive. Most disturbing, each side is often more concerned with proving the other wrong—supporting or subverting the hierarchical difference, but in either case maintaining it—than with finding means of productive engagement.

Conventional wisdom has it that the sciences require clear distinctions between theory and practice, theory and fact, and fact and fiction. Truth is seen, and presented in a literal form stripped of metaphors, ideally mathematics. In this world most scientists are absolutely convinced that those clear distinctions allow them to discover a preformed reality. And they have excellent reasons for holding that conviction. In this world scientists as well as other scholars have made historical and sociological inquiries into the sciences, and are absolutely convinced that one must speak of the constructedness of reality—that the truths of science and nature stem in fact from the worlds of language and its metaphors, politics and its interests, and culture and its values. And they, too, have excellent reasons for thinking as they do.

The Culture of the Third, however, is a world of symbiogenesis, a developmental-evolutionary system vitalized by both nuclear (reason, logic, science) and cytoplasmic (history, politics, culture) forces. It is a world not simply of interactions between these elements, but fusions, confusions, and profusions of them: wonderfully and woefully complex systems of muddy hybrid components that always present challenges and questions along with products and results.

You realize what we're talking about: muddling through. And to see if we can realize muddling through, actualize it through practical trials in this very trying, contradictory world, we offer the following experimental principles.

PURSUE A THIRD TERM

Muddling through isn't about overcoming or moving beyond oppositional thinking or contradictions. Oppositions are necessary for practicing any rationality, especially the sciences; contradictions can't be overcome, and they don't have a transcendent "beyond." We're hobbled/enabled by them. Classical logic excludes the middle or third term because its both/and, neither/nor characteristic is so horrifying—but the third term is almost always present. You have to get in between, *make* the in-between if necessary.

We're certainly not the first people to point out that the sciences—and indeed, Western culture and thought considered broadly—are boxed in by and organized according to such opposed terms. That we're not the first points to the *persistence* of these oppositions—they're a deep gravitational well which continues to capture anything that tries to circle around it or pass by it. Such oppositions shape our "thought-space" like our sun or a neutron star shapes and curves the spacetime around it. Our goal should not be to transcend binary thinking, as some people would put it (apparently not noticing that the recommendation requires its own set of oppositions, transcendence and immanence)—as if we could just shed the intellectual and cultural habits of two millennia like old clothes. Instead, we should pursue thirds, which are not syntheses of the two opposed terms, but destabilizing elements which shake the solidifying and stultifying patterns of hard and fast oppositions.

EXPERIMENTING, ARTICULATING, POWERING/KNOWING, AND JUDGING

are our efforts to approach the worlds of the sciences from this in-between. These slightly off-center focal points provide a more reasonable basis for inquiring into and living within a nature that is less preformed than *per*formed. Our world is one in which the sciences neither discover nor construct a world (and certainly do not reflect one), but where something happens that exceeds our capacity to see, think, or speak it.

Our constructions are geared toward making "really" statements—such as "Our language is geared toward making 'really' statements." It's that little verb "is" that has such powerful effects. The "is" implies equivalences when all you ever really get is a kludge—a force-fit that allows you to speak and think quite well, but never perfectly or finally. Language compels and resists, just as the material world compels and resists. You can't say just anything, and some things are harder to say than others. Likewise, the material world isn't just anything that scientists construct; nor is it ripe for discovery as soon as there is enough funding, lab technicians, and genius-scientists. Like the sciences, language needs to be thought of in terms of possible articulations, imperfect tools, and careful, repeated attempts to find idioms for things that couldn't be spoken of previously. Even though our language is finely tuned to the frequencies of the real, it can't ever really rid itself of metaphor. It is and it isn't "realist," just as practicing scientists are, and aren't.

Of all the third terms pursued in this book, perhaps the most important is re-alitty. This word-invention is a constant reminder, which you have to train yourself to see and think, that what we call "real" changes drastically over historical time, as indicated by the intrusive italicized *t*, physicists' and mathematicians' symbol for time. Remember the double reading that "realitty" was supposed to put in motion: one, a changing, transitory amalgam of inner and outer worlds, natural and cultural, technical and conceptual, and so on—we have a say in what realitty is. Re-alitty is in the middle, which makes it ours, and our responsibility. *And* two, an anagram of alterity, the absolutely other, source of pain and surprise. Realitty has nothing to do with us. Realitty is, in principle, an undecidable both/and—which nevertheless gets decided, in practice, in different ways under different historical and social circumstances.

But wait. Aren't we pulling a move not unlike those made in the name of purity, or in the name of Science—suggesting that realitty is the one, proper way to image and enact the world? What gives us the authority, now that we've undercut the concept of the real outer world, to tell anyone how they need to think about the sciences differently, better than they have?

In the first place, undercutting the concept of the real outer world is exactly what pursuing a third term like realitty *doesn't* do. It does question the concept, shake it, trying to get a feel for what makes it so solid, *testing* what its limits are. We tried to respect the concept of the real outer world as an operation of language, ideas, and thinking, and as the way in which many scientists experience the world and their practices in it. The concept is deeply embedded, and is not something that can be easily undercut. At the same time, we've tried to track down the contradictions that it harbors within itself, or to which it leads.

We invented the word "realitty" to make the concept and experience of the real outer world more complicated and encompassing. And by those criteria of more complex and encompassing—which the sciences themselves have often invoked to narrate their own progress—our account of the sciences is a better, more authoritative one. We've paid attention to the questions that conventional accounts often gloss over. We've examined overlooked details of practice, conceptual gaps, charged metaphors, hidden social relations, linkages to power and interests, necessary but contingent judgments—all things which are unavoidably part of pursuing sciences, but too rarely accounted for in public discussion. The complex, shifting realitty that the sciences both pursue and perform demands better accounting methods than were used in the past—the kinds of methods we provide here.

Philosophical hairsplitting? If we've learned anything over the course of this century, it should be that realitty has an incredibly fine-grained structure, demanding both the utmost precision and sensitivity to nuanced, persistent differences. With any luck, and a good deal of hairsplitting, the next century will be a more subtle one, in which the problems outlined in Section II will meet a public equipped with the ability to think critically about—and neither fear nor idolize— the sciences.

CULTIVATE A LONGING FOR PRECISION

When a critically thinking public pursues third terms, they will also be staying close to questions. This principle is very close in spirit to one of the conventional articulations of the sciences that is well worth preserving: The sciences are at their best when they're brutally skeptical, passionately working to overthrow the old regime of truths and practices. As Nobel laureate François Jacob has noted:

> [P]eople do not kill each other only for material benefit but also for reasons of dogma. Nothing is more dangerous than the certainty that one is right. Nothing is potentially so destructive as the obsession with a truth one considers absolute. All crimes in history have been carried out in the name of virtue, of true religion, of legitimate nationalism, of proper policy, of right ideology; in short, in the name of the fight against somebody else's truth. . . . At the end of the twentieth century, it should be clear to each of us that no single system will ever explain the world in all its aspects and detail. The scientific approach has helped to destroy the idea of an intangible and eternal truth. This is not the least of its claims to fame.

There's a modesty and humility that comes with an attitude of muddling through that we desperately need today, coupled with the pursuit of precision and specificity—a goal of the sciences that has been both praised and misunderstood. Galileo, Copernicus, and Darwin pursued this goal. And, as we've seen, scientists in widely varied fields, from primate behavior to quantum teleportation, continue the pursuit.

What's true for work within the sciences is also true for inquiry into the sciences. All of our examples contain dense ethnographic descriptions of practices in the sciences, tracing and questioning the articulated webs that make meaning and produce facts. Depending on where you are in the sciences, you face the task of analyzing the particular combination of metaphors, matters, observations, instruments, and ideas involved. The demand for precision leads to a complexity to be muddled through, inquisitively: How are gender metaphors involved in our descriptions of biological reality? If they're at work there, are they at work in cosmology, and in the same way? Does a quantum physicist encounter a resisting, Second object in the same way that a hydrogeologist does? Does Thirdness in the form of cultural values get kludged onto the measure of intelligence in the same manner, with the same effects, as it does in the measure of, say, coefficients of expansion in metals? In how many different ways do the habits and trajectories of corporations and funding agencies make their presence felt in the worlds of the sciences? All these things and more will be at work—which is to say, they will be in play—in different ways in different situations. They make generalization difficult if not impossible.

Our involvements with realitty are always a matter of power/knowledge. Our world is not one in which these two terms, power and knowledge, can be easily disassociated, contrary to what either science purists or social engineers might believe. We saw how many scientists are quick to invoke (heavily mythified) historical episodes from Nazi Germany and Stalinist Russia to put down any movements that explicitly link the so-called neutral tools of value-free knowledge and the ever-present world of political, economic, and social interests. And many social constructionists are just as quick to argue, in the name of social responsibility, for the necessity of a science completely attuned and subservient to a world of social values. Once again, we contradict each of these demands as well as conventional standard logic. We want neither of these either/or options, and both of them.

Addressing the contradiction starts with an effort at precision: How is power in fact necessary for the production of rigorous and even "objective" knowledge? How is power/knowledge manifested in laboratory experiments, in Darwinian evolutionary theory, in the collisions and collusions between Copernicus, Galileo, and other scientists who are inevitably caught in the intricate and extensive webs of religion, politics, and culture? How can we hold onto the ideals behind a "pure research" and "free inquiry" that has allowed for an incredibly productive combination of play, chance, commitment, and imagination, while dealing better with the unavoidable dilemmas of power that accompany that privileged space? How can we legitimate difficult and complicated decisions about our social world, if not through either the objective and disinterested reading of a Book of Nature, a grounding in a preformed reality, or raw political muscle?

One of the contradictory effects of pursuing keenness and precision is that you often encounter or create more ambiguity in the process. Most of our precise categories are kludge jobs. In Chapter 2, we looked at the production and use of PET images, and the combination of precise and imprecise articulations that made them useful and informative, within certain limits. Dr. Henry Wagner, retired Director of Positron Emission Tomography and Nuclear Medicine at Johns Hopkins University, reflects on what it means to define a disease:

> I personally believe that putting anything into a category is not because there is some kind of intrinsic truth [to it] but because it is useful. My philosophy is pragmatic. Therefore when you say that a person has disease X, it should be because putting him in that pigeonhole makes a difference in some way. These are manmade categories, abstractions are manmade simplifications of an unbelievably complex external world. . . . Right now, you say you have a simple explanation where serotonin is related to mood and dopamine is related to movement and acetylcholine is related to learning or intelligence. But to say that acetylcholine is intelligence and serotonin is mood and dopamine is movement is a gross and unhelpful simplification, a counterproductive simplification. Although it is true that blocking the dopaminergic system has been one way that it has been found to help some patients with schizophrenia, and blocking the serotonin uptake site, or inhibiting monoamine oxidase has been one way of helping patients become less depressed. If it helps, it helps. It helps solve problems, the world is surrounded with problems, people are surrounded with problems. . . . [T]he best invention of all is language—I think that the most important part of consciousness and memory is language, because it translates the past into the present.[6]

Many scientists and doctors know their truths to be "manmade" and inadequate, but those truths often provide much-needed help nevertheless. Simplification is both an aid and a trap, and this is the unavoidable double-bind of all pursuits of precision in the sciences.

Our passionate longing for keenness and precision sometimes results in only more and more information that no longer informs. A medical anthropologist asks an oncologist about how his field has changed:

> Well, the rules over the last seven, eight years have gotten much more complicated—they're now coming out to say that the older women do

benefit from chemotherapy [for breast cancer], but also lymph-node negative women may benefit, and maybe a combination of chemo and hor-
monal therapy is better than either one, and then they've come out with
all these different prognostic factors that may push you to give chemo in
someone that you would have thought originally would have been in too
good a prognostic group to need the chemotherapy.

So suddenly it becomes very complicated. . . . It's easy to get data, but
what to do with that data once you've got it is the hard part. . . . And it
would be easy if everything were bad prognostic, or everything was good
prognostic. But you're going to have all these women in the middle, and
you just simply don't know.[7]

We have more information than ever, but the algorithms that make straightforward calculation and prescription possible are elusive. The ideal extremes seem
clear, but actual bodies fall into a muddled middle. A visit to the doctor can no
longer be predicated on faith in expertise, the simple passing on of precise information, but is now more a question of how mutual uncertainties will be negotiated
and acted out. In pursuing precision, we end up confronting complexity.

KEEP IT COMPLICATED, STUPID!

Pursuing third terms, cultivating a passion for precision, staying tuned to realitty's
signals that resist will eventually lead you to run afoul of the popular principle favored by management gurus, to KISS: Keep it simple, stupid! Instead, a critically
literate public will have to keep it complicated, remembering that the web of articulations that are the sciences are far denser than can ever be completely compre-
hended, and extend farther than can be fully mapped. The sciences in their
complex fullness will always be incomprehensible, to some degree, and make us
look relatively stupid when it comes to understanding or controlling them perfectly.

Nobel laureate Ilya Prigogine and his coauthor Isabel Stengers have kludged together understandings based on investigations in the natural world with some religious elements drawn from the Judaic tradition to make a great argument for
muddling through: realitty is just too complicated, messy, fluctuating, sensitive,
and self-imbricated to be subject to any grand social scheme or pronouncement.
It's better to stick with small modest changes, that nevertheless can set off a chain
of self-organizing events that might turn into something beautiful:

We know now that societies are immensely complex systems involving a
potentially enormous number of bifurcations exemplified by the variety
of cultures that have evolved in the relatively short span of human history. We know that such systems are highly sensitive to fluctuations. This

leads both to hope and a threat: hope, since even small fluctuations may grow and change the overall structure. As a result, individual activity is not doomed to insignificance. On the other hand, this is also a threat, since in our universe the security of stable, permanent rules seems gone forever. We are living in a dangerous, uncertain world that inspires no blind confidence, but perhaps only the same feeling of qualified hope that some Talmudic texts appear to have attributed to the God of Genesis: "Twenty-six attempts preceded the present genesis, all of which were destined to fail. The world of man has arisen out of the chaotic heart of the preceding debris; he too is exposed to the risk of failure, and the return to nothing. 'Let's hope it works' . . . exclaimed God as he created the World, and this hope, which has accompanied all the subsequent history of the world and mankind, has emphasized right from the outset that this history is branded with the mark of radical uncertainty."[8]

Scientists can be attuned to the need to keep it complicated, particularly when they deal with complex systems in their daily work. That includes not only chaos and complexity theorists like Prigogine, but people in zoology departments or in a School of Fisheries. An article published a few years ago in *Science* by three scientists (Donald Ludwig, Ray Hilborn, and Carl Walters) working in natural resource management challenged the conventional view that scientists like themselves are capable of answering questions scientifically—that is, with absolute certainty and objectivity—regarding the amount of available natural resources, the long-term ecological effects of fishing, logging, mining, or farming operations, and what sustainable practices should be adopted. The lack of proper scientific controls; the complexity of ecological systems and the ways in which their natural variability masks overexploitation until it is too late; and the social and economic interests that aggregate around fish, timber, food, and other resources—all these make for situations in which "assigning causes to past events is problematical, future events cannot be predicted, and even well-meaning attempts to exploit responsibly may lead to disastrous consequences." Even in the cases of such "spectacular failures" as the California sardine industry or the harvesting of the Peruvian anchoveta for cattle feed (where the anchoveta yield plunged from 10 million metric tons to almost zero in a few years), "there is no agreement about the causes of these failures." It's impossible to decide finally if climatological, biological, or social forces played the decisive role. The result, they argued, is that "we shall never attain scientific consensus concerning the systems that are being exploited." And even if we *were* able to reach that kind of stable ground of certainty, history has shown that for reasons that are all too easy to understand, "many practices continue even in cases where there is abundant scientific evidence that they are ultimately destructive."

The five recommendations they make bear repeating here, in abbreviated form. Their "Principles of Effective Management" are very similar to many of the things we've been advocating:

1. Include human motivation and responses as part of the system to be studied and managed. . . .
2. Act before scientific consensus is achieved. . . . Calls for additional research may be mere delaying tactics.
3. Rely on scientists to recognize problems, but not to remedy them. The judgment of scientists is often heavily influenced by their training in their respective disciplines, but the most important issues involving resources and the environment involve interactions whose understanding must involve many disciplines. Scientists and their judgments are subject to political pressure.
4. Distrust claims of sustainability. Because past resource exploitation has seldom been sustainable, any new plan that involves claims of sustainability should be suspect. . . .
5. Confront uncertainty. Once we free ourselves from the illusion that science or technology (if lavishly funded) can provide a solution to resource or conservation problems, appropriate action becomes possible. . . . We must consider a variety of plausible hypotheses about the world; consider a variety of possible strategies; favor actions that are robust to uncertainties; hedge; . . . probe and experiment; . . . and favor actions that are reversible.[9]

These are excellent principles from scientists, the kind who usually don't write bestselling books, but who slog away in unsung jobs and institutions—excellent principles which, of course, will have to be muddled through in practice. In the exceedingly complex webs of the sciences, there's always a need for supplemental judgments, and further inquiry. Sometimes it *will* be better to wait for a greater degree of scientific consensus; in some situations we often *do* need scientists to remedy a problem; *trusting* claims of sustainability might occasionally be what's most urgently called for. And there aren't always going to be reliable guidelines for making those kinds of judgments. As a kind of Firstness, these laudable "Principles" always have to be muddled through the sideroads and detours that Secondness, in all its harsh varieties, will place in our paths.

KLUDGE ANOTHER UNUSUAL ASSEMBLAGE
The sciences themselves are complex, heterogenous power/knowledge assemblages of disparate elements—technical, conceptual, social, and cultural—whose linkages are always contingent: sometimes tight, sometimes loose, always changing and becoming more elaborate. Just trying to stay closer to their questions, in

all their specificity and complexity and elusive Thirdness, requires new patterns of inquiry. You can see something of this in the very materiality of our sentences on these pages throughout the book. They've been strewn with parenthetical remarks, complex subjunctive clauses, the formal and the informal juxtaposed, serious and playful comments, long lists of words to evoke multiplicity and complexity, long phrases kludged with dashes into the middle of an already long sentence—like this one—collected into longer subunits and stitched into a chapter, a section, a book.

There is an extraordinary demand for the kinds of expertise recognized as an inescapable necessity by Musil's fictional Ulrich over half a century ago. Even if they never finish their jobs—indeed, *because* they can never finish their jobs—experts who can undertake the most obsessively focused, singular pursuits of specific thoughts and problems will be of utmost importance. Hydrogeologists, immunologists, physicists, geneticists, and every other type of specialist will continue to need the institutional space that has traditionally been associated with pure science and playful curiosity. It's been a tremendously productive ideal, a "monster of energy," as Musil put it.

At the same time, the contradictory opposite is true: Experts will have to open up their obsessive focus, to understand and engage with the extended articulated webs of meanings and social forces which make their inquiries so productive and energetic. The sciences need more perspectives, from within and from without, to prevent any one view from becoming unduly authoritative; more people need to be speaking and questioning. A plurality of perspectives can help us pursue the precision that will uncover the intellectual, social, and political problems specific to each area of scientific work and inquiry. Sarah Blaffer Hrdy, whose work we discussed in Chapter 2, knows that in primatology, "we would do well to encourage multiple studies, restudies, and challenges to current theories by a broad array of observers."[10] We've seen how multiple explanations are not only possible in quantum physics, but inescapable, and even generative of creative dialogue and experimentation. In these and other cases, what's involved are multiple *disciplined* perspectives: people within the sciences who may have a different or marginalized view on things, but who have nevertheless been enculturated into some professional thought-style or another.

At the very least, the sciences and their pursuers will have to link up with the disciplines that study the sciences from historical, anthropological, philosophical, and literary perspectives, and *their* pursuers. Building such cross-disciplinary knowledge and the collaborations for producing them will meet with both logistical and ideological resistance.

In their polemic *Higher Superstition: The Academic Left and Its Quarrels with Science*, Paul Gross and Norman Levitt appeal to their fellow scientists:

On the whole, it is regrettable that serious students of the exact sciences rarely encounter, in their training, courses in the history of their disciplines that pay close attention to social, cultural and political factors. . . . But . . . the burden of essential preparatory studies is enormous, and is continually growing. Time is precious to a young scientist, and the optimal career path leads to the frontier of the subject as quickly as possible, leaving little opportunity for historical rumination. Nevertheless, much as one might lament the rarity of historically oriented science courses . . . in our judgment their absence is, on the whole, preferable to a hypothetical curriculum that requires such courses but hands responsibility for them over to historians and sociologists of the academic left. . . . The humanities, as traditionally understood, are indispensable to our civilization. . . the indispensability of professional academic humanists, on the other hand, is a less certain proposition. . . . The notion that scientists and engineers will always accept as axiomatic the competence and indispensability for higher education of humanists and social scientists is altogether too smug. Other sentiments are clearly astir. How these matters play out in American intellectual life will depend, to some degree, on the ability of the nonscientists to rein in the most grotesque tendencies in their respective fields.[11]

There's less time than ever, more demands on and for expertise than ever, and more resistance to change than ever. That's a harsh reality. But it's also more imperative than ever that scientists inquire into the history of their disciplines, and the political and cultural webs in which they are (contingently) entwined. There needs to be not only more serious exchanges, but more working *collusions* between and among the sciences and its Others. Those collusions will inevitably be kludge jobs, creaky and noisy assemblages that will require patience, watchful maintenance, and many trials and errors. But we hope this book has shown that these assemblages can be effective and generative of new ideas and questions that are far from being "grotesque," superstitious, or antiscience.

The truly difficult questions about pluralism arise when we start throwing "citizens" into the mix. In kludging these kinds of unusual assemblages, you'll have to follow that other principle of ours, recognizing the need for precision and specificity: different sciences, and the different ways in which they intersect with "community concerns," will require different assemblages for pluralistic, democratic involvement. Our case studies of Section II suggest where and how we think the sciences require the involvement of many members of various communities, if they are to have any chance of solving more social problems than they create.

Ours is not the simple political solution of "community direction of science," which is often either an empty slogan or the beginnings of an authoritarian closure of inquiry. Such solutions often postulate a kind of natural wisdom, that the community just knows better, has some ingrained sense of rightness and limits and ethics. What we're talking about are mechanisms of *community inclusion* in the larger assemblages of the sciences, tailored differently for different tasks and questions. For example, it's neither politically feasible nor scientifically productive to have people diagnosed with MCS directing what biomedical research should be done. It *is* politically possible to include these individuals on NIH funding and review panels, to build their knowledge and experience into clinical trials, and to link citizen-centered advocacy groups into the network of government and corporate institutions that push and guide research on this and other conditions. And that inclusion will produce better sciences of MCS than we would otherwise have.

Our challenge is to continually reinvent and reenact intellectual inquiry as well as practical politics. There is no ready-to-hand accounting mechanism for knowing in advance how and when to value expertise, and how and when to value the pursuit of nonexpert questioning and participation. That's the challenge to be muddled through.

MUDDLE THROUGH

In popular understanding, "muddling" is not the most compelling or attractive word, but we are convinced it is imperative to rescue it from the opprobrium usually associated with it, and to make it the basic metaphor for how science must proceed in the twenty-first century. Recognizing the simultaneous complexities of nature, society, and politics, especially in an era of rapid change, demands that we shun any easy answers. Easy answers will be imprecise, and we cannot afford to sacrifice precision in an age in which uncertainty is the rule, complex and indistinct causalities abound, and every idea or action sends repercussions throughout the tightly interconnected spheres of science, political economy, and the disparate and often conflicting values of a democratic, pluralistic society. Pleas to restore values to value-free rationality, to retain an unfettered faith in pure inquiry, or to renounce technological society and its basis in instrumental reason—all amount to quick fixes, the intellectual equivalent of get-rich-quick schemes.

It could be said that the humility and incrementalism that come with muddling through are too easy for us to advocate, reflecting the fact that we (the authors) are doing relatively well by the system; the status quo is definitely in our favor. We know that muddling through is crisscrossed with potential dangers, that it is an admirable and appropriate, but potentially conservative trope. In its early articulation by Edmund Burke, "muddling through" never challenged the status quo, and left institutions and assumptions of power intact. So we, and others like us who can afford to be patient, should remain aware that we can applaud muddling through because we have time. We're not, now, under a threat that requires fast and definitive response. Many others are.

But pushing the crisscross into highest tension, we would also argue that, when it comes to the sciences, we can't afford to be impatient. The combination of enthusiasm, speed, the sciences, and technologies doesn't have a great track record. Because the sciences are such volatile power/knowledge packages, some form of conservatism might very well be in order.

Muddling through does run the risk of tending toward conservatism. But the qualifying phrases are important—muddling through "tends toward" and "runs the risk" of conservatism, reminding us, again, how different articulations do different things, compelling and resisting our efforts to use them effectively. Muddling through tends toward conservatism, but it also harbors ways of engaging the world which may be uniquely suited for these unstable times.

Muddling through may not be a perfect princple—even we haven't thought it through all the way, because you can never think something through all the way. So this book should be considered as an experiment that involves putting the idea of muddling through into circulation, to better see how it works, what and who it works for, and who responds to it and how. We're committed, in other words, to pursuing what muddling through can become, within assemblages that persistently acknowledge their own limits and contradictions.

To return to our theme of Thirdness: one of the connotations of muddling that turns people off is its supposed grayness, if gray is that muddled, middled third term between black and white. We don't have strictly cognitive judgments about gray; it comes packaged with cultural sensibilities. It's subtly charged: Muddling is gray, and gray is dismal, bleak, featureless. But that's cultural code, not cosmological fact. You can, however, look around for different coding possibilities, as Trinh Minh-ha does in Japan (without glossing over the complexities and contradictions *within* that other culture):

> Grey remains largely (in Japanese as well as in many western contexts) a dull colour within culture's boundaries: one that usually implies a lack of brightness; an unfinished state; a dreary and spiritless outlook (the grey prospects, the grey office routine); a negative intermediate condition or position (that evades for example the spirit of moral and legal control without being overtly immoral and illegal) . . . and last but not least, the polluting of the natural world by ecologically destructive technology (in which modern Japan partakes as one of the most powerful producers). . . . But plain grey . . . in Japanese aesthetics is not so much the result of a mixing of equal parts of black and white as it is "the colour of no colour" in which all colours are canceling each other out. The new hue is a distinct colour of its own, neither black nor white, but somewhere in between—*in the middle* where possibilities are boundless. *Intermezzo.* A midway-between-colour, grey is composed of multiplicities. . . .

Trinh explores grayness in everything from the robes worn in the tea ceremony beginning in the sixteenth century, to the writings and practices of contemporary Japanese architects. In Japan, "Rikyu grey" is a combination of four opposing colors: red, blue, yellow, and white. Grayness is multiple, open-ended; it characterizes a space of transition. Gray is the shrouding fog: "One can say that the fog is a transcultural symbol of that which is indeterminate ('the grey area'); it indicates a phase of (r)evolution, between forms and formlessness, when old forms are disappearing while new ones coming into view are not yet distinguishable. . . . The delicate, diffused, ephemeral and transitory lights of dawn and dusk have always been the lights most sought after in colour photography."[12]

While there's something undeniably beautiful and even comforting about the suffused light of foggy dawns and dusks, there's also something haunting about it. It's the space of shipwrecks, navigation failed; people scream, drown, and die in the fog. It's where we lose our bearings and begin to grope around. And one of the things we grope for is better judgment.

BECOME A RESPONSIBLE HOLE-IST

Scientists often locate the "scientificity" of their project, and consequently their authority, in their methods rather than in their truths. The truths, they can admit, are of course always revisable, but the methods themselves are rock solid, and uniquely so. As we would say, it's the pursuing and not the arriving that makes the sciences "science." What we've done is ask questions about that solidity, located the holes in it, traced the more muddled, unmethodical parts of the scientific method—the many points at which judgments had to be made, or where subtle and not-so-subtle charges shifted the whole enterprise. In our pursuits of the subtle Thirdness that holds together and helps extend the articulated webs of the sciences, we've had to become hole-ists rather than holists.

The more we pursue precision, the more it seems ambiguity and uncertainty spring up somewhere else. The more elements we add to keep our conceptual systems complicated, the more holes emerge in their in-betweens. It's not very reassuring to think that our pursuits of sciences and truths never fully arrive. But at least we might be learning not to expect the sciences to deliver certainty, or to be disillusioned or paralyzed by contradictory expert judgments. *This* study recommends mammograms for all women under the age of fifty, *that* study says such a program offers little preventative value; *this* group of scientists says the increased risk of uterine cancer from estrogen replacement therapy is outweighed by the other benefits, including a decreased risk of breast cancer, *that* group of scientists says exactly the opposite—*why can't they make up their minds?*

In fact, they have simply made up their minds differently, and they've made different realities in the process: kludged together different methodologies, different populations to be studied, different assumptions and cutoff lines, different statistical analyses, different reagents, and many other differences, and all the ambiguities in between. Maybe further studies will help clarify the situation, resolve some of the differences, or fill in some of the holes. Maybe not. Judging among these different articulations and monstrously complicated assemblages is no easy task, and certainly not a science. We have to recognize that such decisions, often life and death ones, will almost always come down to someone's judgment call. We've also seen how the terms of those judgments can be specified and questioned as necessary. That's not nearly as comforting as absolute certainty—but it's not nothing.

Many people will say that the social and intellectual challenge we face in the next millennium is to figure out how science and values "go together," or "interact." These people, who are often highly critical of the "ideology of control" represented by modern scientific thought and processes, always have at least one eye turned toward how we might better control this imprecise combination of these vague terms, values and sciences. They presume that science and values are specifiable, determinate things and that their linkages are like a plumbing system: an input here, an elbow joint there, and here's where we suggest you apply the wrench or add a new valve, or value.

The "add-values-and-tighten-it-all-up" proposal displays an understandable longing for mechanism, but we need a new, less mechanical metaphor. Values and sciences can neither be simply opposed nor combined; we have to be able to see, question, and work the tensions and gaps between responsibility and experimenting. A straightforward organic metaphor won't do either, so we'll need another of our cyborg lobsters, another assemblage.

Every day, many times a day, people watch the performance of just such a science-assemblage, the weather report: a strange and wonderful combination of satellite technology, government institutions, local advertising, scientific expertise, show biz, beautiful pictures, precise measurements, history, folksy humor, "Super Doppler Radar," and helpful, often vital, predictions. The object of all this attention: the atmosphere. Oliver Wendell Holmes likened the atmosphere to the worlds of medicine and the sciences. He suggested that "medicine, professedly founded on observation, is as sensitive to outside influence, political, religious, philosophical, imaginative, as is the barometer to the atmospheric density."[13]

Think of sciences and values (or politics, cultures, etc.) as atmospheric events. Start with one enormous, chaotic, life-giving and life-deriving system. Perform an arbitrary separation for purposes of analysis: The sciences are an enormous cold air mass descending from the North, values a highly saturated warm front moving in from the South. Every winter here in New England, at least one major event catches all the local weather forecasters off-guard. Will it snow? Where? How much? Run the computer models of historical remembrance and cross your fingers as the commercial break ends and the red light of the camera flicks on and you're forced to make your prediction. Even if you get lucky, and the contours on your map the next day correspond in some manner to the vast stretches of blanketed territory, what exactly will you have accomplished? "We're already tracking this new possible storm system for the weekend, but it's too early to tell. Stay tuned. . . ."

Sciences/values, responsibility/experimenting is that fantastic world which we inhabit, not the problem we try to solve. We can predict probabilities, we can build new technological systems to sharpen those probabilities, we can take all kinds of protective measures from grabbing an umbrella on the way out the door to evacuating cities, but to think you can control the complex confluences is a big mistake.

If muddling through the sciences responsibly is sort of like predicting the weather, it's also sort of like the law. Jacques Derrida provides a helpful articulation:

> For a decision to be just and responsible, it must, in its proper moment, if there is one, be both regulated and without regulation: it must conserve the law and also destroy it or suspend it enough to have to reinvent it in each case, rejustify it, at least reinvent it in the reaffirmation and the new and free confirmation of its principle. Each case is other, each decision is different and requires an absolutely unique interpretation, which no existing, coded rule can or ought to guarantee absolutely. At least, if the rule guarantees it in no uncertain terms, so that the judge is a calculating machine—which happens—we will not say that he is just, free and responsible. But we also won't say it if he doesn't refer to any law, to any rule or if, because he doesn't take any rules for granted beyond his own interpretation, he suspends his decision, stops short before the undecidable or if he improvises and leaves aside all rules, all principles.[14]

Responsibility and justice are defined here in terms of competing demands. The law must be upheld; the law must be overturned. The sciences must be upheld; the sciences must be overturned. A good decision, a good science happens in the middle, in that space tracked on our radar screens where currents cross and storms brew. The stakes can be as high as life and death, requiring precise calculations and the most vigilant technologies, but these will be no guarantees.

Being a good, creative, responsible scientist is very similar to being a good, creative, responsible judge. You have to be aware that you're caught in several contradictions at once. You have to abide by existing rules, but those rules are no guarantee; in addition, your own actions will establish new rules to replace the old, which will then be seen as quaint and lacking any authority. You have to conserve existing truths *and* destroy them. You have to improvise new arrangements *and* obey old principles. You have to experiment, *and* you have to be responsible. You can't do both, and you must.

You must recognize that legal decisions necessarily adjudicate, resolving the complexity of a dispute into a final judgment. Similarly, the truths proclaimed by science necessarily silence the excess complexity of the data, resolving all ambiguity into a formula or argument that can be published and circulated. Since "each case is other, each decision is different," responsible experimenting in the sciences—in the laboratory, and in the social world—can come only from intellectual and social engagement with issues and questions on the ground.

Or on the coastline. Return again to Mark Tansey's painting *Coastline Measure* on the cover. In our chart, opposite terms meet like colliding air masses, or like furious sea and craggy rock. The front, the coastline, and the middle are all forms of Thirdness, where the pursuits of the sciences always happen. A big part of pursuing them more responsibly will involve the difficult, ceaseless measure of the fractal contours where the opposing forces meet. Triangulating on the holes—the ineradicable limits of any analysis, scientific or social—is crucial. It's a collaborative enterprise, to be undertaken by diverse teams with a variety of instruments. It's irreducibly messy and reliable, clumsy and precise, exciting and dangerous.

Our final metaphors here have shifted rapidly, from atmospheres to grounds to fractal coastlines. It says something about the simultaneous uncertainties and opportunities presented by the challenge of muddling through. But we have to turn muddling through into the hottest pursuit of the sciences, or else the next century will be even colder, less pluralistic, more dogmatic, and more obsessed with exterminating all forms of impurity than this past one. To pursue sciences and truths more responsibly, we're going to need every medium we can get, because neither realism nor cultural analysis nor studying history nor doing philosophy nor having the right politics nor being socially responsible will work on its own. But putting all of these into a new assemblage, one that threatens to fly apart any minute, will allow us to continue experimenting. There's no return to the safety of any kind of pure space, whether of science or of ethics. The first demand of responsibility in the sciences is simply to be willing and able to respond, and the first things demanding response are the ever-present holes in the sciences, ethics, and politics alike.

As with all the sciences, the reader has to take *Muddling Through* into realitty, and see how it holds up, how it works, and what it works for. Read the references we've provided, think carefully and critically about the ideas here, and, most important, run some more experiments. Do those experiments in public, where you'll be forced to account for your articulations over and over again, and where the outcomes can be judged from many perspectives. Crack this book, rip out pages and insert new ones of your own, kludge it into other sets of practices and communities.

And?

Let's hope it works.

Appendix

The EPR-Bohm Spin-State

In this appendix, we briefly explain the physics behind three important effects, recently demonstrated in laboratories: the Einstein-Podolsky-Rosen (EPR) correlations, the Greenberger-Horne-Zeilinger (GHZ) refutation of EPR's definition of realism, and quantum teleportation.

There are many interpretations of quantum theory, which we have no intention (or space) to go into here. Indeed, physicists agree far more on the mathematical manipulations and rules of translation into laboratory practice than they do on what their equations mean. So we should give our readers a simplified mathematical description and let them formulate their own interpretations—which is precisely what Bernstein does in his teaching. Here, however, we cannot assume the mathematical interest—nor provide the quantitative skills support—shown in a classroom. Instead, we must present verbal descriptions, adopting a framework where information about the world and experimental intra-action with materials are emphasized to make rather remarkable quantum effects understandable and to bring out their salient features.

As discussed in Chapter 8, the 1935 Einstein-Podolsky-Rosen thought experiment was designed to test the limits of quantum mechanics, in the form of the Copenhagen interpretation associated with Niels Bohr. The EPR thought experiment hinged on the measurement of two properties, P and Q, of two correlated or entangled particles. P and Q were the position and momentum of these particles.

David Bohm put on a new twist on this thought experiment with his proposal to measure the spins of the particles, which have discrete values rather than the range of continuous values that position and momentum can have. Figure 1 schematizes Bohm's thought experiment in three stages: before, during, and after. The two particles start from an initial state of zero total spin (as indicated by $S = 0$ in the "before" diagram).

With two spinning particles coming from an initial spin-zero state, we can see why the experimental results for one particle seem to depend on the other's reality,

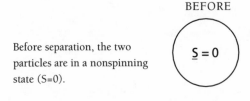

BEFORE

Before separation, the two
particles are in a nonspinning
state (S=0).

$\underline{S} = 0$

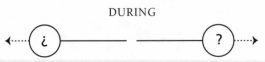

DURING

The two particles are ONE system, so they have opposite spin even when each is
inderterminate.

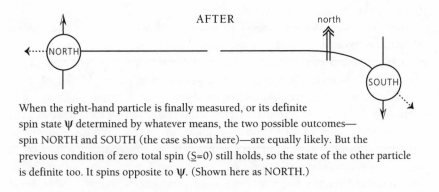

AFTER

north

NORTH

SOUTH

When the right-hand particle is finally measured, or its definite
spin state ψ determined by whatever means, the two possible outcomes—
spin NORTH and SOUTH (the case shown here)—are equally likely. But the
previous condition of zero total spin ($\underline{S}$=0) still holds, so the state of the other particle
is definite too. It spins opposite to ψ. (Shown here as NORTH.)

Figure 1. The Einstein–Podolksy–Rosen–Bohm state.

and why this isn't so "spooky" after all. Because they are created in this condition,
they always have opposite spins in every direction, e.g., if one has spin NORTH,
the other will have spin SOUTH. This is true at any point *during* their flight from
each other, even when they haven't been measured by a physicist's apparatus and
their individual states are indeterminate, unknown, or nonexistent (as indicated
by ¿ and ? in the "during" diagram).

Separating the particles doesn't change the circumstance that the two must be
spinning in opposite directions for the total spin to add up to zero. Any measurement will have two possible outcomes. Each outcome will happen at random, fifty
percent of the time. The figure shows a measurement of spin along the north direction on the rightward traveling particle, with possible outcomes of NORTH or
SOUTH. By measuring the right-going particle's spin in some direction—which re-

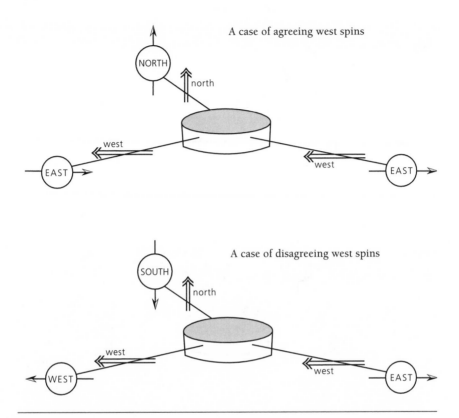

Figure 2a. The GHZ theorem. The setup: mixed experiment measuring one north and two west spins.

quires the magnetic field indicated by the north-pointing arrow—it is shown, known, or made to be in state **Ψ**—in this case, it turned out to be a spin SOUTH state. The left-going particle must be in the exact *opposite* state, spin NORTH, whether someone chooses to measure it or not, and even if it's far across the laboratory. The same opposition occurs for any chosen direction in space: two spinning particles emerging from a zero total spin state have opposite luck. While each individual outcome is random, every pair's spins along the *same* direction are opposite.

The Greenberger-Horne-Zeilinger (GHZ) Theorem

Figure 2 describes the GHZ theorem, again in three stages: the setup, the prediction, and the results. First, a machine is set up that produces three fully entangled particles. A source produces a three-particle state which can be written as UP UP UP - DOWN DOWN DOWN. (The up-down axis is perpendicular to the north-south and east-west axes. Think of the corner of a box: Its height is the up-

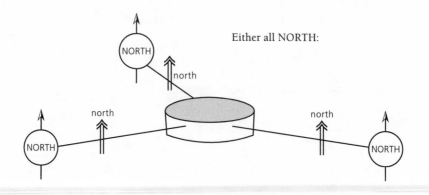

Either all NORTH:

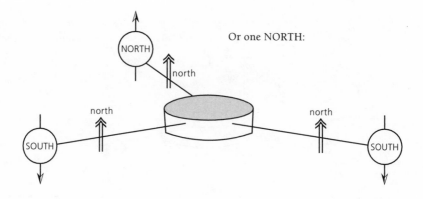

Or one NORTH:

Never zero or two NORTHS!

Figure 2b. The prediction: measuring north spins only.

down axis, length is east-west, and width is north-south.) This state of the ensemble of the three particles is actually a superposition of two states of equal probability: one state in which all three particles have spin parallel to the up direction, and one state in which all three particles have spin in the opposite direction, down. It is usually called the GHZ state, after Bernstein's three coinvestigators, a name given by Cornell University physicist David Mermin, who first investigated it.

In this state, if you know (or make, or show, or are told) the values for the properties of any two spins, you know the value for the third. The mixed experiment of two west magnetic fields and one north field (indicated by the double arrows) shows why. This set of north and west magnetic spin detectors define those spins as Einsteinian "elements of reality."

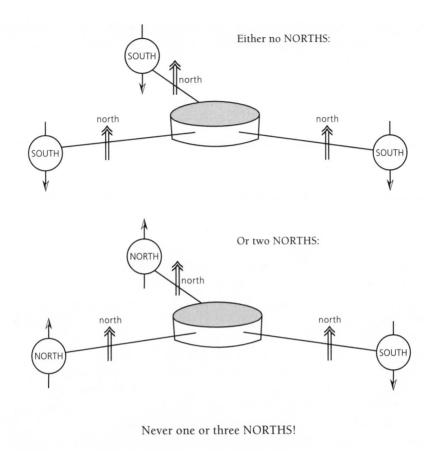

Figure 2c. The actual results: measuring north spin only.

Without in any way disturbing or measuring particle one, you can predict its north spin with absolute certainty by measuring the west states of the other two. There are two classes of outcomes, as shown in Figure 2a. When the magnetic spin detectors for the other two particles are set west, those particles either agree in their spin (i.e., are spinning in the same direction, say EAST), or disagree, with one spin oriented EAST and the other WEST. If they agree, the spin of particle one will *always* be oriented NORTH when it passes through its magnetic spin detector aligned along the north-south spin axis. If the other two particles disagree, then particle one *always* comes through with its spin aligned SOUTH. The property of north-spin, then, is real in the Einsteinian sense—and should exist independently of our acts of measuring or knowing.

In this setup west-spin is also "real." If we measure the north-spin of particle one and the west-spin of one of the other particles, we can predict the value (WEST or EAST) of west-spin of the third particle without in any way disturbing or touching it. You run a million sets of particles through this system, and they always come out in the predicted state, EAST or WEST.

But this setup actually becomes a powerful disproof of Einsteinian elements of reality, and precisely because it uses only one hundred percent, perfectly predictable outcomes, and not the probabilistic results so frequently associated with the physics and philosophy of quantum mechanics. Here's why.

Without realizing it, we have just made possible a prediction about what should happen if the magnetic spin detectors are now all set to point north. This prediction is illustrated in Figure 2b. One would predict there must be either one NORTH spin, or three. Those are the numbers predicted because, since the west spins are "real" in the Einsteinian sense—that is, existing independent of our acts of measuring and knowing, they must either agree completely (e.g., all WEST-WEST-WEST) or have one agreement and two disagreements among the possible pairs of the three particles (e.g., the set WEST-WEST-EAST is composed of one WEST-WEST pair, one WEST-EAST pair, and one EAST-WEST pair).

You can understand the logic by looking case by case at the outcomes from Figure 2a. As we saw in this figure, if any two particles agree in their west-spin, the third particle will have NORTH spin. So if particles one and two are both WEST, particle three will be NORTH after it passes through a north magnetic detector; if particles two and three are WEST, particle one is NORTH; and if particles one and three agree in west-spin, then particle two will be NORTH. So with all three particles agreeing in their west spins, then, if Einsteinian reality holds, the particles would all be spinning NORTH when they passed through detectors oriented along that axis, and the physicist would get three NORTH signals.

The other general possibility is that there is one agreeing pair and two disagreeing pairs of west spins. Here's the case-by-case logic for each pair: particle one and particle two are both WEST, and particle three will then come out NORTH; particle two is WEST and particle three is EAST, so particle one will come out SOUTH; and particle one is WEST and particle three is EAST, making particle two come out SOUTH. The physicist gets one NORTH signal in this situation.

In other words, since these west spins are real (as established in the setup experiment), there should be an odd number of clicks on the NORTH detectors for every group of entangled particles which passes through the apparatus.

But when you actually do that experiment, *every* single time the prediction is wrong! Only the incorrect values ever occur. Each time you measure you get only two NORTH spins or none, never an odd number, as illustrated in Figure 2c.

Why does the north-spin "element of reality" have contradictory values when measured directly and indirectly? Because the prediction uses the "reality" property of west-spin as if it still exists even if it is not measured. Drawing conclusions from west-spin measurements that we have never made—but only *could* have made—is just not possible. There are at least two ways to articulate why this is so.

The mathematical reason is that the different possibilities of two terms in the GHZ-Mermin state actually interfere with each other to create the probability for west-spin. The terms have mathematical signs that interfere destructively like waves that are 180° out-of-phase, exactly where Einstein's definition would need them to add up instead of canceling. But another way to express why the outcome differs so dramatically from the prediction is in terms of experimenting, since in quantum physics, as Israeli physicist Asher Peres once put it, "unperformed experiments have no results." In other words, the west-spins of the particles aren't "real" until they are constructed, or imposed, or extracted, in the work of experimenting. The west-spins can't be used to make predictions before an experiment is done, because it's only after the experiment that they will have a "local reality."

Quantum Teleportation

In quantum teleportation, we again have a way to picture the geometry of the spins, as we did in the EPR-Bohm and GHZ cases. The experiment is schematized in Figure 3. Like the GHZ theorem, it requires quantum states of three particles. Particle A is in a state Ψ (*psi*) to be teleported, and particles a and b are the two shared by Alice and Bob which are entangled in one of those perfectly correlated, exactly opposite-pointing singlet states that we saw in the first EPR-Bohm example above. In the course of teleportation, the first two (particles A and a) are measured at Alice's station, while particle b is transformed into B—an exact replica of the original state Ψ of particle A—wherever Bob happens to be. The "mirror" doesn't change any properties of the particle, but is only there to bounce particle b back to give Bob the time to receive Alice's two-bit message (indicated by the vertical double arrow) telling him what to do, and then pick and set up the proper treatment to convert particle b's state into Ψ.

Here are the details of how this happens. The diagram should be read from bottom to top as a schematic picture of what happens as time goes on. Alice and Bob begin their experimenting by arranging to have the pair of entangled particles a and b (the dashed lines) fly to them from a source, such as a down-converting crystal. Someone—we call her Carol—sends the teleportee particle A in the state Ψ to Alice, timing it to arrive exactly when particle a does, so she can experiment with *both of them* together as a pair. (In practice this is done by generating photon A in a second independent down-conversion from the same pulse of a powerful ultraviolet laser.)

At the shaded ellipse Alice does her part by measuring the combined state of the particles A and a, as if they were fully entangled (i.e., perfectly correlated as in the two previous examples). These two particles have in fact never interacted before. But a *has* interacted with b. Indeed, the spins of particles a and b are exactly opposite because they are an EPR-Bohm pair. By experimenting on A and a as if *they too* were fully entangled, Alice produces a realitty that enables her to signal Bob how to change the state of particle b to an exact replica of A.

There are four states of combined spin for Alice's particles (that is, two times two) since each particle has exactly two outcomes for any spin measurement. In the experiment Alice chooses to perform, the particular four states she seeks are fully entangled, called Bell states in honor of John Bell. In each of them the spin of particle a is perfectly correlated with the spin of particle A. But the relationship is different in the different cases. All four Bell states have equal probability; each occurs exactly one-quarter of the time. One of them is the EPR state itself, in which the particles have opposite spins.

The simplest case occurs when Alice finds A and a in this EPR state of opposite spins. Consider this state, the one used in the first actual teleportation experiment, performed in Anton Zeilinger's laboratory in Austria. In this case, the spin of particle A is exactly opposite that of particle a. She doesn't need to know which particle has which particular spin; she just needs to know that the two are opposite each other. If Alice is lucky enough to find this spin-zero combination, all she has to do is signal Bob to do nothing. Particle a and particle b were formed as a spin-zero state by design; their spins are always already opposite each other. So when Bob does nothing and simply passes b on to a fourth party (called Ted in the figure), Alice and Bob together have managed to make a perfect replica of Carol's state Ψ: that state has been teleported. (Ted could check the teleportation by seeing that he received the correct state by a series of tests. In the Innsbruck experiment, the spins used were actually those of polarized photons; the role of Ted was played by a polarization analysis of the teleported photons.)

The other three possible outcomes call for Bob to give particle b just a half-turn around one of three different directions, north, west, or up. In principle, these other three Bell states are somewhat harder to teleport. In practice, implementing Bob's teleporting procedures for these cases has yet to be done in a laboratory. But conceptually they are related to the easiest first case. In each of these three Bell states the two particles combine to have a total spin of one unit, but with zero spin in a particular direction. As a pair, they have one axis along which a spins exactly opposite to A.[1] Alice uses the complete outcome of her measurement in the gray circle to tell which of the three possible axes (up-down, north-south, or east-west) has opposite spins. Alice's Bell state measurement enables her to tell Bob which axis he should use to turn particle b into state Ψ. She signals that information to Bob, sending one of four possible messages, which we call Zero, Up, North, West.[2] Thus for example, if Alice finds particles a and A in a state with spins opposite along the west axis, she sends the message "West." This signal is indicated in Figure 3 by the vertical double arrow, since it requires two bits of information to convey four equally likely messages.

Bob gives b exactly one-half turn around the axis Alice signals him: a half-turn about the up-down axis for Up, about the north pole for North; about the east-west line for West; and no axis at all for Zero (the easiest case we treated first). We can picture the result correctly: when Alice signals "Up" her measurement has indicated that particle a spins opposite to particle A along the up-down axis. Bob's half-turn about this direction won't change b's up-spin. Which is good, because opposites of opposites are the same, and we want to make that state of b come out Ψ. Guaranteeing that up-down spins are the same is half the struggle. Fortunately, a half turn about the up-down axis is just what we need to turn *both* north- and west-spins to their opposites. (Picture yourself facing north with your arms out; your right arm points east. If you turn 180 degrees, your face will be south and your right arm will point to the west.)

We have emphasized measuring the "combined" Bell states, because Alice does *not* measure the spin of each particle separately. That would simply destroy the unknown state Ψ without telling her how it relates to Bob's particle. Instead, she measures the combined state to find the spin of particle A *relative* to particle a. This in turn lets her know the relationship to Bob's particle b because it spins fully opposite to a.

Knowing the outcome of the relative spin measurement, Alice then signals to Bob which of the four operations to make on his EPR particle b to turn it into a perfect copy of the original state Ψ. In quantum teleportation, these instructions from Alice to Bob go at ordinary speed over ordinary media, certainly no faster than radio. They can even be published in your weekly newspaper. Instead of going faster than light to a preset coordinate as in *Star Trek*, the message for quantum teleportation goes at a leisurely pace.

And so does Bob: He merely reads the weekly paper wherever he is, opens his spinning-particle storage, and performs one of four operations whenever he is ready. If the message from Alice says Zero, he does nothing. If it is one of the three direction names, he calmly and carefully turns the particle spin by exactly half a turn—180 degrees—about the axis mentioned, north, west, or up.

Of course Alice's quantum mechanical measurement still destroys the original state of particle A, or rather turns it (together with a, her half of the EPR pair) into a state characteristic of whichever Bell state came out of the measurement, no matter what the original state Ψ.

The original state destroyed, then re-created at a distance, and at the expense of certain resources of information sharing—it certainly sounds like teleportation. Those resources are a simple signal to instruct Bob's action, and a previously established quantum information link in the form of perfect EPR correlations.

It's all quite remarkable. Alice's two-bit message—one of four possibilities—is enough for Bob to transform his half of the EPR state into an exact copy of the original spin state, ψ. Remember, this is a state which can have its spin pointing exactly toward any of hundreds of thousands of different directions, to within a fraction of a degree of precision. Somehow those hundreds of thousands of possibilities are summed up into just four operations. Even more remarkable is the fact that the EPR particle which Bob transforms into the exact right possibility is, by itself, completely without information. It is not spinning about any particular axis direction whatever. Indeed by itself its behavior is as unpredictable and useless as if its spin axis had been picked totally at random. Whenever we measure an EPR particle's spin by itself, in whatever direction we may chose, the probability for either of the two possible outcomes is exactly fifty-fifty.

Note where Alice's choice was crucial. She chooses not to do the simplest thing, i.e., to measure total spin in a given single direction.[3] Instead her measurement looks at the two particles as totally entangled. Alice knows that her two particles, A and a, are totally unrelated. There is no a priori reason for Alice to think her particle a is entangled with Carol's particle A. Quite the contrary: they were prepared in procedures performed independently, by different people for different purposes; Carol gave her particle A in state Ψ to send it far away, to deliver it secretly to Ted. Yet Alice chooses to measure four possible outcomes that are each totally entangled. Alice does her part by insisting on viewing the unknown state as totally entangled with that of particle a, her half of the EPR pair.

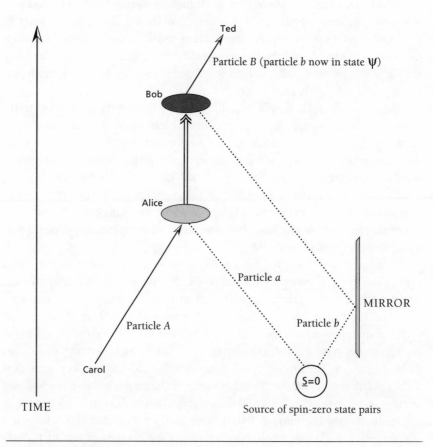

Figure 3. Quantum teleportation.

Chapter 2

Notes

Prologue

1. Evelyn Fox Keller, *Secrets of Life, Secrets of Death: Essays on Language, Gender, and Scien e* (London: Routledge, 1992), p. 181.
2. François Jacob, *The Logic of Life: A History of Heredity* (New York: Random House, 1982), p. ix.
3. Avital Ronell, quoted in *Angry Women*, ed. Andrea Juno and V. Vale (San Francisco: Re/Search Publications, 1991), p. 127.
4. C.E. Lindblom, "The Science of Muddling Through," *Public Administration Review*, Vol. 19, No. 2(1959); reprinted in *Organization Theory*, ed. D.S. Pugh (New York: Penguin, 1990), pp. 278–294.

Chapter 1

1. John Dewey, *Essays in Experimental Logic* (University of Chicago Press, 1916), p. 35
2. Ibid., p. 67
3. Ibid., pp. 36–37. For more on Dewey's philosophy of science and technology, see Larry A. Hickman, *John Dewey's Pragmatic Technology* (Bloomington: Indiana University Press, 1990). Cornel West's *The American Evasion of Philosophy: A Genealogy of Pragmatism* (Madison: University of Wisconsin Press, 1989) gives an excellent reading of the full scope of the pragmatist tradition, while Joseph Rouse's *Knowledge and Power: Towad a Political Philosophy of Science* (Ithaca: Cornell University Press, 1987) focuses more on the sciences.
4. "Falsifiability" is the criterion invented by Sir Karl Popper, the philosopher of science who has probably done more to shape our culture's conventional notions of science and scientists than almost anyone else in the twentieth century. Forced to the realization that scientific theories could not be verified in the strictest sense (and philosophers of science like Popper are a very strict bunch), Popper proposed that theories could nevertheless be falsified or disproven, and that this openness to empirical or theoretical refutation was the chief quality by which science should be demarcated from less reliable, less rational pursuits.

5. The best guide to why none of these standard explanations hold is Imre Lakatos, a philosopher of science who nevertheless was most committed to preserving a complete rationality at the core of science (against such "cultural relativists" as Thomas Kuhn and Paul Feyerabend). See Imre Lakatos and Elie Zahar, "Why Did Copernicus's Research Program Supersede Ptolemy's?" in *The Copernican Achievement*, ed. Robert Westman (University of California Press, 1975), pp. 354–383.

6. Indeed, the greatest improvement in calculation lies in the "latitude problem," as historians of science have only very recently recognized. This problem is the tendency of planets to wander north and south of the equitorial plane in which the sun seems to move around the earth. The problem is caused by slight differences between the orientation of the earth's orbital plane, and those of other planets, about the sun. [David Pingree, personal communication, 1998.]

7. Steven Shapin and Simon Schaffer, *Leviathan and the Air-Pump: Hobbes, Boyle, and the Experimental Life* (Princeton University Press, 1985).

8. Shapin and Schaffer, p. 67.

9. Shapin and Schaffer, pp. 45–46.

10. Shapin and Schaffer, p. 71, quoting Boyle's *The Sceptical Chymist*.

11. Quoted in Silvan S. Schweber, *QED and the Men Who Made It: Dyson, Feynman, Schwinger, and Tomonaga* (Princeton: Princeton University Press, 1994), pp. 465–466.

12. Gerald Holton, "Subelectrons, Presuppositions, and the Millikan-Ehrenhaft Dispute," in *The Scientific Imagination: Case Studies* (Cambridge: Cambridge University Press, 1978), pp. 25–83.

13. Peter Galison, *How Experiments End* (Chicago: University of Chicago Press, 1987), p. 88. Galison goes on to describe how Millikan's commitments did not serve him so well in his subsequent work on cosmic rays.

14. Quoted in Abraham Pais, *Inward Bound: Of Matter and Forces in the Physical World* (New York: Oxford University Press, 1986), p. 71.

15. Thomas Kuhn's foreword to Ludwik Fleck, *Genesis and Development of a Scientific Fact* (Chicago: University of Chicago Press, 1979), p. viii.

16. Fleck, p. 89.

17. Fleck, p. 92.

18. Fleck, pp. 95, 98, 102.

19. Among the different kinds of historians, anthropologists, sociologists, feminist theorists, ethnomethodologists, rhetoricians, philosophers, and others who are badly thrown together under the label "social constructionists," Fleck is often considered by all of them to be a common intellectual ancestor. We decided that defending the entire field of the "social construction of science" would be a thankless, fruitless, and humorless task, and so have focused here on some of the shared themes first introduced by Fleck. There is much work in this diverse body of scholarship called "social constructionism" that is valuable and provocative, some that is bland and poorly thought through, and some that is a mix. In addition to the work which we draw from and cite over the course of this book, the following books are well worth reading for insights into how the sciences and scientists actually work: Adele E. Clarke and Joan H. Fujimura, eds., *The Right Tools for the Job: At Work*

in *Twentieth-Century Life Sciences* (Princeton: Princeton University Press, 1992); Anne Fausto-Sterling, *Myths of Gender: Biological Theories About Women and Men* (New York: Basic Books, 1985); Joan Fujimura, *Crafting Science* (Cambridge, MA: Harvard University Press, 1997); Peter Galison, *Image and Logic: A Material Culture of Microphysics* (Chicago: University of Chicago Press, 1997); Ian Hacking, *Representing and Intervening* (Cambridge: Cambridge University Press, 1983); Evelyn Fox Keller, *Secrets of Life, Secrets of Death: Essays on Language, Gender, and Science* (London: Routledge, 1992); Karin Knorr-Cetina, *The Manufacture of Knowledge: An Essay on the Constructivist and Contextual Nature of Science* (New York: Pergamon Press, 1981); Bruno Latour and Steve Woolgar, *Laboratory Life: The Construction of Scientific Facts* (Princeton: Princeton University Press, 1986); Helen Longino, *Science as Social Knowledge: Values and Objectivity in Scientific Inquiry* (Princeton University Press, 1990); Michael Lynch, *Art and Artifact in Laboratory Science: A Study of Shop Work and Shop Talk in a Research Laboratory* (London: Routledge and Kegan Paul, 1985); Andrew Pickering, *The Mangle of Practice* (Chicago: University of Chicago Press, 1996); Brian Rotman, *Ad Infinitum . . . : The Ghost in Turing's Machine: Taking God Out of Mathematics and Putting the Body Back In* (Palo Alto: Stanford University Press, 1993); Sharon Traweek, *Beamtimes and Lifetimes: The World of High- Energy Physicists* (Cambridge, MA: Harvard University Press, 1988).

20. Max Planck, *Where Is Science Going?* (New York: Norton, 1932), pp. 64–65.

21. For those readers lucky enough to have missed the "Science Wars" that preoccupied many scientists, other academics, a few reporters and pundits, and even the authors as we were writing this book, a few reference points: the initiating event was the publication of Paul Gross (a biologist) and Norman Levitt's (a mathematician) book *Higher Superstition: The Academic Left and Its Quarrels with Science* (Baltimore: Johns Hopkins University Press, 1994). The titles of the articles discussing their book and the debates it spawned suggest the fundamentalist and purist tenor of the events: Anthony Flint, "Science Isn't Immune to Cultural Critique," *The Boston Globe*, November 15, 1994, pp. 1, 28; Christina Hoff Sommers, "The Flight from Science and Reason," *The Wall Street Journal*, July 10, 1995, p. 24; Franklin Hoke, "Scientists See Broad Attack Against Research and Reason," *The Scientist*, Vol. 9, no. 14 (July 10, 1995), p. 1ff ; Robert L. Park, "The Danger of Voodoo Science," *The New York Times*, July 9, 1995, p. A31; and Malcolm W. Browne, "Scientists Deplore Flight From Reason," *The New York Times*, June 6, 1995, p. C1. Gross and Levitt's most recent literary effort is an edited volume, named after the conference which they organized under the auspices of the New York Academy of Sciences in 1995: *The Flight From Science and Reason* (Annals of the New York Academy of Sciences, Vol. 775, 1996).

   Then New York University physicist Alan Sokal published two articles in the same week in the spring of 1996, touching off a media flurry that also pitted the forces of science and reason against their opposites, antiscience and irrationality. The cultural studies journal *Social Text* (No. 46/47, Spring/Summer 1996, pp. 217–52) published Sokal's article, "Transgressing the Boundaries: Toward a Transformative Hermeneutics of Quantum Gravity" in its special issue devoted to the "Science Wars" in the academy. In a confessional exposé prepared simultaneously for *Lingua Franca* (July/August 1996, pp. 54–64), a journal that covers the beat of university culture, Sokal revealed that his *Social Text* article was a hoax, a parody built on extensive quotes from academic theorists writing about science, stitched together by what he considered Leftist ideological cant and nonsensical "jargon." The Associated Press picked up this story, prompting National Public Radio to interview Sokal, and things snowballed from there. In his own words, Sokal was intent on showing "the proliferation . . . of a particular kind of nonsense and sloppy thinking: one that denies the existence of objective realities, or (when challenged) admits their existence but downplays their practical relevance;" and how the "subjectivist thinking" and "epistemic relativism" infecting and polluting the "self-perpetuating subculture" of science studies is both "false (when not simply meaningless)" and "undermines the already fragile prospects for progressive social critique." "The dispute over the article," suggested the front-page article in *The New York Times* (May 18, 1996, pp. 1, 22), "goes to the heart of the public debate over left-wing scholarship, and particularly over the belief that social, cultural, and political conditions influence and may even determine knowledge and ideas about what is truth."

We haven't relegated the "Science Wars" to this footnote because we don't think they're serious. The "Science Wars" are quite serious: people have lost their jobs over these issues, careers have been stunted, and an intellectual culture of respect and open inquiry has been poisoned. Our hope is that this entire book of ours will serve as a more indirect, and hopefully more effective, intervention in these issues.

22. John Wheeler, "Information, Physics, Quantum: The Search for Links," in *Complexity, Entropy, and the Physics of Information: Santa Fe Institute Studies in the Sciences of Complexity,* Vol. VIII, ed. W.H. Zurek (New York: Addison-Wesley, 1990), p. 4.

23. Planck, *Where is Science Going?* p. 82.

24. Planck, pp. 214–15.

25. Dewey, *Essays*, p. 44.

26. Dewey, *Essays*, p. 45.

Chapter 2

1. Steve Heims, a physicist turned historian of science, situates Bateson (along with Norbert Wiener, John von Neumann, Margaret Mead, and others) in the changing social and scientific contexts of post-World War II America, where interdisciplinary conferences sponsored by the Macy Foundation pushed the concepts and techniques of cybernetics in multiple directions. Steve Joshua Heims, *Constructing a Social Science for Postwar America: The Cybernetics Group 1946–1953* (Cambridge, MA: MIT Press, 1993).

2. Gregory Bateson, *Steps to an Ecology of Mind* (San Francisco: Chandler Publishing Company, 1972), pp. 35–37.

3. Shapin and Schaffer, *Leviathan and the Air-Pump*, p. 291.

4. Quoted in Robert Proctor, *Value-Free Science?: Purity and Power in Modern Knowledge* (Cambridge, MA: Harvard University Press, 1991), p. 33

5. Quoted in Shapin and Schaffer, p. 306.

6. Quoted in Shapin and Schaffer, p. 307.

7. We are treating Hobbes's contrary position in these debates in a condensed and simplified form and recommend a full reading of Shapin and Schafer's book for the important subtleties and complexities not found in our discussion. Also, for work that puts questions of gender in their proper position in these events, see Evelyn Fox Keller, *Reflections on Gender and Science* (New Haven: Yale University Press, 1985); Donna J. Haraway, *Modest Witness@Second Millenium: FemaleMan Meets OncoMouse* (New York and London: Routledge, 1997), esp. pp. 24–31; and David Noble, *A World Without Women: The Christian Clerical Culture of Western Science* (New York: Oxford University Press, 1992).

8. Quoted in Shapin and Schaffer, p. 106.

9. Paul W. Knoll, "The Arts Faculty at the University of Cracow at the end of the Fifteenth Century," in *The Copernican Achievement*, ed. Robert Westman (University of California Press, 1975), p. 156.

10. Robert Westman, "Proof, Poetics, and Patronage," *Reappraisals of the Scientific Revolution*, ed. David Lindberg and Robert Westman (Cambridge: Cambridge University Press, 1990), p. 178.

11. Westman, pp. 187–8, 192, 194.

12. Quoted in Brockman, *The Third Culture*, p. 23.

13. Brockman, p. 91.

14. Richard Dawkins, *The Blind Watchmaker: Why the Evidence of Evolution Reveals a Universe Without Design* (New York: W.W. Norton, 1987), p. 111.

15. Richard Doyle, *On Beyond Living: Rhetorical Transformations of the Life Sciences* (Stanford: Stanford University Press, 1997), p. 127.

16. Henri Atlan, *Enlightenment to Enlightenment: Intercritique of Science and Myth*, translated by Lenn J. Schramm (Albany: State University of New York Press, 1993), pp. 315–6. Atlan makes some of the same arguments in a more conventionally scientific venue in his article (with Moshe Koppel), "The Cellular Computer DNA: Program or Data?," *Bulletin of Mathematical Biology* 52:3 (1990), pp. 335–48.

17. Donna Haraway, "Situated Knowledges: The Science Question in Feminism and the Privilege of Partial Perspective," in *Simians, Cyborgs, and Women: The Reinvention of Nature* (London: Routledge: 1991), p. 187.

18. Quoted in Sarah Blaffer Hrdy, "Empathy, Polyandry, and the Myth of the Coy Female," in Ruth Bleier, ed., *Feminist Approaches to Science* (New York: Pergamon Press, 1986), p. 119.

19. As historian of science Robert Kohler argues in his recent book *Lords of the Fly* (Chicago: University of Chicago Press, 1994), famed geneticist T.H. Morgan and his colleagues at Columbia had — through intensive and extensive labor of breeding, record-keeping, statistical analysis, anatomical observation, and original theorization — turned *Drosophila* into a kind of finely tuned, productive scientific instrument. They had, in effect, machined fruit flies into a tool for producing the science of genetics, like the grinding of microscope lenses to aid in the work of microbiology.

20. Hrdy, p. 121.

21. Hrdy, p. 127.

22. Hrdy, p. 141.

23. Her "About the Author" description, in Bleier, p. 208

24. Pamela J. Asquith, "Japanese Science and Western Hegemonies: Primatology and the Limits Set to Questions," in *Naked Science: Anthropological Inquiry Into Boundaries, Knowledge, and Power*, ed. Laura Nader (London: Routledge, 1996), pp. 258–9.

25. Sharon Traweek, "Unity, Dyads, Triads, Quads, and Complexity: Cultural Choreographies of Science," *Social Text*, No. 46/47 (Spring/Summer 1996), pp.129–139, on p. 136.

26. All the quotes here are taken from Karen Barad, "A Feminist Approach to Teaching Quantum Physics," in *Teaching the Majority: Breaking the Gender Barrier in Science, Mathematics, and Engineering*, ed. Sue V. Rosser (New York: Teachers College Press, 1995), pp. 43–75. The quotes from Niels Bohr are taken by Barad from *The philosophical writings of Niels Bohr*, Vols. 3 and 1, respectively (Woodbridge, CT: Ox Bow Press, 1963).

27. Donna J. Haraway, "Reading Buchi Emecheta: Contests for 'Women's Experience' in Women's Studies," in *Simians, Cyborgs, and Women: The Reinvention of Nature* (London: Routledge, 1991), p. 110–1.

28. This quote and all following quotes on PET are from: Joseph Dumit, "Twenty-First Century PET: Looking for Mind and Morality Through the Eye of Technology," in George Marcus, ed., *Technoscientific Imaginaries* (Chicago: University of Chicago, 1995), 87–128.

29. This discussion of Peirce's system of signs has been drawn from E. Valentine Daniel, *Fluid Signs: Being a Person the Tamil Way* (Berkeley: University of California Press, 1984). All the quotes, including those from T.L. Short, Isabel Stearns, and Peirce himself, can be found on pp. 15, 16, 18–19, and 241–2.

Chapter 3

1. Thomas Mallon, "Galileo, Phone Home," *The New York Times Magazine*, December 3, 1995, pp. 57–9.

2. Peter Huber, *Galileo's Revenge: Junk Science in the Courtroom* (New York: Basic Books, 1991), p. 16.

3. Mario Biagioli, *Galileo Courtier: The Practice of Science in the Culture of Absolutism* (Chicago: University of Chicago Press, 1993), pp. 94–5.

4. Biagioli, pp. 99–100.

5. All quotes are from Biagioli, pp. 301–5.

6. Biagioli, p. 331.

7. Biagioli, p. 336

8. Biagioli, p. 342.

9. Quoted in Biagioli, p. 342.

10. Michel Foucault, "Two Lectures," in *Power/Knowledge. Selected Interviews and Other Writings, 1972–1977*, ed. C. Gordon (New York: Pantheon, 1980), p. 93.

11. Simon Jackman, "Liberalism, Public Opinion, and Their Critics: Some Lessons for Defending Science," in *The Flight from Science and Reason*, ed. Paul R. Gross, Norman

Levitt, and Martin W. Lewis (New York Academy of Sciences 1996), p. 355; quoting Foucault, *Discipline and Punish*, p. 27.

12. Michel Foucault, "Truth and Power," in *Power/Knowledge: Selected Interviews and Other Writings 1972–1977* (New York: Pantheon 1980), p. 132.

13. This material is taken from Gina Kolata, "The Many Myths About Sex Offenders," *The New York Times*, September 1, 1996, p. E10.

14. James H. Jones, *Bad Blood: The Tuskegee Syphilis Experiment* (New York: The Free Press, 1981), p. 25.

15. Jones, p. 8

16. Adrian Desmond and James Moore, *Darwin: The Life of a Tormented Evolutionist* (New York: W.W. Norton, 1991), pp. xx-xxi.

17. Desmond and Moore, p. 191.

18. Desmond and Moore, p. 222.

19. Desmond and Moore, p. 199.

20. Desmond and Moore, p. 294.

21. Richard Bernstein, *The New Constellation: The Ethical-Political Horizons of Modernity/Postmodernity*, (Cambridge, MA: MIT Press, 1991), p.201.

22. See, for example, Gillian Beer, *Darwin's Plots: Evolutionary Narrative in Darwin, George Eliot, and Nineteenth-Century Fiction* (London: Routledge and Kegan Paul, 1983).

23. Desmond and Moore, p. xxi.

24. The project that we're describing here, and undertaking, is akin to what Donna Haraway calls "situated knowledges" and Sandra Harding calls, in a re-appropriation of conventional categories, "strong objectivity." In Harding's terms, such a project involves "extending the notion of scientific research to include systematic examination of such powerful background beliefs," that come from "cultural agendas" and personal experience. "If the goal is to make available for critical scrutiny *all* the evidence marshaled for or against a scientific hypothesis, then this evidence too requires critical examination *within* scientific research processes." Sandra Harding, *Whose Science? Whose Knowledge?: Thinking from Women's Lives* (Ithaca: Cornell University Press, 1991), p. 149. Haraway argues for "a doctrine and practice of objectivity that privileges contestation, deconstruction, passionate construction, webbed connections, and hope for transformations of systems of knowledge and ways of seeing." Donna Haraway, "Situated Knowledges: The Science Question in Feminism and the Privilege of Partial Perspective," in *Simians, Cyborgs, and Women: The Reinvention of Nature* (New York and London: Routledge, 1991), pp. 191–2.

Chapter 4

1. The story is recounted by James T. Kloppenberg in his *Uncertain Victory: Social Democracy and Progressivism in European and American Thought, 1870–1920* (New York: Oxford University Press, 1986), p. 247.

2. Kloppenberg, p. 11.

3. Silvan S. Schweber, "Physics, Community, and the Crisis in Physical Theory," *Physics Today*, November 1993, pp. 34–40.

4. Paul Forman, "Recent Science: Late Modern and Post-Modern," *The Historiography of Contemporary Science and Technology*, ed. Thomas Söderqvist (Harwood Academic Publishers, 1997), pp. 197–8.

5. Quoted in Forman, p. 201.

6. Mark B. Adams, "Towards a Comparative History of Eugenics," in *The Wellborn Science: Eugenics in Germany, France, Brazil and Russia*, ed. Mark B. Adams (New York: Oxford University Press, 1990), pp. 217–31.

7. See Diane Paul, *Controlling Human Heredity: 1865 to the Present* (New York: Humanities Press, 1995).

8. Peter W. Huber, *Galileo's Revenge: Junk Science in the Courtroom* (New York: Basic Books, 1991), p. 217.

9. Richard Levins and Richard Lewontin, "The Problem of Lysenkoism," in *The Dialectical Biologist* (Harvard University Press, 1985), p. 163.

10. Wes Jackson, *Becoming Native To This Place* (Washington, DC: Counterpoint Press, 1996), p. 35.

11. See David Joravsky, *The Lysenko Affair* (Cambridge, MA: Harvard University Press, 1970) and Zhores Medvedev, *The Rise and Fall of T.D. Lysenko* (New York: Columbia University Press, 1969).

12. Levins and Lewontin, pp. 170–1.

13. Quoted in Mark B. Adams, "Eugenics in Russia 1900–1940," in *The Wellborn Science*, ed. Adams, p. 195.

14. Adams, pp. 196–7.

15. Levins and Lewontin, p. 178.

16. Oxford University Press, 1987.

17. Gregg Mitman and Anne Fausto-Sterling, "Whatever Happened to *Planaria*? C. M. Child and the Physiology of Inheritance," in *The Right Tools for the Job: At Work in Twentieth-Century Life Sciences*, ed. Adele E. Clarke and Joan H. Fujimura (Princeton: Princeton University Press, 1992), pp. 198–232.

18. Evelyn Fox Keller, "A World of Difference," in *Reflections on Gender and Science* (Yale University Press, 1985), p. 171.

19. Quoted in Mitman and Fausto-Sterling, p. 191.

20. See Garland E. Allen, "The Eugenics Record Office at Cold Spring Harbor, 1910–1940: An Essay in Institutional History," *Osiris* (2nd series) Vol.2, pp. 225–64.

21. All quotes are from Diane Paul and Barbara Kimmelman, "Mendel in America: Theory and Practice," in *The American Development of Biology*, ed. Ronald Rainger, Keith R. Benson, and Jane Maienschein (Philadelphia: University of Pennsylvania Press, 1988), pp. 296–301.

22. Gross and Levitt, p. 226.

23. Shana M. Solomon and Edward J. Hackett, "Setting Boundaries Between Science and Law: Lessons from *Daubert v. Merrell Dow Pharmaceuticals Inc.*," *Science Technology and Human Values* Vol. 21 (1996):131–156.

24. P.C.W. Davies and J.R. Brown, eds., *The Ghost in the Atom* (New York: Oxford University Press, 1986), p. 134.
25. Hon. Mark I. Bernstein, Opinion, Court of Common Pleas of Philadelphia County, Civil Trial Division, *Blum v. Merrell Dow Pharmaceuticals, Inc.*, September Term 1982, No. 1027 (Opinion delivered December 13, 1996).
26. An appellate court has reversed the judgment; at this writing, the plaintiffs are taking that ruling to the highest Pennsylvania court.
27. See Daniel J. Kevles, "The Assault on David Baltimore," *The New Yorker*, May 27, 1996, pp. 94–109.
28. We have not provided page numbers for these quotes, since the pagination on our downloaded version will obviously differ from others. The full report is available at http://www.os.dhhs.gov/progorg/dab/dab1582.txt.
29. Frederick Grinnell, "Ambiguity in the Practice of Science," *Science* Vol. 272 (April 19, 1996), p. 333.
30. Donald E. Buzzelli, "Misconduct: Judgment Called For," *Science* Vol. 272 (May 17, 1996), pp. 935–9.
31. Paul de Man, "The Rhetoric of Temporality," in *Blindness and Insight: Essays in the Rhetoric of Contemporary Criticism* (Minneapolis: University of Minnesota Press, 1983), p. 214.
32. Henri Atlan, *Enlightenment to Enlightenment: Intercritique of Science and Myth*, translated by Lenn J. Schramm (Albany: State University of New York Press, 1993), pp. 290–92.
33. Atlan, p. 376.
34. Atlan, p. 11.

Chapter 5

1. *Estimating the Cold War Mortgage: The 1995 Baseline Environmental Management Report. Executive Summary, March 1995* (Washington, DC: U.S. Department of Energy, Office of Environmental Management), p. ix.
2. For an overview of the problems on Cape Cod, and for further examples of the social, technical, and political processes involved in the cleanup of federal facilities, see their Web page at www.mmr.org.
3. *Interim Report of the Federal Facilities Environmental Restoration Dialogue Committee* (Washington, DC: U.S. Environmental Protection Agency and The Keystone Center, February 1993), p. v. The *Final Report* was released in April 1996.
4. There were some interesting results from these studies, but mostly in the form of anomalies that aren't explained by the current theories about what's happening underground. For example, monitoring well 2 is just east of the landfill, and is the most contaminated; the groundwater there contains chlorinated solvents. But well 9, which is further east and thus "downstream" from well 2, shows no chlorinated solvents but does test positive for petroleum products. So the data from these two wells can't be fitted into the model of a continuous plume. The terrain conductivity studies added to the anomalies: There were "blobs" of high conductivity around wells 2 and 9, suggesting high concen-

trations of ions and hence contaminants, but no one had very good data for what was happening in between these wells. One possible explanation is that the contamination near well 9 might be coming from the fire training area, which would mean that groundwater flow models would have to be revised. Or it could be coming from a source that is still undiscovered. In any case, all we had were some nagging questions, and not much interest or willingness on the part of the base to pursue them.

Chapter 6

1. Mark R. Cullen, "The Worker with Multiple Chemical Sensitivities: An Overview," in M.R. Cullen, ed., *Workers with Multiple Chemical Sensitivities, Occupational Medicine: State of the Art Reviews* (Philadelphia: Hanley and Belfus, 1987), p. 655.
2. Ronald E. Gots, "Multiple Chemical Sensitivities—Public Policy," *Clinical Toxicology*, Vol. 33, No. 2 (1995), p. 111.
3. Lynn Lawson, *Staying Well in a Toxic World: Understanding Environmental Illness, Multiple Chemical Sensitivities, Chemical Injuries, and Sick Building Syndrome* (Chicago: The Noble Press, 1993) pp. 15–6.
4. Lawson, p. 9.
5. Roy L. DeHart, "Multiple Chemical Sensitivities — What Is It?," from *Multiple Chemical Sensitivities: A Workshop*, Board on Environmental Studies and Toxicology, Commission on Life Sciences, National Research Council; reprinted in *Multiple Chemical Sensitivity: A Scientific Overview*, ed. Frank L. Mitchell, U.S. Department of Health and Human Services, Public Health Service, Agency for Toxic Substance and Disease Registry (Princeton: Princeton Scientific Publishing Co., 1995), pp. 35–8.
6. See the summary of these and other definitions of MCS in Claudia S. Miller, "White Paper: Chemical Sensitivity: History and Phenomenology," *Toxicology and Environmental Health*, Vol. 10, No. 4/5 (1994), pp. 274–5.
7. Press Release, March 26, 1996, MCS Referral and Resources.
8. The material in this section is drawn from Eric Nelson and Mark Worth, "It's All In Your Head," *Washington Free Press*, No. 8, February-March 1994, pp. 10*ff*. This reporting on the Boeing plants and MCS was supported by the Fund for Investigative Journalism. Readers can find it on the World Wide Web at http://www.speakeasy.org/wfp/08/Boeing1.html.
9. Gregory E. Simon, "Epidemic Multiple Chemical Sensitivity in an Industrial Setting" in *Multiple Chemical Sensitivity: A Scientific Overview*, ed. Frank L. Mitchell, U.S. Department of Health and Human Services, Public Health Service, Agency for Toxic Substance and Disease Registry (Princeton: Princeton Scientific Publishing Co., 1995), p. 42.
10. Ann Davidoff and Linda Fogarty, for example, have documented and analyzed (and published in a peer-review journal) the "serious methodologic flaws regarding sample selection, measurement, and study design" in ten studies claiming to show that MCS is "psychogenic" in origin. See their article, "Psychogenic Origins of Multiple Chemical Sensitivities Syndrome: A Critical Review of the Research Literature," *Archives of Environmental Health* Vol. 49, No. 5 (September/October 1994), pp. 316–25.

11. American Council on Science and Health, *Multiple Chemical Sensitivity* brochure (1994), pp. 2–3, 26, available at their web site at www.acsh.org.
12. *Our Toxic Times,* July 1996, published by the Chemical Injury Information Network (P.O.Box 301, White Sulphur Springs, MT 59645).
13. Quoted in *Multiple Chemical Sensitivities: A Scientific Overview* p. 650 [note 5].
14. Peter W. Huber, *Galileo's Revenge: Junk Science in the Courtroom* (New York: Basic Books, 1991).
15. *Science and Technology in Judicial Decision Making: Creating Opportunities and Meeting Challenges,* Report of the Carnegie Commission on Science, Technology, and Government, March 1993, pp. 26–8.
16. Norman E. Rosenthal, "Multiple Chemical Sensitivity: Lessons From Seasonal Affective Disorder," *Toxicology and Industrial Health,* Vol. 10 (1994), reprinted in *Multiple Chemical Sensitivities: A Scientific Overview* [note 5], pp. 623–32.
17. Miller, "White Paper," p. 275 [note 6].
18. Claudia S. Miller, "Toxicant-Induced Loss of Tolerance — An Emerging Theory of Disease," *Environmental Health Perspectives,* Vol. 105 (1997, Supplement 2), pp. 445–53.
19. Miller, "White Paper," p. 275 [note 6].

Chapter 7

1. J.B.S. Haldane, "The Biology of Inequality," Chapter 1 of *Heredity and Politics* (London: Allen & Unwin, 1938), reprinted in *On Being the Right Size and Other Essays* (Delhi: Oxford University Press, 1992), pp. 128–9.
2. François Jacob, *The Logic of Life,* p. 16.
3. Jacob, p. 11.
4. The quotes in this section are all taken from Evelyn Fox Keller, *Refiguring Life: Metaphors of Twentieth-Century Biology* (New York: Columbia University Press, 1995).
5. Robert Kohler's *Lords of the Fly: Drosophila Genetics and the Experimental Life* (Chicago: University of Chicago Press, 1994) will reward readers with an excellent account of how *Drosophila* was crafted and kludged into one of the most important pieces of "experimental equipment" in twentieth-century biology.
6. U.S. Senate, Subcommittee on Energy Research and Development, Committee on Energy and Natural Resources, *The Human Genome Project,* July 11, 1990 (101st Congress, 1st Session, S. Hrg. 101–894), pp. 91–2.
7. Eric Lander, Testimony before the Subcommittee on Labor, Health and Human Services, Education and Related Agencies of the House Appropriations Committee, April 23, 1990.
8. Quoted in Michael Fortun, "Making and Mapping Genes and Histories: The Genomics Project in the United States, 1980–1990," Ph.D. dissertation, Department of the History of Science, Harvard University, 1993.
9. See Michael Fortun, "The Human Genome Project and the Acceleration of Biotechnology," in Arnold Thackeray, ed., *Private Science: Biotechnology and the Rise of the Molecular Sciences* (Philadelphia: University of Pennsylvania Press, 1998), pp. 182–201.

10. Vicki Glaser, "Lilly and Millennium Cut the Genome Cake a New Way," *Bio/Technology*, Vol. 13 (November 1995), pp. 1149–50.

11. See Paul Rabinow, *Making PCR: A Story of Biotechnology* (Chicago: University of Chicago Press, 1996); Robert Cook-Deegan, *Science, Politics, and the Human Genome* (New York: W.W. Norton, 1994); and Joan Fujimura, *Crafting Science: A Sociohistory of the Quest for the Genetics of Cancer* (Cambridge, MA: Harvard University Press, 1996).

12. Keller, p. 22 [note 4].

13. Keller, pp. 85–6 [note 4].

14. Wray Herbert, "Politics of Biology," *U.S. News and World Report*, April 21, 1997, pp. 72–80.

15. Arthur Allen, "Policing the Gene Machine," *Lingua Franca* (March 1997), p. 31.

16. Stephen J. Gould, "Curveball," in *The Bell Curve Wars*, ed. Steven Fraser (New York: Basic Books, 1995), p. 11.

17. See note 16.

18. Quoted in Allen, pp. 32–33, 34.

19. Stephen J. Gould, *The Mismeasure of Man* (New York: W.W. Norton, 1981).

20. Howard Gardner, "Cracking Open the IQ Box," in *The Bell Curve Wars*, p. 29.

21. Dean Hamer and Peter Copeland, *The Science of Desire: The Search for the Gay Gene and the Biology of Behavior* (New York: Simon and Schuster, 1994), pp. 18–22.

22. Ibid, p. 25.

23. Dean Hamer et al., "A Linkage Between DNA Markers on the X Chromosome and Male Sexual Orientation," *Science*, Vol. 261 (July 16, 1993), pp. 321–7.

24. As quoted in Arthur Allen, "Policing the Gene Machine," *Lingua Franca*, March 1997, p. 34.

25. Simon LeVay, *The Sexual Brain* (Cambridge, MA: MIT Press, 1993), pp. 137–8.

26. Vernon A. Rosario, "Homosexual Bio-Histories: Genetic Nostalgias and the Quest for Paternity," in V.A. Rosario, ed., *Science and Homosexualities* (New York: Routledge, 1997), pp. 12, 18.

27. Paul D. Markel et al., "Confirmation of Quantitative Trait Loci for Ethanol Sensitivity in Long-Sleep and Short-Sleep Mice," *Genome Research*, Vol. 7 (1997), pp. 92–9.

28. Michael Lynch, "Sacrifice and Transformation of the Animal Body Into a Scientific Object: Laboratory Culture and Ritual Practice in the Neurosciences," *Social Studies of Science*, Vol. 18 (1988), pp. 265–289.

29. James Boyle, *Shamans, Software, and Spleens: Law and the Construction of the Information Society* (Cambridge, MA: Harvard University Press, 1996).

Chapter 8

1. We've adapted this image from James Clerk Maxwell, the British physicist who transformed electromagnetic theory in the late nineteenth century. In describing the work of the physicist, Maxwell used the metaphoric image of the church bell ringer, who tugged on (and was tugged by) various ropes passing through the ceiling to the unseen belfry beyond, thereby learning about the heavenly bells and their mechanisms. The image is

almost entirely Kantian, but with less of an emphasis on the inaccessibility of the "thing-in-itself." It's an uneasy combination of direct and indirect connection.

2. A. Einstein, B. Podolsky, and N. Rosen, "Can Quantum-Mechanical Description of Reality Be Considered Complete?," *Physical Review*, Vol. 47, (May 15, 1935), p. 777.

3. Ibid., emphasis in original.

4. The work was by GHZ—Greenberger, Horne, and Zeilinger, Bernstein's three long-term collaborating principal investigators on the Hampshire College National Science Foundation grant "Quantum interferometry." See the Appendix for a brief description.

5. Einstein, Podolsky, and Rosen, p. 780.

6. David Bohm, "On Bohr's Views Concerning the Quantum Theory," in A.P. French and P. J. Kennedy, eds., *Niels Bohr: A Centenary Volume* (Cambridge, MA: Harvard University Press, 1985), p. 153.

7. Quoted in N. David Mermin, "A Bolt from the Blue: The E-P-R Paradox," in A.P. French and P. J. Kennedy, eds., *Niels Bohr: A Centenary Volume* (Cambridge, MA: Harvard University Press, 1985), p. 142.

8. Niels Bohr, "Can Quantum-Mechanical Description of Physical Reality Be Considered Complete?," *Physical Review*, Vol. 48 (1935), pp. 696–702.

9. So unsatisfied that he did physics about it, for Einstein's commitment was no mere battle of wills. The EPR paper implies there should be completely deterministic explanation for the quantum phenomena including those that seem probabilistic. In the same year, in the very same journal as the EPR paper, he and Rosen published a work showing how two very remote places could be connected by a topological structure: the Einstein-Rosen bridge. Today it's famous for introducing "wormholes" to general relativity. But in a remarkable passage, they suggest it could be a structure of quantum particles, taking the model so seriously as to remark that inside of protons each would need at least two bridges to account for their electrical repulsion, since like charges repel. Positive mass bridges—as gravitational structures—can only attract. A. Einstein and N. Rosen, "The Particle Problem in the General Theory of Relativity," *Physical Review*, Vol. 48 (1935), pp. 73–6. The great final revenge would come if a nonlocal but geometric theory of General Relativity actually *entailed* quantum mechanics. The current string theories or M-theories of quantum field theory may someday produce just such a history bending, time-twisting outcome. Only time will tell.

10. The cat apparatus is not the best example of how "consciousness" or at least an act of observing or interfering is "responsible" for the "creation" of reality. We've highlighted it here because of its dramatic illustration of the situation, because of its historical significance, and because it persists as a point of reference in popular treatments of quantum physics.

11. See Trevor Pinch, "What Does a Proof Prove If It Does Not Prove?" in *The Social Production of Scientific Knowledge*, ed. E. Mendelsohn, P. Weingart and R. Whitley (Dordrecht: D. Reidel, 1977), pp. 171–215.

12. David Bohm, "Hidden Variables and the Implicate Order," in *Quantum Implications: Essays in Honour of David Bohm*, ed. B.J. Hiley and F. David Peat (London: Routledge and Kegan Paul, 1987), pp. 33–45; on 35.

13. Ibid., p.36.
14. Bohm, p. 39.
15. Quoted in John Gribbin, *Schrödinger's Kittens and the Search for Reality: Solving the Quantum Mysteries* (Boston: Little, Brown) p. 155.
16. Paul Forman, "Weimar Culture, Causality, and Quantum Theory, 1918–1927," *Historical Studies in the Physical Sciences*, Vol. 3 (1971), pp. 1–116.
17. Quoted in Pinch, p. 188 [note 11].
18. Or add a second differently trained set of bloodhounds. Some physicists *had* followed Einstein, eschewing quantum mechanics to become General Relativists. Abner Shimony reports a conversation between Peter Bergman and Einstein in Valentine Bargmann's presence where Einstein pointed at von Neumann's Postulate B', a corollary of axiom E (derived from it and B and two others) saying "it needn't be like that, the dispersion-free state mustn't obey this!" (*Search for a Naturalistic Worldview* [New York: Cambridge University Press, 1993], p. 89.) By 1966 at latest, Bernstein was aware that the von Neumann proof was wrong because it had assumed for the submicroscopic Hidden Variables something that needed only be true for the quantum variables.
19. Thanks to David Edgar for teaching us about hounds.
20. See Pinch, p. 186 [note 11].
21. J.S. Bell, *Speakable and Unspeakable in Quantum Mechanics* (Cambridge: Cambridge University Press, 1987), p. 171.
22. One modern physicist, punning with Einstein's "spooky action at a distance" called EPR correlations "passion at a distance."
23. Actually the inequality is *obeyed* by quantum physics for the deterministic 100 percent–0 percent cases which EPR had discussed. Only by extending the reasoning to alternative pairs of measurements that do not have definite results does quantum mechanics fail the inequality.
24. An unfortunate, apparently totally accidental coincidence diminishes the snappiness of the punchline to Aspect's famous experiment: The time it took for the photons to reach the apparatus at each end was exactly the switching time for the measurement choice; if the source somehow conspired to give exactly the right quantum results for their settings at time of emission it would also match the theory upon arrival. Anton Zeilinger, "Testing Bell's Inequality with Periodic Switching, *Physics Letters*, Vol. 118 (1986), p. 1 An experiment at Zeilinger's labratory has just been completed which not only chose measurements at random, but also recorded all counts at either end and only *later* compared the records [to be published in *PRL* 1998].
25. See Daniel M. Greenberger, Michael A. Horne, and Anton Zeilinger, "Multiparticle Interferometry and the Superposition Principle," *Physics Today* (August 1993), pp. 22–9.
26. Charles H. Bennett, Gilles Brassard, Claude Crépeau, Richard Jozsa, Asher Peres, and William K. Wootters, "Teleporting an Unknown Quantum State via Dual Classical and Einstein-Podolsky-Rosen Channels," *Physical Review Letters*, Vol. 70, No. 13 (March 29, 1993), pp. 1895–9.
27. The A-bomb has had remarkably deep personal effects on at least two generations of Americans. Every baby boomer remembers growing up with duck-and-cover drills un-

der their grade-school desks, and it often gets much more personal and pointed than that. Bernstein's career (like that of Michio Kaku and several other currently active physicists) was strongly influenced by his parents' wartime and Manhattan Project experiences. The story of Bernstein's father—and his joy at narrowly having declined Edward Teller's call to join the project in Oak Ridge—appears in *New Ways of Knowing* by Marcus Raskin and Herbert Bernstein (Totowa, NJ: Rowman & Littlefield, 1987), Chapter 2. Available from the authors.

28. Paul Forman, "Behind Quantum Electronics: National Security as Basis for Physical Research in the United States, 1940–1960," *Historical Studies in the Physical Sciences*, Vol. 18, No. 1 (1987), pp. 149–229.

29. Norbert Wiener, *I Am a Mathematician* (Cambridge, MA: MIT Press, 1956), p. 308.

Chapter 9

1. *Science, Technology, and Democracy: Research on Issues of Governance and Change*, National Science Foundation, no date, p. 1. Copies of the report are available from Rachelle Hollander, (703) 306–1743; rholland@nsf.gov.

2. Quoted in Kurt Kleiner, "Fear and Loathing at the Smithsonian Institution," *New Scientist*, April 8, 1995, p. 42.

3. See Langdon Winner, "The Gloves Come Off: Shattered Alliances in Science and Technology Studies," *Social Text* Vol. 46/47 (Spring/Summer 1996), pp. 89–90.

4. John Brockman, ed., *The Third Culture: Beyond the Scientific Revolution* (New York: Simon and Schuster, 1995), p. 17.

5. This quote and subsequent ones are from Robert Musil, *The Man Without Qualities*, translated by Sophie Wilkins and Burton Pike (New York: Alfred A. Knopf, 1995), pp. 231–5.

6. Quoted in Joseph Dumit, "Twenty-First Century PET: Looking for Mind and Morality Through the Eye of Technology," in George Marcus, ed., *Technoscientific Imaginaries* (Chicago: University of Chicago Press, 1995), pp. 118–9.

7. Quoted in Mary-Jo Delvecchio Good, et al., "Medicine on the Edge: Conversations With Oncologists," in George Marcus, ed., *Technoscientific Imaginaries* (Chicago: University of Chicago Press, 1995), pp. 140–1.

8. Ilya Prigogine and Isabel Stengers, *Order Out of Chaos: Man's New Dialogue With Nature* (New York: Bantam Books, 1984), pp. 312–3.

9. Donald Ludwig, Ray Hilborn, and Carl Walters, "Uncertainty, Resource Exploitation, and Conservation: Lessons from History," *Science*, Vol. 260 (April 2, 1993), pp. 17, 36.

10. Sarah Blaffer Hrdy, "Empathy, Polyandry, and the Myth of the Coy Female," in Ruth Bleier, ed., *Feminist Approaches to Science* (New York: Pergamon Press, 1986), p. 141.

11. Paul Gross and Norman Levitt, *Higher Superstition: The Academic Left and its Quarrels with Science* (Baltimore: Johns Hopkins Press, 1994), pp. 242–3.

12. Trinh T. Minh-ha, "Nature's r: A Musical Swoon," in George Robertson et al., eds., *FutureNatural: Nature, Science, Culture* (New York: Routledge, 1996), pp. 86–104; pp. 100, 101.

13. Quoted in James H. Jones, *Bad Blood: The Tuskegee Syphilis Experiment* (New York: The Free Press, 1981), p. 16.

14. Jacques Derrida, "Force of Law: The 'Mystical Foundation of Authority'," *Cardozo Law Review*, Vol. 11, Nos. 5–6 (1990), p. 961.

Appendix

1. For example, if A spins along an axis pointing skyward at 45° above the horizon in the southeast direction, then a particle which is opposite in the north direction but parallel in the other two must spin along an axis 45° above the northeast horizon. A particle that's spinning opposite along the west direction would point 45° above the southwest.

2. We capitalize words for message tokens to distinguish them from the names of directions north, west, up (lowercase) *and* from the state names for particles spinning in a given direction NORTH, WEST, and UP (uppercase).

3. That would not work because the state of maximum vertical spin, for example, is Up Up. Because this is a simple product it represents two independent states—definitely not entangled since measurements in any direction other than vertical gives no definitive correlation with the second particle's outcome, even for measuring both particles along the same axis.

Index

values, 76–77; in sciences, 113
Vavilov, Nikolai, 119
via media, 109–10
Victorian culture, 53, 92–93, 102
Vienna Circle, 4
visualization. *See* manner of thought

W

Wadsworth, Marian, 161
Wagner, Henry, 281
Wallace, Alfred Russell, 98
Walters, Carl, 283–84
Watson, James, 204
Weaver, Warren, 209
Weimar culture, 242
Westman, Robert, 43, 45–46
Westover Air Reserve Base, 153–57, 313n, 314n; geology of, 159–60; history of,

160–64; Landfill B, 164–71; and RAB, 169–71
Wheeler, John, 27–28
Widmanstetter, Johann Albrecht, 44
Wiener, Norbert, 254–55
willies, 248–50, 252
Wilson, E. O., 51
Wittgenstein, Ludwig, 37, 66, 151
women: biotechnology and, 208; and eugenics, 115; in science, 12
wormholes, 317n

Y

Yeats, William Butler, 257

Z

Zeilinger, Anton, 246–47, 297–300
Ziem, Grace, 188, 318n